CONCEPTS OF

DATABASE MANAGEMENT

Fifth Edition

Philip J. Pratt
Grand Valley State University

Joseph J. Adamski
Grand Valley State University

THOMSON
COURSE TECHNOLOGY ™

Australia • Canada • Mexico • Singapore • Spain

D1361484

THOMSON

COURSE TECHNOLOGY

Concepts of Database Management, Fifth Edition

by Philip J. Pratt and Joseph J. Adamski

Executive Vice President, Academic and Technology Publishing:
Kristen Duerr

Executive Editor:
Mac Mendelsohn

Acquisitions Editor:
Maureen Martin

Product Manager:
Janet Aras

Developmental Editor:
Jessica Evans

Production Editor:
Melissa Panagos

Associate Product Manager:
Mirella Misiaszek

Senior Marketing Manager:
Karen Seitz

Editorial Assistant:
Amanda Piantedosi

Text Designer:
Books By Design

Cover Designer:
Betsy Young

Cover Art Designer:
Rakefet Kenaan

Senior Manufacturing Coordinator:
Laura Burns

For permission to use material from this text or product, submit a request online at www.thomsonrights.com
Any additional questions about permissions can be submitted by email to thomsonrights@thomson.com.

Disclaimer
Course Technology reserves the right to revise this publication and make changes from time to time in its content without notice.

The Web addresses in this book are subject to change from time to time as necessary without notice.

This publication is designed to provide accurate and authoritative information regarding the subject matter covered. It is sold with the understanding that neither the publisher nor the author are engaged in rendering legal, accounting, or other professional services. If legal, accounting, or other expert advice is required, the services of a competent professional person should be sought.

Some of the product names and company names used in this book have been used for identification purposes only and may be trademarks or registered trademarks of their respective manufacturers and sellers.

Microsoft is a registered trademark of Microsoft Corporation in the United States and/or other countries. Thomson Learning, Inc. is an independent entity from the Microsoft Corporation, and not affiliated with Microsoft in any manner.

Oracle is a registered trademark, and Oracle10g is a trademark of Oracle Corporation.

ISBN 0-619-21529-1

CONTENTS

9 Database Management Approaches 285

A Appendix A: Comprehensive Design Example: Marvel College 327

A Appendix B: Additional Case Problem: Alexamara Marina Group 363

PREFACE

The advent of database management systems for personal computers in the 1980s moved database management beyond the realm of database professionals and into the hands of everyday users from all segments of the population. A field once limited to highly trained users of large, mainframe, database-oriented application systems became an essential productivity tool for such diverse groups as home computer owners, owners of small businesses, and end-users in large organizations.

The major PC-based database software systems have continually added features to increase their ease of use, allowing users to enjoy the benefits of database tools relatively quickly. Truly effective use of such a product, however, requires more than just knowledge of the product itself, although that knowledge is obviously important. It requires a general knowledge of the database area, including such topics as database design, database administration, and application development using these systems. While the depth of understanding required is certainly not as great for the majority of users as it is for the data processing professional, a lack of any understanding in these areas precludes effective use of the product in all but the most limited applications.

About This Book

This book is intended for anyone who is interested in gaining some familiarity with database management. It is appropriate for students in introductory database classes in computer science or information systems programs. It is appropriate for students in database courses in related disciplines, such as business, at either the undergraduate or graduate level. Such students require a general understanding of the database environment. In addition, courses introducing students of any discipline to database management have become increasingly popular over the past few years, and this book is ideal for such courses. It is also appropriate for individuals considering purchasing a PC-based database package and who want to make effective use of such a package.

This book assumes that you have some familiarity with computers; a single introductory course is all the background that is required. While you need not have any background in programming to use this text effectively, there are certain areas where a little programming background will enable you to explore topics in more depth.

Changes to the Fifth Edition

The Fifth Edition includes the following new features and content:

- As in the Fourth Edition, the SQL material is covered using Microsoft Access. Also included are generic forms of all examples that students can use on a variety of platforms, including Oracle. The Fifth Edition also includes two new appendices that provide a useful reference for anyone wanting to use SQL effectively. Appendix C includes a command reference of all the SQL commands and operators that are taught in the chapters. Students can use this appendix as a quick resource when constructing commands. Each command includes a short description, a table that shows the required and optional clauses and operators, and an example and its results. Appendix D provides students with an opportunity to ask a question, such as "How do I delete rows?," and to identify the appropriate section in Appendix C to use to find the answer. Appendix D is extremely valuable when students know what they want to accomplish, but can't remember the exact SQL command they need.

- In addition to the section of Review Questions, the end of each chapter includes two sets of exercises—one featuring the Premiere Products database and the other featuring the Henry Books database—that give students "hands-on" experiences with the concepts found in the chapter. The Fifth Edition includes additional exercises in each chapter that are not included in previous editions.

- As in the previous edition, the Fifth Edition covers entity-relationship diagrams. The database design material in the Fifth Edition has been expanded to include a discussion of the entity-relationship model as a database model. It also includes a discussion of a characterization of various types of primary keys.

- Appendix B in the Fifth Edition contains an additional case problem, Alexamara Marina Group. This case problem introduces the database for Alexamara Marina Group and also includes additional exercises that can be assigned along with chapters 1 through 6.

- The Premiere Products, Henry Books, and Alexamara Marina Group databases are available in Access 2000 format (the default database format for Access 2000, Access 2002, and Access 2003) on the Instructor's Resource CD-ROM and at *www.course.com*. For those students using database management systems that run scripts (such as Oracle), the data files also include the script files that create the tables and add the data to the tables in the databases used in the book.

- For instructors who want to use an Access or SQL text as a companion to the Fifth Edition, the Instructor's Manual for this book includes detailed tips on integrating the Fifth Edition with other books from Course Technology that cover Microsoft Access 2003 and SQL (for more information, see the "Instructor's Resource CD-ROM" section in this preface).

Special Features

Detailed Coverage of the Relational Model, including Query-By-Example (QBE) and SQL

The text features detailed coverage of the important aspects of the relational model, including comprehensive coverage of SQL. It also covers QBE and relational algebra as well as advanced aspects of the model, such as views, the use of indexes, the catalog, and the relational integrity rules.

Normalization Coverage

The Fifth Edition covers first normal form, second normal form, third normal form (Boyce-Codd normal form), and fourth normal form. The text describes in detail the update anomalies associated with lower normal forms as part of the motivation for the need for higher normal forms. Finally, there are right ways and wrong ways to normalize tables. This text specifically addresses this by showing students some of the mistakes people can make in the normalization process, explaining why the approach is incorrect, demonstrating the problems that would result from incorrect normalizations, and, most importantly, identifying how to avoid these mistakes.

Views Coverage

This text covers the important topic of views. It discusses the creation and use of views as well as the advantages of using views.

Database Design

The important process of database design is given detailed treatment. A highly useful methodology for designing databases is presented and illustrated through a variety of examples. In addition to the methodology, this text includes important design topics such as the use of survey forms, obtaining information by reviewing existing documents, special relationship considerations, and entity subtypes. Appendix A contains a comprehensive design example that illustrates how to apply the complete design process to a large and complex set of requirements. After mastering the design methodology presented in this text, students should be able to produce correct database designs for future database requirements they encounter.

Functions Provided by a Database Management System

With such a wide range of features included in current PC-based database management systems, it is important for students to know the functions that such systems should provide. These functions are presented and discussed in detail.

Database Administration

While database administration (DBA) is absolutely essential in the mainframe environment, it is also important in a personal computer environment, especially if the database is to be shared among several users. Thus, this text includes a detailed discussion of the database administration function.

Database Management System Selection

The process of selecting a database management system is important considering the number of available systems from which to choose. Unfortunately, selecting the correct database management system is not an easy task. To prepare students to be able to do an effective job in this area, the text includes a detailed discussion of the selection process together with a comprehensive checklist that greatly assists in making such a selection.

Advanced Topics

The text also covers distributed database management systems, client/server systems, data warehouses, object-oriented database management systems, and Web access to databases. Each of these topics encompasses an enormous amount of complex information, but the goal is to introduce students to these important topics.

Glossary

The glossary included at the end of this text contains definitions to the key terms in the chapters.

Numerous Realistic Examples

The book contains numerous examples illustrating each of the concepts. A running "case" example—Premiere Products—is used throughout the book to illustrate concepts. The examples are realistic and representative of the kinds of problems students will encounter in the design, manipulation, and administration of databases. Exercises that use the Premiere Products case are included at the end of each chapter. In addition, there is another complete set of exercises at the end of each chapter that feature a second case—Henry Books—giving students a chance to apply what they have learned to a database that they have not seen in the chapter material.

Review Material

This text contains a wide variety of questions. At key points within the chapters, students are asked questions to reinforce their understanding of the material before proceeding. The answers to these questions follow the questions. A summary and a list of key terms appear at the end of each chapter, followed by review questions that test the students' knowledge of the important points in the chapter and that occasionally test their ability to apply what they have learned. The answers to the odd-numbered review questions are provided in Appendix E. Each chapter also contains hands-on exercises related to the Premiere Products and Henry Books case examples.

Teaching Tools

When this book is used in an academic setting, instructors may obtain the following teaching tools from Course Technology through their sales representative or by visiting *www.course.com*:

- **Instructor's Manual** The Instructor's Manual has been carefully prepared and tested to ensure its accuracy and dependability. The Instructor's Manual includes suggestions and strategies for using this text, including the incorporation of companion texts on Access or SQL for those instructors who desire to do so. For instructors who want to use an Access or SQL text as a companion to the Fifth Edition, the Instructor's Manual for this book includes detailed tips on integrating the Fifth Edition with the following books, also published by Course Technology: *New Perspectives on Microsoft Office Access 2003—Comprehensive, New Perspectives on Microsoft Office Access 2003—Introductory,* and *New Perspectives on Microsoft Office Access 2003—Brief,* by Adamski and Finnegan; *Microsoft Office Access 2003: Introductory Concepts and Techniques, Microsoft Office Access 2003: Complete Concepts and Techniques,* and *Microsoft Office Access 2003: Comprehensive Concepts and Techniques,* by Shelly, Cashman, Pratt, and Last; and *A Guide to SQL, Sixth Edition,* by Pratt.

- **Data and Solution Files** Data and solution files are available at *www.course.com* and on the Instructor's Resource CD-ROM. Data files consist of copies of the Premiere Products, Henry Books, and Alexamara Marina Group databases in Access 2000 format and script files to create the tables and data in these databases in other systems, such as Oracle.

- **ExamView®** This text is accompanied by ExamView, a powerful testing software package that allows instructors to create and administer printed, computer (LAN-based), and Internet exams. ExamView includes hundreds of questions that correspond to the topics covered in this text, enabling students to generate detailed study guides that include page references for further review. The computer-based and Internet testing components allow students to take exams at their computers, and also save the instructor time by grading each exam automatically.

- **PowerPoint Presentations** Microsoft PowerPoint slides are included for each chapter as a teaching aid for classroom presentations, to make available to students on a network for chapter review, or to be printed for classroom distribution. Instructors can add their own slides for additional topics they introduce to the class. The presentations are included on the Instructor's CD.

- **Figure Files** Figure files are included so that instructors can create their own presentations using figures appearing in the text.

- **Distance Learning** Course Technology is proud to present online content in WebCT and Blackboard to provide the most complete and dynamic learning experience possible. For more information on how to bring distance learning to your course, contact your local Course Technology sales representative.

ORGANIZATION OF THE TEXTBOOK

This text includes nine chapters covering general database topics that are relevant to any database management system. A brief description of the organization of topics in the chapters and an overview of updates from the Fourth Edition follows.

Introduction

Chapter 1 provides a general introduction to the field of database management.

The Relational Model

The relational model is covered in detail in Chapters 2, 3, and 4. Chapter 2 covers the data definition and manipulation aspects of the model using QBE and relational algebra. The text uses Microsoft Access 2003 to illustrate the QBE material. The relational algebra section includes the entire relational algebra.

Note: The extra material on relational algebra is optional and can be omitted if desired.

Chapter 3 is devoted exclusively to SQL. The SQL material is illustrated using Access, but the chapter also includes generic versions of all examples that can be used with a variety of platforms, including Oracle.

Chapter 4 covers some advanced aspects of the relational model such as views, the use of indexes, the catalog, and relational integrity rules.

Database Design

Chapters 5 and 6 are devoted to database design. Chapter 5 covers the normalization process, which enables students to identify and correct bad designs. This chapter discusses and illustrates the use of first, second, third, and fourth normal forms.

Note: The material on fourth normal form is optional and can be omitted if desired.

Chapter 6 presents a methodology for database design using many examples. The material includes entity-relationship diagrams and their role in database design. It also includes discussions of several special design issues as well as the use of survey forms, obtaining information by reviewing existing documents, special relationship considerations, and entity subtypes. After completing Chapter 6, students can further challenge themselves by completing Appendix A, which includes a comprehensive design example that illustrates the application of the complete design process to a large and complex set of requirements.

Note: Chapters 5 and 6 can be covered immediately after Chapter 2 if desired.

Database Management System Functions

Chapter 7 discusses the features that should be provided by a full-functioned PC-based database management system. This chapter includes coverage of journaling, forward recovery, backward recovery, authentication, and authorizations.

Database Administration

Chapter 8 is devoted to the role of database administration. Also included in this chapter is a discussion of the process of selecting a database management system.

Database Management Approaches

Chapter 9 provides an overview of several advanced topics: distributed databases, client/server systems, data warehouses, object-oriented databases, and Web access to databases. This chapter has been expanded from the previous edition to include more details on client/server systems and data warehouses and a new section on UML.

Additional Case Problem

Appendix B contains an additional case problem, Alexamara Marina Group. The appendix introduces the database for Alexamara Marina Group and also includes additional exercises that can be assigned along with chapters 1 through 6.

GENERAL NOTES TO THE STUDENT

Embedded Questions

There are a number of places in the text where special questions have been embedded. Sometimes the purpose of these questions is to ensure that you understand some crucial material before you proceed. In other cases, the questions are designed to give you the chance to consider some special concept in advance of its actual presentation. In all cases, the answers to these questions follow each question. You could simply read the question and its answer. You will receive maximum benefit from the text, however, if you take the time to work out the answers to the questions and then check your answer against the one provided before continuing.

End-of-Chapter Material

The end-of-chapter material consists of a summary, a list of key terms, review questions, and exercises for the Premiere Products and Henry Books databases. The summary briefly describes the material covered in the chapter. The review questions require you to recall and apply the important material in the chapter. The answers to the odd-numbered review questions appear in Appendix E. The Premiere Products and Henry Books exercises test your knowledge of the chapter material; your instructor will assign one or more of these exercises for you to complete.

ACKNOWLEDGMENTS

We would like to acknowledge the following individuals who all made contributions during the preparation of this book. We appreciate the following individuals who reviewed the text and made many helpful suggestions: Shahed Mustafa, Long Island Business Institute—Flushing; Holly M. Blaskiewicz, Ivy Tech State College; Diane E. Del Monte, Briarcliffe College, Bethpage Campus; Aurea Janine Loveless, Eastern Iowa Community College; Dave Braunschweig, Harper College; Irene Zadnik, Saskatchewan Institute of Applied Science and Technology—Moose Jaw, Saskatchewan; James P. Verdon, Briarcliffe College; Barbara J. Nicolai, Purdue University—Calumet; Michael Chilton, Kansas State University—Manhattan; Angela Mattia, J. Sargeant Reynolds Community College—Parham Road; Risa Blair, Champlain College; Phil Chmielewski, Milwaukee Area Technical College; Q. Chung, Villanova University; Chris Davis, Baker College of Muskegon; Barbara Klein, University of Michigan—Dearborn; Charles Lamb, County College of Morris; Donna Meyerholz, Trinidad State Junior College; E. Jean Peterson,

Tillamook Bay Community College; Patty Santoianni, Sinclair Community College; Cherie Ann Sherman, Ramapo College of New Jersey; Rahul Tikekar, Southern Oregon University; Peter Van der Goes, Rose State College; William Wagner, Villanova University; and Amanda Whaley, Austin Community College—Rio Grande. The efforts of the following individuals have been invaluable: Kristen Duerr, Executive Vice President, Academic and Technology Publishing; Maureen Martin, Acquisitions Editor; Janet Aras, Product Manager; Nicole Ashton and Serge Palladino, Quality Assurance testers; Jessica Evans, Developmental Editor; Melissa Panagos, Production Editor; and Mirella Misiaszek, Associate Product Manager.

We have again been privileged to work with Jessica Evans as our Developmental Editor. Thank you, Jess, for your knowledge, skill, guidance, and energy, and for your hard work and positive influence on our work. We appreciate your many contributions to this book and value you as a friend.

CHAPTER 1

Introduction to Database Management

OBJECTIVES

- Introduce Premiere Products, the company that is used as the basis for many of the examples throughout the text.

- Introduce basic database terminology.

- Describe database management systems.

- Explain the advantages and disadvantages of database processing.

- Introduce Henry Books, the company that is used in the case that runs throughout the text.

Introduction

You begin this chapter by examining the requirements of Premiere Products, a company that will be used in many examples in this chapter and in the rest of the text. You will learn how Premiere Products initially stored its data, what problems employees encountered with this storage method, and why management decided to use a database management system. Then you will study the basic terminology and concepts of databases and database management systems, and learn the advantages and disadvantages of database processing. Finally, you will examine the database requirements for Henry Books, the company featured in the case that is used throughout the text.

▇ Premiere Products Background

Premiere Products is a distributor of appliances, housewares, and sporting goods. Since its inception, the company has used spreadsheet software to maintain customer, order, inventory, and sales representative (sales rep) data. Management has determined that the company's recent growth means that it is no longer feasible to use spreadsheets to maintain its data.

What has led the managers at Premiere Products to this decision? One of the company's spreadsheets, shown in Figure 1.1, displays sample order data and illustrates the company's problems with the spreadsheet approach. For each order, the spreadsheet displays the number and name of the customer placing the order; the number and date of the order; the number, description, number ordered, quoted price, and warehouse of the item ordered; and the number of the sales rep assigned to the customer. Note that Ferguson's order, order number 21610, and the first Johnson's Department Store order, order number 21617, appear in two rows because these customers purchased two different items in those orders.

FIGURE 1.1
....
Sample orders
spreadsheet

Customer Number	Customer Name	Order Number	Order Date	Part Number	Part Description	Number Ordered	Quoted Price	Warehouse	Rep Number
148	Al's Appliance and Sport	21608	10/20/2007	AT94	Iron	11	$21.95	3	20
148	Al's Appliance and Sport	21619	10/23/2007	DR93	Gas Range	1	$495.00	2	20
282	Brookings Direct	21614	10/21/2007	KT03	Dishwasher	2	$595.00	3	35
356	Ferguson's	21610	10/20/2007	DR93	Gas Range	1	$495.00	2	65
356	Ferguson's	21610	10/20/2007	DW11	Washer	1	$399.99	3	65
408	The Everything Shop	21613	10/21/2007	KL62	Dryer	4	$329.95	1	35
608	Johnson's Department Store	21617	10/23/2007	BV06	Home Gym	2	$794.95	2	65
608	Johnson's Department Store	21617	10/23/2007	CD52	Microwave Oven	4	$150.00	1	65
608	Johnson's Department Store	21623	10/23/2007	KV29	Treadmill	2	$1,290.00	2	65

Orders requiring more than one spreadsheet row

Redundancy is one problem that employees have with the orders spreadsheet. **Redundancy** is the duplication of data or the storing of the same data in more than one place. In the orders spreadsheet, redundancy occurs in the Customer Name column because the name of a customer is stored in more than one place. All three rows for customer number 608, for example, store "Johnson's Department Store" as the customer name. In the orders spreadsheet, redundancy also occurs in other columns, such as the Order Date and Part Description columns.

....

Q UESTION

ANSWER

What problems does redundancy cause?

Redundancy wastes space because you're storing the same data in multiple places. The extra space results in larger spreadsheets that require more space in memory and on disk and that take longer to save and open.

When you need to change data, redundancy also makes your changes more cumbersome and time-consuming. For example, if you incorrectly entered "Johnson's Department Store" in the Customer Name column, you'd need to correct it in three places. Even if you use the global find-and-replace feature, multiple changes require more computer time than does a single change.

Finally, redundancy can lead to inconsistencies. For example, you might enter "Johnson's Department Store," "Johnsons Department Store," and "Johnsons' Department Store" in the Customer Name column, and then not be sure which is the correct spelling. Further, if that customer's name is spelled three different ways and you use the search feature with one of the three values, you'd find a single match instead of three matches.

....

Difficulty accessing related data is another problem that employees at Premiere Products encounter with their spreadsheets. For example, if you want to see a customer's address and a part's standard price, you must open and search other spreadsheets that contain this data.

Spreadsheets also have limited security features to protect data from access by unauthorized users. A spreadsheet's data-sharing features also prevent multiple employees from updating data in one spreadsheet at the same time. Finally, if the increase in sales and growth at Premiere Products continues at its planned rate, spreadsheets have inherent size limitations that will eventually force the company to split its order data into multiple spreadsheets. Splitting the spreadsheets would create further redundancy.

Having decided to replace its spreadsheet software, management has determined that Premiere Products must maintain the following information about its sales reps, customers, and parts inventory:

- Premiere Products must store the sales rep number, last name, first name, address, total commission, and commission rate for each sales rep.

- The company must store the customer number, name, address, current balance, and credit limit for each customer, as well as the number of the sales rep who represents the customer.

- The company must store the part number, description, number of units on hand, item class, number of the warehouse where the item is stored, and unit price for each part in inventory.

Premiere Products must also store information about orders. Figure 1.2 shows a sample order.

FIGURE 1.2
....
Sample order

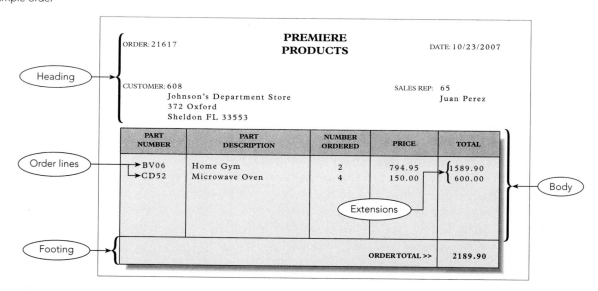

The sample order has three components:

- The heading (top) of the order contains the order number and date; the customer's number, name, and address; and the sales rep number and name.

- The body of the order contains one or more order lines, sometimes called line items. Each order line contains a part number, a part description, the number of units of the part ordered, and the quoted price for the part. Each order line also contains a total, usually called an extension, which is the result of multiplying the number ordered by the quoted price.

- Finally, the footing (bottom) of the order contains the order total.

Premiere Products must also store the following items for each customer's order:

- For each order, it must store the order number, the date the order was placed, and the number of the customer that placed the order. The customer's name and address and the number of the sales rep who represents the customer are stored with customer information. The name of the sales rep is stored with the sales rep information.

- For each order line, it must store the order number, the part number, the number of units ordered, and the quoted price. Remember that the part description is stored with the information about parts. The result of multiplying the number of units ordered by the quoted price is not stored because the computer can calculate it when necessary.

- The overall order total is not stored. Instead, the computer calculates the total whenever an order is printed or displayed on the screen.

The problem facing Premiere Products is common to many businesses and individuals that need to store and retrieve data in an efficient and organized way. Furthermore, most organizations are interested in more than one category of information. For example, Premiere Products is interested in categories such as sales reps, customers, orders, and parts. A school is interested in students, faculty, and classes; a real estate agency is interested in clients, houses, and agents; and a used car dealership is interested in customers, vehicles, and manufacturers.

Besides wanting to store data that pertains to more than one category, Premiere Products is also interested in the relationships between the categories. For example, company employees want to be able to associate orders with the customers that ordered them, the sales reps who coordinated them, and the parts that the customers requested. Likewise, a real estate agency wants to know not only about clients, houses, and agents, but also about the relationships between clients and houses (which clients have listed which houses and which clients have expressed interest in which houses). It also wants to know about the relationships between agents and houses (which agent sold which house) and so on.

Database Background

After studying the alternatives to using spreadsheet software, Premiere Products decided to switch to a database system. A database is a structure that contains information about many different categories of information and about the relationships between these categories. The Premiere Products database, for example, will contain information about sales reps, customers, orders, and parts. It will also provide facts that relate sales reps to the customers they represent and customers to the orders they currently have placed.

With the use of a database, employees can enter the number of a particular order and find out which customer placed the order, as well as which parts were ordered. Alternately, employees can start with a customer and find all the orders that the customer placed, together with which parts were ordered and the amount of the commission earned by the customer's sales rep. Using a database, Premiere Products can not only maintain its data better, but it can use the data in the database to produce a variety of reports and to answer a variety of questions.

There are some terms and concepts that are important for you to know in the database environment. For instance, the terms *entity*, *attribute*, and *relationship* are fundamental when discussing databases. An **entity** is a person, place, object, event, or idea for which you want to store and process data. The entities of interest to Premiere Products, for example, are sales reps, customers, orders, and parts.

An **attribute** is a characteristic or property of an entity. The term is used in this text exactly as it is used in everyday English. For the entity person, for example, the list of attributes might include such things as eye color and height. For Premiere Products, the attributes of interest for the entity customer are such things as customer name, street, city, and so on. An attribute is also called a **field** or **column** in many database systems.

Figure 1.3 shows two entities, Rep (short for Sales Rep) and Customer, along with the attributes for each entity. The Rep entity has nine attributes: RepNum, LastName, FirstName, Street, City, State, Zip, Commission, and Rate. The attributes are the same as the columns in a spreadsheet. The Customer entity has nine attributes: CustomerNum, CustomerName, Street, City, State, Zip, Balance, CreditLimit, and RepNum.

FIGURE 1.3
· · · · · · · · · · · · · · · · · ·
Entities and
attributes

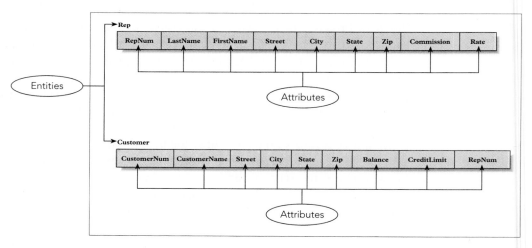

The final key term is relationship. A **relationship** is an association between entities. There is an association between reps and customers, for example, at Premiere Products. A rep is associated with all of his or her customers, and a customer is associated with its rep. Technically, you say that a rep is *related to* all of his or her customers, and a customer is *related to* its rep.

6

This particular relationship is called a **one-to-many relationship** because each rep is associated with *many* customers, but each customer is associated with only *one* rep. In this type of relationship, the word *many* is used differently than in everyday English; it might not always indicate a large number. In this context, for example, the term *many* means that a rep can be associated with *any* number of customers. That is, a given rep can be associated with zero, one, or more customers.

A one-to-many relationship often is represented visually in the manner shown in Figure 1.4. In such a diagram, entities and attributes are represented in precisely the same way as they are shown in Figure 1.3. A line connecting the entities represents the relationship. The *one* part of the relationship, in this case Rep, does not have an arrow on its end of the line, and the *many* part of the relationship, in this case Customer, is indicated by a single-headed arrow.

FIGURE 1.4
••••••••••••••••••
One-to-many
relationship

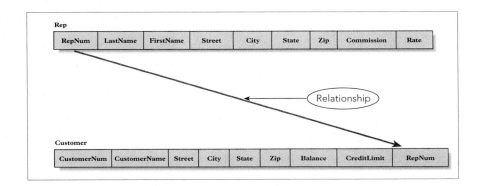

Spreadsheets, word-processed documents, Web pages, and other computer information sources are stored in files. Basically, a file that is used to store data, which is often called a **data file**, is the computer counterpart to an ordinary paper file you might keep in a file cabinet, an accounting ledger, or other place. A database, however, is much more than a file. Unlike a typical data file, a database can store information about multiple entities. There is also another difference. A database holds information about the relationships among the various entities. Not only will the Premiere Products database have information about both reps and customers, it also will hold information relating reps to the customers they service, customers to orders, parts to orders, and so on. Formally, the definition of a database is as follows:

Definition: A **database** is a structure that can store information about multiple types of entities, the attributes of those entities, and the relationships among the entities.

How does a database handle entities, attributes of entities, and relationships between entities? Entities and attributes are fairly simple. Each entity has its own table. In the Premiere Products database, for example, there will be one table for reps, one table for customers, and so on. The attributes of an entity become the columns in the table. In the table for reps, for example, there will be a column for the rep number, a column for the rep last name, and so on.

What about relationships between entities? At Premiere Products, there is a one-to-many relationship between reps and customers (each rep is related to the many customers that he or she represents, and each customer is related to the one rep who represents the customer). How is this relationship handled in a database system? It is handled by using common columns in the two tables. Consider Figure 1.4 again. The RepNum column in the Rep table and the RepNum column in the Customer table are used to implement the relationship between reps and customers. Given a rep, you can use these columns to determine all the customers that he or she represents; given a customer, you can use these columns to find the rep who represents the customer.

How will Premiere Products store its data in a database? Figure 1.5 shows sample data for Premiere Products.

FIGURE 1.5
••••••••••••••••
Sample data for
Premiere Products

Rep

RepNum	LastName	FirstName	Street	City	State	Zip	Commission	Rate
20	Kaiser	Valerie	624 Randall	Grove	FL	33321	$20,542.50	0.05
35	Hull	Richard	532 Jackson	Sheldon	FL	33553	$39,216.00	0.07
65	Perez	Juan	1626 Taylor	Fillmore	FL	33336	$23,487.00	0.05

Customer

CustomerNum	CustomerName	Street	City	State	Zip	Balance	CreditLimit	RepNum
148	Al's Appliance and Sport	2837 Greenway	Fillmore	FL	33336	$6,550.00	$7,500.00	20
282	Brookings Direct	3827 Devon	Grove	FL	33321	$431.50	$10,000.00	35
356	Ferguson's	382 Wildwood	Northfield	FL	33146	$5,785.00	$7,500.00	65
408	The Everything Shop	1828 Raven	Crystal	FL	33503	$5,285.25	$5,000.00	35
462	Bargains Galore	3829 Central	Grove	FL	33321	$3,412.00	$10,000.00	65
524	Kline's	838 Ridgeland	Fillmore	FL	33336	$12,762.00	$15,000.00	20
608	Johnson's Department Store	372 Oxford	Sheldon	FL	33553	$2,106.00	$10,000.00	65
687	Lee's Sport and Appliance	282 Evergreen	Altonville	FL	32543	$2,851.00	$5,000.00	35
725	Deerfield's Four Seasons	282 Columbia	Sheldon	FL	33553	$248.00	$7,500.00	35
842	All Season	28 Lakeview	Grove	FL	33321	$8,221.00	$7,500.00	20

Orders

OrderNum	OrderDate	CustomerNum
21608	10/20/2007	148
21610	10/20/2007	356
21613	10/21/2007	408
21614	10/21/2007	282
21617	10/23/2007	608
21619	10/23/2007	148
21623	10/23/2007	608

(continued)

OrderLine

OrderNum	PartNum	NumOrdered	QuotedPrice
21608	AT94	11	$21.95
21610	DR93	1	$495.00
21610	DW11	1	$399.99
21613	KL62	4	$329.95
21614	KT03	2	$595.00
21617	BV06	2	$794.95
21617	CD52	4	$150.00
21619	DR93	1	$495.00
21623	KV29	2	$1,290.00

Part

PartNum	Description	OnHand	Class	Warehouse	Price
AT94	Iron	50	HW	3	$24.95
BV06	Home Gym	45	SG	2	$794.95
CD52	Microwave Oven	32	AP	1	$165.00
DL71	Cordless Drill	21	HW	3	$129.95
DR93	Gas Range	8	AP	2	$495.00
DW11	Washer	12	AP	3	$399.99
FD21	Stand Mixer	22	HW	3	$159.95
KL62	Dryer	12	AP	1	$349.95
KT03	Dishwasher	8	AP	3	$595.00
KV29	Treadmill	9	SG	2	$1,390.00

In the Rep table, you see that there are three reps whose numbers are 20, 35, and 65. The name of sales rep 20 is Valerie Kaiser. Her street address is 624 Randall. She lives in Grove, FL, and her zip code is 33321. Her total commission is $20,542.50, and her commission rate is 5% (0.05).

Premiere Products has 10 customers that are identified with the numbers 148, 282, 356, 408, 462, 524, 608, 687, 725, and 842. The name of customer number 148 is Al's Appliance and Sport. This customer's address is 2837 Greenway in Fillmore, FL, with a zip code of 33336. The customer's current balance is $6,550.00, and its credit limit is $7,500. The number 20 in the RepNum column indicates that Al's Appliance and Sport is represented by sales rep 20 (Valerie Kaiser).

Skipping to the table named Part, you see that there are 10 parts, whose part numbers are AT94, BV06, CD52, DL71, DR93, DW11, FD21, KL62, KT03, and KV29. Part AT94 is an iron and the company has 50 units of this part on hand. Irons are in item class HW (housewares) and are stored in warehouse 3. The price of an iron is $24.95. Other item classes are AP (appliances) and SG (sporting goods).

Moving back to the table named Orders, you see that there are seven orders, which are identified with the numbers 21608, 21610, 21613, 21614, 21617, 21619, and 21623. Order number 21608 was placed on October 20, 2007 by customer 148 (Al's Appliance and Sport).

Note: In some database systems, the word "Order" has a special purpose. Having a table named Order could cause problems in such systems. For this reason, Premiere Products uses the table name Orders instead of Order.

The table named OrderLine might seem strange at first glance. Why do you need a separate table for the order lines? Couldn't the order lines be included in the Orders table? The answer is yes. The Orders table could be structured as shown in Figure 1.6. Notice that this table contains the same orders as shown in Figure 1.5, with the same dates and customers. In addition, each table row in Figure 1.6 contains all the order lines for a given order. Examining the fifth row, for example, you see that order 21617 has two order lines. One of these order lines is for two BV06 parts at $794.95 each, and the other order line is for four CD52 parts at $150.00 each.

FIGURE 1.6
••••••••••••••••••
Alternative Orders
table structure

Orders

OrderNum	OrderDate	CustomerNum	PartNum	NumOrdered	QuotedPrice
21608	10/20/2007	148	AT94	11	$21.95
21610	10/20/2007	356	DR93	1	$495.00
			DW11	1	$399.99
21613	10/21/2007	408	KL62	4	$329.95
21614	10/21/2007	282	KT03	2	$595.00
21617	10/23/2007	608	BV06	2	$794.95
			CD52	4	$150.00
21619	10/23/2007	148	DR93	1	$495.00
21623	10/23/2007	608	KV29	2	$1,290.00

QUESTION

ANSWER

How is the information from Figure 1.5 represented in Figure 1.6?

Examine the OrderLine table shown in Figure 1.5 and note the sixth and seventh rows. The sixth row indicates that there is an order line in order 21617 for two BV06 parts at $794.95 each. The seventh row indicates that there is an order line in order 21617 for four CD52 parts at $150.00 each. Thus, the information that you find in Figure 1.6 is represented in Figure 1.5 in two separate rows rather than in one row.

It might seem inefficient to use two rows to store information that can be represented in one row. There is a problem, however, with the arrangement shown in Figure 1.6—the table is more complicated. In Figure 1.5, there is a single entry at each position in the table. In Figure 1.6, some of the individual positions within the table contain multiple entries, thus making it difficult to track the information between columns. In the row for order number 21617, for example, it is crucial to know that the BV06 corresponds to the 2 in the NumOrdered column (not the 4) and that it corresponds to the $794.95 in the QuotedPrice column (not the $150.00). In addition, having a more complex table means that there are practical issues to worry about, such as:

- How much room do you allow for these multiple entries?

- What if an order has more order lines than you have allowed room for?

- Given a part, how do you determine which orders contain order lines for that part?

Certainly, none of these problems is unsolvable. These problems do add a level of complexity, however, that is not present in the arrangement shown in Figure 1.5. In Figure 1.5, there are no multiple entries to worry about, it doesn't matter how many order lines exist for any order, and finding every order that contains an order line for a given part is easy (just look for all order lines with the given part number in the PartNum column). In general, this simpler structure is preferable, and that is why order lines appear in a separate table.

To test your understanding of the Premiere Products data, use the data shown in Figure 1.5 to answer the following questions.

QUESTION What are the numbers of the customers represented by Valerie Kaiser?

ANSWER 148, 524, and 842. (Look up the RepNum value for Valerie Kaiser in the Rep table and obtain the number 20. Then find all customers in the Customer table that have the number 20 in the RepNum column.)

QUESTION What is the name of the customer that placed order 21610, and what is the name of the rep who represents this customer?

ANSWER Ferguson's is the customer, and Juan Perez is the rep. (Look up the CustomerNum value in the Orders table for order number 21610 and obtain the number 356. Then find the customer in the Customer table with a CustomerNum value of 356. Using this customer's RepNum value, which is 65, find the name of the rep in the Rep table.)

QUESTION List all the parts that appear in order 21610. For each part, give the description, number ordered, and quoted price.

ANSWER Part number: DR93; part description: Gas Range; number ordered: 1; and quoted price: $495.00. Also, part number: DW11; part description: Washer; number ordered: 1; and quoted price: $399.99. (Look up each OrderLine table row in which the order number is 21610. Each of these rows contains a part number, the number ordered, and the quoted price. Use the part number to look up the corresponding description in the Part table.)

QUESTION Why is the QuotedPrice column part of the OrderLine table? Can't you just use the part number to look up the price in the Part table?

ANSWER If the QuotedPrice column didn't appear in the OrderLine table, you would need to obtain the price for a part on an order line by looking up the price in the Part table. Although this might not be a bad practice, it prevents Premiere Products from charging different prices to different customers for the same part. Because Premiere Products wants the flexibility to quote and charge different prices to different customers, the QuotedPrice column is included in the OrderLine table. If you examine the OrderLine table, you will see cases in which the quoted price matches the actual price in the Part table and cases in which it differs. For example, in order number 21608, Al's Appliance and Sport bought 11 irons, and Premiere Products charged only $21.95 per iron, rather than the regular price of $24.95.

A visual way to represent a database is with an **entity-relationship (E-R) diagram**. In an E-R diagram, rectangles represent entities, and lines represent relationships between connected entities. The E-R diagram for the Premiere Products database appears in Figure 1.7.

FIGURE 1.7
· · · · · · · · · · · · · · · · · · · ·
E-R diagram for the
Premiere Products
database

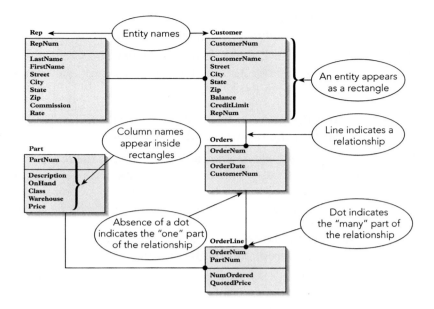

Each of the five entities in the Premiere Products database appears as a rectangle in the E-R diagram shown in Figure 1.7. The name of each entity appears above the rectangle. The columns that comprise each entity appear within the rectangle. Because the Rep and Customer entities have a one-to-many relationship, a line connects these two entities; similarly, a line connects the Customer and Orders entities, the Orders and OrderLine entities, and the Part and OrderLine entities. The dot at the end of a line, such as the dot at the Customer end of the line that connects the Rep and Customer entities, indicates the "many" part of the one-to-many relationship between two entities. You will learn more about E-R diagrams in Chapter 6.

■ Database Management Systems

Managing a database is inherently a complicated task. Fortunately, software packages, called database management systems, can do the job of manipulating databases for us. A **database management system (DBMS)** is a program, or a collection of programs, through which users interact with a database. The actual manipulation of the underlying database is handled by the DBMS. In some cases, users may interact with the DBMS directly, as shown in Figure 1.8.

FIGURE 1.8
· · · · · · · · · · · · · · · · · · ·
Using a DBMS
directly

In other cases, users may instead interact with programs, such as those created with Visual Basic, Java, or C++; these programs in turn interact with the DBMS, as shown in Figure 1.9. In either case, it is only the DBMS that actually accesses the database.

FIGURE 1.9
••••••••••••••••
Using a DBMS
through another
program

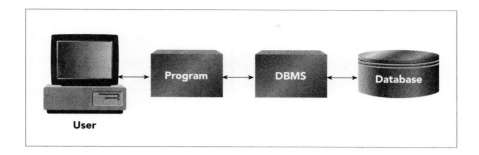

With a DBMS, for example, users at Premiere Products can ask the system to find data about part KV29, and the system will either locate this part and provide the data or display a message that no such part exists in the database. All the work involved in this task is performed by the DBMS. If part KV29 is in the database, users then can ask for the order lines that contain this part, and again the system will perform all the work involved in locating the order lines. Likewise, when users add data about a new customer to the database, the DBMS performs all the tasks necessary to ensure that the customer data is added and that the customer is related to the appropriate rep.

Popular DBMSs include Access, Oracle, DB2, MySQL, and SQL Server. Because Premiere Products uses the Microsoft Office suite of programs, which includes Access, management elects to use Access initially as its DBMS. Using the tables shown in Figure 1.5 as the starting point, a computer expert at Premiere Products determines the structure of the required database—this process is called **database design**. Then this person enters the design in the DBMS and creates several **forms** (which are screen objects used to maintain, view, and print data from a database). Employees then begin to enter data.

The form that employees use to process part data is shown in Figure 1.10. Using this form, employees can enter a new part; view, change, or delete an existing part; and print the information for a part. No one at Premiere Products needs to write a program to create this form; instead, the DBMS creates the form based on answers provided in response to the DBMS's questions about the form's content and appearance.

FIGURE 1.10
••••••••••••••••
Part form

The DBMS creates the other forms Premiere Products needs in this same way. A more complicated form for processing order data is shown in Figure 1.11. This form displays data about an order and its order lines, using data from the Orders table and related data from the OrderLine table.

FIGURE 1.11
....
Orders form

Premiere Products can create the reports it needs in a similar way—the DBMS asks questions about the desired content and appearance of each report and then creates these reports automatically based on the answers. The Parts report, listing each part in stock, is shown in Figure 1.12.

FIGURE 1.12
....
Parts report

Parts

PartNum	Description	OnHand	Class	Warehouse	Price
AT94	Iron	50	HW	3	$24.95
BV06	Home Gym	45	SG	2	$794.95
CD52	Microwave Oven	32	AP	1	$165.00
DL71	Cordless Drill	21	HW	3	$129.95
DR93	Gas Range	8	AP	2	$495.00
DW11	Washer	12	AP	3	$399.99
FD21	Stand Mixer	22	HW	3	$159.95
KL62	Dryer	12	AP	1	$349.95
KT03	Dishwasher	8	AP	3	$595.00
KV29	Treadmill	9	SG	2	$1,390.00

Finally, Premiere Products creates a **switchboard** system—a set of special forms used to provide controlled access to the data, forms, reports, and other objects in a database. The switchboard system contains a main switchboard, shown in Figure 1.13, which appears when users open the database, and two other switchboards: "Maintain Data" and "Produce Reports."

FIGURE 1.13
••••••••••••••••••
Main Switchboard
form

When a Premiere Products user chooses the "Maintain Data" option on the main switchboard, the switchboard shown in Figure 1.14 appears. Depending on which data the user wants to maintain in the database, he or she chooses the corresponding option on this switchboard.

FIGURE 1.14
••••••••••••••••••
Maintain Data
switchboard

■ Advantages of Database Processing

The database approach to processing offers nine clear advantages over alternate data management methods. These advantages are listed in Figure 1.15 and are discussed on the following pages.

FIGURE 1.15
••••••••••••••••••
Advantages of
database
processing

1. **Getting more information from the same amount of data**
2. **Sharing data**
3. **Balancing conflicting requirements**
4. **Controlling redundancy**
5. **Facilitating consistency**
6. **Improving integrity**
7. **Expanding security**
8. **Increasing productivity**
9. **Providing data independence**

1. *Getting more information from the same amount of data.* The primary goal of a computer system is to turn data (recorded facts) into information (the knowledge gained by processing those facts). In a nondatabase, file-oriented environment, data often is partitioned into several disjointed systems, with each system having its own collection of files. Any request for information that necessitates accessing data from more than one of these collections can be extremely difficult. In some cases, for all practical purposes, it is impossible. Thus, the desired information is unavailable—it has been stored in the computer but it is scattered across multiple files. When all the data for the various systems is stored in a single database, however, the information becomes available. Given the power of a DBMS, the information is available, and the process of getting it is quick and easy.

2. *Sharing data.* The data of various users can be combined and shared among authorized users, allowing all users access to a greater pool of data. Several users can have access to the same piece of data—for example, a customer's address—and still use it in a variety of ways. When one user changes a customer's address, the new address immediately becomes available to all users. In addition, the existing data can be used in new ways, such as generating new types of reports, without having to create additional data files, as is the case in the nondatabase approach.

3. *Balancing conflicting requirements.* For the database approach to function adequately within an organization, a person or group should be in charge of the database itself, especially if the database will serve many users. This person or group is often called the **database administrator** or **database administration (DBA)**. By keeping the overall needs of the organization in mind, a DBA can structure the database in such a way that it benefits the entire organization, not just a single group. Although this approach might mean that an individual user group is served less well than it would have been if it had its own isolated system, the organization as a whole is better off. Ultimately, when the organization benefits, so do the individual groups of users.

4. *Controlling redundancy.* With database processing, data that formerly was kept separate in a nondatabase, file-oriented system is integrated into a single database, so multiple copies of the same data no longer exist. With the nondatabase approach, each user group at Premiere Products has its own copy of each customer's address. With the database approach, each customer's address would occur only once, thus eliminating redundancy.

16

Eliminating redundancy not only saves space, but also makes the process of updating data much simpler. With the database approach, changing a customer's address means making one change. With the nondatabase approach, in which data for each customer might be stored in three different places, the same address change means that three changes have to be made.

Although eliminating redundancy is the ideal, it is not always possible. Sometimes, for reasons having to do with performance, you might choose to introduce a limited amount of redundancy into a database. However, even in these cases, you would be able to keep the redundancy under tight control, thus obtaining the same advantages. This is why it is better to say that you *control* redundancy rather than *eliminate* it.

5. *Facilitating consistency.* Suppose an individual customer's address appears in more than one place. Customer 148, for example, might be listed at 2837 Greenway in one place and at 2856 Wisner in another place. In this case, the data for this customer is inconsistent. Because the potential for this sort of problem is a direct result of redundancy, and because the database approach reduces redundancy, there is much less potential for the occurrence of this sort of inconsistency with the database approach.

6. *Improving integrity.* An **integrity constraint** is a rule that must be followed by data in the database. For example, the rep number given for any customer must be one that is already in the database. In other words, users cannot enter an incorrect or nonexistent rep number for a customer. A database has **integrity** if the data in it satisfies all established integrity constraints. A good DBMS should provide an opportunity for users to incorporate these integrity constraints when they design the database. The DBMS then should ensure that these constraints are not violated. According to the integrity constraint about customers, the DBMS should *not allow* you to store data about a given customer if the rep number that you enter is not the number of a rep that already is in the database.

7. *Expanding security.* **Security** is the prevention of access to the database by unauthorized users. A DBMS has many features that help ensure the enforcement of security measures. For example, a DBA can assign passwords to authorized users, and only those users who enter an acceptable password can gain access to the data in the database. Further, a DBMS lets you assign users to groups, with some groups permitted to view and update data in the database and other groups permitted to view only certain data in the database. With the nondatabase approach, you have limited security features and are more vulnerable to intentional and accidental access and changes to data.

8. *Increasing productivity.* A DBMS frees the programmers who are writing database access programs from having to engage in mundane data manipulation activities, such as adding new data and deleting existing data, thus making the programmers more productive. A good DBMS has many features that allow users to gain access to data in a database without having to do any programming at all. These features increase the productivity both of programmers, who may not need to write complex programs in order to perform certain tasks, and of nonprogrammers, who may be able to get the results they seek from the data in a database without waiting for a program to be written for them.

9. **Providing data independence.** The structure of a database often needs to be changed. For example, changing user requirements might necessitate the addition of an entity, an attribute, or a relationship, or a change might be required to improve performance. A good DBMS provides **data independence**, which is a property that lets you change the structure of a database without requiring you to change the programs that access the database; examples of these programs are the forms you use to interact with the database and the reports that provide information from the database. Without data independence, a lot of effort can be expended in changing programs to match the new database structure. The presence of many programs in the system may make this effort so prohibitive that management might decide to avoid changing the database, even though the change might improve the database's performance or add valuable data. With data independence, management is more likely to make the decision to change the database.

■ Disadvantages of Database Processing

As you would expect, if there are advantages to doing something in a certain way, there are also disadvantages. Database processing is no exception. In terms of numbers alone, the advantages outweigh the disadvantages; the latter are listed in Figure 1.16 and explained next.

FIGURE 1.16
••••••••••••••••••
Disadvantages of
database
processing

1. **Larger file size**
2. **Increased complexity**
3. **Greater impact of failure**
4. **More difficult recovery**

1. **Larger file size.** In order to support all the complex functions that it provides to users, a DBMS must be a large program that occupies a great amount of disk space as well as a substantial amount of internal memory. In addition, because all the data that the database manages for you is stored in one file, the database file itself requires a large amount of disk space and internal memory.

2. **Increased complexity.** The complexity and breadth of the functions provided by a DBMS make it a complex product. Users of the DBMS must learn a great deal to understand the features of the system in order to take full advantage of it. In the design and implementation of a new system that uses a DBMS, many choices have to be made; it is possible to make incorrect choices, especially with an insufficient understanding of the system. Unfortunately, a few incorrect choices can spell disaster for the whole project. A sound database design is critical to the successful use of a DBMS.

3. **Greater impact of failure.** In a nondatabase, file-oriented system, each user has a completely separate system; the failure of any single user's system does not necessarily affect any other user. On the other hand, if several users are sharing the

18

same database, a failure on the part of any one user that damages the database in some way might affect all the other users.

4. ***More difficult recovery.*** Because a database inherently is more complex than a simple file, the process of recovering it in the event of a catastrophe also is more complicated. This is particularly true if the database is being updated by many users at the same time. The database must first be restored to the condition it was in when it was last known to be correct; any updates made by users since that time must be redone. The greater the number of users involved in updating the database, the more complicated this task becomes.

■ Introduction to the Henry Books Database Case

Similar to the management of Premiere Products, Ray Henry, the owner of a bookstore chain named Henry Books, has decided to store his data in a database. He wants to achieve the same benefits; that is, he wants to ensure that his data is current and accurate. He also needs to create forms to interact with the data and to produce reports from that data. In addition, he wants to be able to ask questions concerning the data and to obtain answers to these questions easily and quickly.

In running his chain of bookstores, Ray gathers and organizes information about branches, publishers, authors, and books. Figure 1.17 shows sample branch and publisher data for Henry Books. Each branch has a number that uniquely identifies the branch. In addition, Ray tracks the branch's name, location, and number of employees. Each publisher has a code that uniquely identifies the publisher. In addition, Ray tracks the publisher's name and city.

FIGURE 1.17
• • • • • • • • • • • • • • • • •
Sample branch and
publisher data for
Henry Books

Branch

BranchNum	BranchName	BranchLocation	NumEmployees
1	Henry Downtown	16 Riverview	10
2	Henry On The Hill	1289 Bedford	6
3	Henry Brentwood	Brentwood Mall	15
4	Henry Eastshore	Eastshore Mall	9

Publisher

PublisherCode	PublisherName	City
AH	Arkham House	Sauk City WI
AP	Arcade Publishing	New York
BA	Basic Books	Boulder CO
BP	Berkley Publishing	Boston
BY	Back Bay Books	New York
CT	Course Technology	Boston
FA	Fawcett Books	New York
FS	Farrar Straus & Giroux	New York
HC	HarperCollins Publishers	New York
JP	Jove Publications	New York
JT	Jeremy P. Tarcher	Los Angeles
LB	Lb Books	New York
MP	McPherson and Co.	Kingston
PE	Penguin USA	New York
PL	Plume	New York
PU	Putnam Publishing Group	New York
RH	Random House	New York
SB	Schoken Books	New York
SC	Scribner	New York
SS	Simon & Schuster	New York
ST	Scholastic Trade	New York
TA	Taunton Press	Newtown CT
TB	Tor Books	New York
TH	Thames and Hudson	New York
TO	Touchstone Books	Westport CT
VB	Vintage Books	New York
WN	W.W. Norton	New York
WP	Westview Press	Boulder CO

Figure 1.18 shows sample author data for Henry Books. Each author has a number that uniquely identifies the author. In addition, Ray records each author's last and first names.

FIGURE 1.18
· · · · · · · · · · · · · · · · · ·
Sample author data
for Henry Books

Author

AuthorNum	AuthorLast	AuthorFirst
1	Morrison	Toni
2	Solotaroff	Paul
3	Vintage	Vernor
4	Francis	Dick
5	Straub	Peter
6	King	Stephen
7	Pratt	Philip
8	Chase	Truddi
9	Collins	Bradley
10	Heller	Joseph
11	Wills	Gary
12	Hofstadter	Douglas R.
13	Lee	Harper
14	Ambrose	Stephen E.
15	Rowling	J.K.
16	Salinger	J.D.
17	Heaney	Seamus
18	Camus	Albert
19	Collins, Jr.	Bradley
20	Steinbeck	John
21	Castelman	Riva
22	Owen	Barbara
23	O'Rourke	Randy
24	Kidder	Tracy
25	Schleining	Lon

Figure 1.19 shows sample book data for Henry Books. Each book has a code that uniquely identifies the book. For each book, Ray also tracks the title, publisher, type of book, price, and whether the book is a paperback.

FIGURE 1.19
••••••••••••••••••
Sample book data
for Henry Books

Book

BookCode	Title	PublisherCode	Type	Price	Paperback
0180	A Deepness in the Sky	TB	SFI	7.19	Yes
0189	Magic Terror	FA	HOR	7.99	Yes
0200	The Stranger	VB	FIC	8.00	Yes
0378	Venice	SS	ART	24.50	No
079X	Second Wind	PU	MYS	24.95	No
0808	The Edge	JP	MYS	6.99	Yes
1351	Dreamcatcher: A Novel	SC	HOR	19.60	No
1382	Treasure Chests	TA	ART	24.46	No
138X	Beloved	PL	FIC	12.95	Yes
2226	Harry Potter and the Prisoner of Azkaban	ST	SFI	13.96	No
2281	Van Gogh and Gauguin	WP	ART	21.00	No
2766	Of Mice and Men	PE	FIC	6.95	Yes
2908	Electric Light	FS	POE	14.00	No
3350	Group: Six People in Search of a Life	BP	PSY	10.40	Yes
3743	Nine Stories	LB	FIC	5.99	Yes
3906	The Soul of a New Machine	BY	SCI	11.16	Yes
5163	Travels with Charley	PE	TRA	7.95	Yes
5790	Catch-22	SC	FIC	12.00	Yes
6128	Jazz	PL	FIC	12.95	Yes
6328	Band of Brothers	TO	HIS	9.60	Yes
669X	A Guide to SQL	CT	CMP	37.95	Yes
6908	Franny and Zooey	LB	FIC	5.99	Yes
7405	East of Eden	PE	FIC	12.95	Yes
7443	Harry Potter and the Goblet of Fire	ST	SFI	18.16	No
7559	The Fall	VB	FIC	8.00	Yes
8092	Godel, Escher, Bach	BA	PHI	14.00	Yes
8720	When Rabbit Howls	JP	PSY	6.29	Yes
9611	Black House	RH	HOR	18.81	No
9627	Song of Solomon	PL	FIC	14.00	Yes
9701	The Grapes of Wrath	PE	FIC	13.00	Yes
9882	Slay Ride	JP	MYS	6.99	Yes
9883	The Catcher in the Rye	LB	FIC	5.99	Yes
9931	To Kill a Mockingbird	HC	FIC	18.00	No

QUESTION To check your understanding of the relationship between publishers and books, answer the following questions: Who published *Jazz*? Which books did Jove Publications publish?

ANSWER Plume published *Jazz*. In the row in the Book table for *Jazz* (see Figure 1.19), find the publisher code PL. Examining the Publisher table (see Figure 1.17), you see that PL is the code assigned to Plume.

Jove Publications published *The Edge*, *When Rabbit Howls*, and *Slay Ride*. To find the books published by Jove Publications, find its code (JP) in the Publisher table. Next, find all records in the Book table for which the publisher code is JP.

The table named Wrote, shown in Figure 1.20, is used to relate books and authors. The Sequence field indicates the order in which the authors of a particular book are listed on the cover. The table named Inventory in the same figure is used to indicate the number of copies of a particular book that are currently on hand at a particular branch of Henry Books. The first row, for example, indicates that there are two copies of the book with the number 0180; they are on hand at branch 1.

FIGURE 1.20

Sample data that relates books to authors and books to branches for Henry Books

Wrote

BookCode	AuthorNum	Sequence
0180	3	1
0189	5	1
0200	18	1
0378	11	1
079X	4	1
0808	4	1
1351	6	1
1382	23	2
1382	25	1
138X	1	1
2226	15	1
2281	9	2
2281	19	1
2766	20	1
2908	17	1
3350	2	1
3743	16	1
3906	24	1
5163	20	1
5790	10	1
6128	1	1

(continued)

Inventory

BookCode	BranchNum	OnHand
0180	1	2
0189	2	2
0200	1	1
0200	2	3
0378	3	2
079X	2	1
079X	3	2
079X	4	3
0808	2	1
1351	2	4
1351	3	2
1382	2	1
138X	2	3
2226	1	3
2226	3	2
2226	4	1
2281	4	3
2766	3	2
2908	1	3
2908	4	1
3350	1	2

Wrote				Inventory		
BookCode	**AuthorNum**	**Sequence**		**BookCode**	**BranchNum**	**OnHand**
6328	14	1		3743	2	1
669x	7	1		3906	2	1
6908	16	1		3906	3	2
7405	20	1		5163	1	1
7443	15	1		5790	4	2
7559	18	1		6128	2	4
8092	12	1		6128	3	3
8720	8	1		6328	2	2
9611	5	2		669X	1	1
9611	6	1		6908	2	2
9627	1	1		7405	3	2
9701	20	1		7443	4	1
9882	4	1		7559	2	2
9883	16	1		8092	3	1
9931	13	1		8720	1	3
				9611	1	2
				9627	3	5
				9627	4	2
				9701	1	2
				9701	2	1
				9701	3	3
				9701	4	2
				9882	3	3
				9883	2	3
				9883	4	2
				9931	1	2

QUESTION

To check your understanding of the relationship between authors and books, answer the following questions: Who wrote _Black House_? (Make sure to list the authors in the correct order.) Which books did Toni Morrison write?

ANSWER

Stephen King and Peter Straub wrote _Black House_. To determine who wrote _Black House_, first examine the Book table (see Figure 1.19) to find its book code (9611). Next, look for all rows in the Wrote table in which the book code is 9611. There are two such rows. In one row, the author number is 5, and in the other, it is 6. Then, look in the Author table to find the authors who have been assigned the numbers 5 and 6. The answer is Peter Straub (5) and Stephen King (6). The sequence number for author number 5 is 2 and the sequence number for author number 6 is 1. Thus, listing the authors in the proper order results in Stephen King and Peter Straub.

Toni Morrison wrote *Beloved*, *Jazz*, and *Song of Solomon*. To find the books written by Toni Morrison, look up her author number in the Author table and find that it is 1. Then look for all rows in the Wrote table for which the author number is 1. There are three such rows. The corresponding book codes are 138X, 6128, and 9627. Looking up these codes in the Book table, you find that Toni Morrison wrote *Beloved*, *Jazz*, and *Song of Solomon*.

QUESTION **A customer in branch 1 wishes to purchase** *The Soul of a New Machine*. **Is it currently in stock at branch 1?**

ANSWER No. Looking up the code for *The Soul of a New Machine* in the Book table, you find it is 3906. To find out how many copies are in stock at branch 1, look for a row in the Inventory table with 3906 in the BookCode column and 1 in the BranchNum column. Because there is no such row, branch 1 doesn't have any copies of *The Soul of a New Machine*.

QUESTION **You would like to obtain a copy of** *The Soul of a New Machine* **for this customer. Which other branches currently have it in stock, and how many copies does each branch have?**

ANSWER Branch 2 has one copy, and branch 3 has two copies. You already know that the code for *The Soul of a New Machine* is 3906. (If you didn't know the book code, you would look it up in the Book table.) To find out which branches currently have copies, look for rows in the Inventory table with 3906 in the BookCode column. There are two such rows. The first row indicates that branch number 2 currently has one copy. The second row indicates that branch number 3 currently has two copies.

The E-R diagram for the Henry Books database appears in Figure 1.21.

FIGURE 1.21

E-R diagram for the Henry Books database

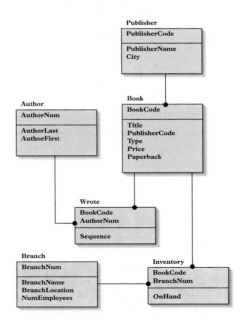

■ SUMMARY

- Problems with nondatabase approaches to data management include redundancy, difficulties accessing related data, limited security features, limited data-sharing features, and potential size limitations.

- An entity is a person, place, object, event, or idea for which you want to store and process data. An attribute, field, or column is a characteristic or property of an entity. A relationship is an association between entities.

- A one-to-many relationship between two entities exists when each occurrence of the first entity is related to many occurrences of the second entity and each occurrence of the second entity is related to only one occurrence of the first entity.

- A database is a structure that can store information about multiple types of entities, the attributes of the entities, and the relationships among the entities.

- Premiere Products is an organization whose requirements include information about the following entities: reps, customers, parts, orders, and order lines.

- An entity-relationship (E-R) diagram represents a database pictorially by using a rectangle for each entity that includes the entity's name above the rectangle and the entity's columns inside the rectangle, using a line to connect two entities that have a relationship, and placing a dot at the end of a line to indicate the "many" part of a one-to-many relationship.

- A database management system (DBMS) is a program, or a collection of programs, through which users interact with a database. DBMSs let you create forms, reports, and switchboards quickly and easily, as well as obtain answers to questions about the data stored in a database.

- Database processing offers the following advantages: getting more information from the same amount of data, sharing data, balancing conflicting requirements, controlling redundancy, facilitating consistency, improving integrity, expanding security, increasing productivity, and providing data independence. The disadvantages of database processing include the following: larger file size, increased complexity, greater impact of failure, and more difficult recovery.

- Henry Books is an organization whose requirements include information about the following entities: branches, publishers, authors, books, inventory, and author sequence.

■ KEY TERMS

attribute
column
data file
data independence
database
database administration (DBA)
database administrator
database design
database management system (DBMS)
entity

entity-relationship (E-R) diagram
field
form
integrity
integrity constraint
one-to-many relationship
redundancy
relationship
security
switchboard

■ REVIEW QUESTIONS

1. What is redundancy? What are the problems associated with redundancy?

2. Besides redundancy, what are the other problems associated with the nondatabase approach to processing data?

3. What is an entity? An attribute?

4. What is a relationship? A one-to-many relationship?

5. What is a database?

6. How do you create a one-to-many relationship in a database system?

7. What is an E-R diagram?

8. What is a DBMS?

9. What is database design?

10. What is a form?

11. What is a switchboard system?

12. How is it possible to get more information from the same amount of data by using a database approach as opposed to a nondatabase approach?

13. What is meant by the sharing of data?

14. What is a DBA? What kinds of responsibilities does a DBA have in a database environment?

15. How does consistency result from controlling redundancy?

16. What is an integrity constraint? When does a database have integrity?

17. What is security? How does a DBMS provide security?

18. What is data independence? Why is it desirable?

19. How is file size a disadvantage in a database environment?

20. How can the complexity of a DBMS be a disadvantage?

21. Why can a failure in a database environment be more serious than one in a nondatabase environment?

22. Why might recovery of data be more difficult in a database environment?

■ Premiere Products Exercises

Answer each of the following questions using the Premiere Products data shown in Figure 1.5. No computer work is required.

1. List the names of all customers that have a credit limit less than $10,000.

2. List the descriptions of all parts in item class AP and located in warehouse number 3.

3. List the order numbers for orders placed by customer number 608 on October 23, 2007.

4. List the part number, part description, and on-hand value (OnHand * Price) for each part in item class SG.

5. List the name of each customer that placed an order for two different parts in the same order.

6. List the name of each customer that has a credit limit of $5,000 and is represented by Richard Hull.

7. Find the sum of the balances for all customers represented by Juan Perez.

8. For each order, list the order number, order date, customer number, and customer name.

9. For each order placed on October 21, 2007, list the order number, order date, customer number, and customer name.

10. For each order placed on October 20, 2007, list the order number and customer name, along with the name of the rep who represents the customer.

▦ Henry Books Case

In later chapters, you will be asked to perform many tasks to help Ray Henry manage and manipulate his database. In order to familiarize yourself with this database in preparation for these tasks, answer each of the following questions using the Henry Books data shown in Figures 1.17 through 1.20. No computer work is required.

1. How many total employees work at all branches of Henry Books?

2. List the name of each publisher that's not located in New York.

3. List the title of each book published by Jove Publications.

4. List the title of each book that has the type PSY.

5. List the title of each book that has the type MYS and that is in paperback.

6. List the title of each book that has the type HOR or whose publisher code is SC.

7. List the titles of paperback books that have the type FIC and a price of less than $8.00.

8. List the title of each book that has the type HIS or TRA.

9. List the title and publisher name for all books with a price greater than $20.00.

10. For each book with co-authors, list the title, price, and author names (in the order listed on the cover).

11. For each book with more than six total copies on hand across all branches, list the book's title and price.

12. List the title of each book that has the type FIC and that was written by John Steinbeck.

CHAPTER 2

The Relational Model 1: Introduction, QBE, and Relational Algebra

OBJECTIVES

- Describe the relational model.

- Understand Query-By-Example (QBE).

- Use criteria in QBE.

- Create calculated columns in QBE.

- Calculate statistics in QBE.

- Sort data in QBE.

- Join tables in QBE.

- Update data using QBE.

- Understand relational algebra.

Introduction

The database management approach implemented by most personal computer database management systems (DBMSs) (and many mainframe systems as well) is the relational model. In this chapter, you will study the relational model and examine a method of retrieving data from relational databases called Query-By-Example (QBE). Finally, you will learn about relational algebra, which is one of the original ways of manipulating a relational database.

■ Relational Databases

A relational database is a collection of tables like the ones you viewed for Premiere Products in Chapter 1. These tables also appear in Figure 2.1. You might wonder why this type of database is not called a "table" database, or something similar, if a database is really just a collection of tables. Formally, these tables are called relations, and this is where this type of database gets its name.

FIGURE 2.1
....
Sample data for
Premiere Products

Rep

RepNum	LastName	FirstName	Street	City	State	Zip	Commission	Rate
20	Kaiser	Valerie	624 Randall	Grove	FL	33321	$20,542.50	0.05
35	Hull	Richard	532 Jackson	Sheldon	FL	33553	$39,216.00	0.07
65	Perez	Juan	1626 Taylor	Fillmore	FL	33336	$23,487.00	0.05

Customer

CustomerNum	CustomerName	Street	City	State	Zip	Balance	CreditLimit	RepNum
148	Al's Appliance and Sport	2837 Greenway	Fillmore	FL	33336	$6,550.00	$7,500.00	20
282	Brookings Direct	3827 Devon	Grove	FL	33321	$431.50	$10,000.00	35
356	Ferguson's	382 Wildwood	Northfield	FL	33146	$5,785.00	$7,500.00	65
408	The Everything Shop	1828 Raven	Crystal	FL	33503	$5,285.25	$5,000.00	35
462	Bargains Galore	3829 Central	Grove	FL	33321	$3,412.00	$10,000.00	65
524	Kline's	838 Ridgeland	Fillmore	FL	33336	$12,762.00	$15,000.00	20
608	Johnson's Department Store	372 Oxford	Sheldon	FL	33553	$2,106.00	$10,000.00	65
687	Lee's Sport and Appliance	282 Evergreen	Altonville	FL	32543	$2,851.00	$5,000.00	35
725	Deerfield's Four Seasons	282 Columbia	Sheldon	FL	33553	$248.00	$7,500.00	35
842	All Season	28 Lakeview	Grove	FL	33321	$8,221.00	$7,500.00	20

(continued)

Orders

OrderNum	OrderDate	CustomerNum
21608	10/20/2007	148
21610	10/20/2007	356
21613	10/21/2007	408
21614	10/21/2007	282
21617	10/23/2007	608
21619	10/23/2007	148
21623	10/23/2007	608

OrderLine

OrderNum	PartNum	NumOrdered	QuotedPrice
21608	AT94	11	$21.95
21610	DR93	1	$495.00
21610	DW11	1	$399.99
21613	KL62	4	$329.95
21614	KT03	2	$595.00
21617	BV06	2	$794.95
21617	CD52	4	$150.00
21619	DR93	1	$495.00
21623	KV29	2	$1,290.00

Part

PartNum	Description	OnHand	Class	Warehouse	Price
AT94	Iron	50	HW	3	$24.95
BV06	Home Gym	45	SG	2	$794.95
CD52	Microwave Oven	32	AP	1	$165.00
DL71	Cordless Drill	21	HW	3	$129.95
DR93	Gas Range	8	AP	2	$495.00
DW11	Washer	12	AP	3	$399.99
FD21	Stand Mixer	22	HW	3	$159.95
KL62	Dryer	12	AP	1	$349.95
KT03	Dishwasher	8	AP	3	$595.00
KV29	Treadmill	9	SG	2	$1,390.00

How does a relational database handle entities, attributes of entities, and relationships between entities? Each entity is stored in its own table. For example, in the Premiere Products database, there is a table for sales reps, a table for customers, and so on. The attributes of an entity become the fields or columns in the table. In the table for sales reps, for example, there is a column for the rep number, a column for the rep's last name, and so on.

What about relationships? At Premiere Products, there is a one-to-many relationship between sales reps and customers (each sales rep is related to the *many* customers he or she represents, and each customer is related to the *one* sales rep who represents it). How is this relationship implemented in a relational database? The answer is through common columns in two or more tables. Consider Figure 2.1 again. The RepNum columns in the Rep and Customer tables implement the relationship between sales reps and customers; that is, given a sales rep, you can use these columns to determine all the customers he or she represents, and, given a customer, you can use these columns to find the sales rep who represents the customer. If the Customer table did not include the sales rep number, you would not be able to identify the sales rep for a given customer and the customers for a given sales rep.

More formally, a relation is essentially just a two-dimensional table. If you consider the tables shown in Figure 2.1, however, you might see that there are certain restrictions you can place on relations. Each column in a table should have a unique name, and entries in each column should all "match" this column name. For example, in the CreditLimit column, all entries should in fact *be* credit limits. In addition, each row should be unique. After all, if two rows in a table contain identical data, the second row doesn't provide any information that you don't already have. In addition, for maximum flexibility, the order in which columns and rows appear in a table should be immaterial. Finally, a table's design is less complex if you restrict each position in the table to a single entry; that is, you should not permit multiple entries (often called **repeating groups**) in the table. These ideas lead to the following definitions:

Definition: A **relation** is a two-dimensional table in which:

1. The entries in the table are single-valued; that is, each location in the table contains a single entry.

2. Each column has a distinct name (technically called the attribute name).

3. All values in a column are values of the same attribute (that is, all entries must match the column name).

4. The order of columns is immaterial.

5. Each row is distinct.

6. The order of rows is immaterial.

Definition: A **relational database** is a collection of relations.

Later in this text, you will encounter situations in which a structure satisfies all the properties of a relation *except for the first item*; that is, some of the entries contain repeating groups and thus are not single-valued. Such a structure is called an **unnormalized relation**. This jargon is certainly a little strange in that an unnormalized relation is really not a relation at all. This term is used for such a structure, however. The table shown in Figure 2.2 is an example of an unnormalized relation.

FIGURE 2.2
....
Sample structure of
an unnormalized
relation

Orders

OrderNum	OrderDate	CustomerNum	PartNum	NumOrdered	QuotedPrice
21608	10/20/2007	148	AT94	11	$21.95
21610	10/20/2007	356	DR93 DW11	1 1	$495.00 $399.99
21613	10/21/2007	408	KL62	4	$329.95
21614	10/21/2007	282	KT03	2	$595.00
21617	10/23/2007	608	BV06 CD52	2 4	$794.95 $150.00
21619	10/23/2007	148	DR93	1	$495.00
21623	10/23/2007	608	KV29	2	$1,290.00

Note: Rows in a table (relation) are often called **records** or **tuples**. Columns in a table (relation) are often called **fields** or **attributes**.

There is a commonly accepted shorthand representation to show the structure of a relational database: You write the name of the table and then, within parentheses, list all the columns in the table. In addition, each table should appear on its own line. Using this method, you would write the Premiere Products database as follows:

```
Rep (RepNum, LastName, FirstName, Street, City, State,
     Zip, Commission, Rate)
Customer (CustomerNum, CustomerName, Street, City, State,
     Zip, Balance, CreditLimit, RepNum)
Orders (OrderNum, OrderDate, CustomerNum)
OrderLine (OrderNum, PartNum, NumOrdered, QuotedPrice)
Part (PartNum, Description, OnHand, Class, Warehouse, Price)
```

The Premiere Products database contains some duplicate column names. For example, the RepNum column appears in *both* the Rep table *and* the Customer table. Suppose a situation exists wherein the two columns might be confused. If you write RepNum, how would the computer or another programmer know which RepNum column in which table you intend to use? That could be a problem.

When duplicate column names exist in a database, you need a way to indicate the column to which you are referring. One common approach to this problem is to write both the table name and the column name, separated by a period. Thus, you would write the RepNum column in the Customer table as Customer.RepNum and the RepNum column in the Rep table as Rep.RepNum. Technically, when you combine a column name with a table name, you say that you **qualify** the column names. It is *always* acceptable to qualify column names, even if there is no possibility of confusion. If confusion may arise, however, it is *essential* to qualify column names.

The **primary key** of a table (relation) is the column or collection of columns that uniquely identifies a given row in that table. In the Rep table, the sales rep's number uniquely identifies a given row. For example, rep 35 occurs in only one row of the table. Thus, RepNum is the primary key.

The primary key provides an important way of distinguishing one row in a table from another. Primary keys are usually represented by underlining the column or collection of columns that comprises the primary key for each table in the database. Thus, the complete representation for the Premiere Products database is as follows:

```
Rep (RepNum, LastName, FirstName, Street, City, State,
     Zip, Commission, Rate)
Customer (CustomerNum, CustomerName, Street, City, State,
     Zip, Balance, CreditLimit, RepNum)
Orders (OrderNum, OrderDate, CustomerNum)
OrderLine (OrderNum, PartNum, NumOrdered, QuotedPrice)
Part (PartNum, Description, OnHand, Class, Warehouse, Price)
```

• •

QUESTION

Why does the primary key of the OrderLine table consist of two columns, not just one?

ANSWER

No single column uniquely identifies a given row in the OrderLine table. It requires a combination of two columns: OrderNum and PartNum.

• •

■ Query-By-Example (QBE)

When you ask Access or any other DBMS a question about the data in a database, the question is called a query. A **query** is simply a question represented in a way that the DBMS can recognize and process. In this section, you will investigate **Query-By-Example (QBE)**, an approach to writing queries that is very visual. With QBE, users enter their questions by entering column names and other criteria using an on-screen grid, and data appears on the screen in tabular form.

Note: This chapter features a specific version of QBE to illustrate its use. The examples used in this chapter are based on the Premiere Products database and are only for illustration purposes; you will not need to do any computer work at this point. The QBE version used in the examples is Microsoft Office Access 2003. Although the various versions of QBE are not identical, the differences are relatively minor. After you have mastered one version of QBE, you can apply your skills to learn another version of QBE.

In Access, you create queries using a special window with two panes. The upper pane (portion) of the window contains a field list for each table you want to query. The lower pane contains the **design grid**, the area where you specify the format of your output, the fields to be included in the query output, a sort order for the query results, and any criteria the records you are looking for must satisfy.

The following figures and examples will show you how to retrieve data using the Access version of QBE.

Note: If you plan to work through the examples in this chapter using a computer, you should use a copy of the Premiere Products database because the version of the database used in subsequent chapters does not include the changes you will make.

■ Simple Queries

To include a field in the results of a query, you place it in the design grid.

EXAMPLE 1 : List the number, name, balance, and credit limit of all customers in the database.

To include a field in an Access query, double-click the field in the field list to place it in the design grid, as shown in Figure 2.3. The check marks in the Show check boxes indicate the fields that will appear in the query results.

FIGURE 2.3
· · · · · · · · · · · · · · · · · · ·
Fields added to the design grid

34

Clicking the Run button on the toolbar runs the query and displays the query results, as shown in Figure 2.4.

EXAMPLE 2   :   List all fields and all rows in the Orders table.

To display all fields in the Orders table, you could add each field to the design grid. There is a shortcut, however. In Access, you can add all fields from a table to the design grid by double-clicking the asterisk in the table's field list. As shown in Figure 2.5, the asterisk appears in the design grid, indicating that all fields will be included in the query results.

The results of executing this query appear in Figure 2.6.

■ Simple Criteria

When the records that you want to display in a query's results must satisfy a condition, you enter the condition in the appropriate column in the design grid. Conditions that data must satisfy are also called **criteria**. (A single condition is called a **criterion**.) The following example illustrates the use of a criterion to select data.

E XAMPLE 3   :   Find the name of customer 148.

To enter a criterion for a field, include the field in the design grid and then enter the criterion in the row labeled Criteria for that field, as shown in Figure 2.7.

FIGURE 2.7
••••••••••••••••
Query to find
the name of
customer 148

Note: Access automatically adds quotation marks around values in the design grid that are formatted as Text fields when you run the query or move the insertion point to another cell in the design grid. Typing these quotation marks directly is optional.

The query results shown in Figure 2.8 show an exact match; the query selects a record only when CustomerNum equals 148.

FIGURE 2.8
••••••••••••••••
Query results

If you want something other than an exact match, you must enter the appropriate **comparison operator**, also called a **relational operator**, as you will see in the next example. The comparison operators are = (equal to), > (greater than), < (less than), >= (greater than or equal to), <= (less than or equal to), and NOT (not equal to).

Note: It is common in QBE to omit the = symbol in "equal to" comparisons, although you can use it every time.

36

Compound Criteria

You can use the comparison operators by themselves to create conditions. You can also combine criteria to create **compound criteria**, or **compound conditions**. In many query languages, you create compound criteria by including the word AND or OR between the separate criteria. In an **AND criterion**, both criteria must be true for the compound criterion to be true. In an **OR criterion**, the overall criterion is true if either of the individual criteria is true.

In QBE, to create an AND criterion, place the criteria on the same Criteria row in the design grid; to create an OR criterion, place the criteria on different Criteria rows in the design grid.

EXAMPLE 4

List the description of all parts for which there are more than 10 units on hand *and* that are located in warehouse 3.

To indicate that two criteria must both be true to select a record, place the criteria on the same Criteria row, as shown in Figure 2.9. In this case, you want the query to select those parts where the value in the OnHand field is greater than 10 (which requires the use of the > comparison operator) and the value in the Warehouse field is 3. Normally, if a field is included in the design grid, it will appear in the query results. In Example 4, the request is for the part description only, so you'll need to omit the OnHand and Warehouse fields from the query results. To omit a field from the query results, remove the check mark from the field's Show check box.

The query results appear in Figure 2.10.

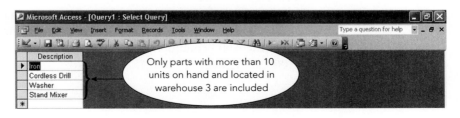

EXAMPLE 5 : List the description of all parts for which there are more than 10 units on hand *or* that are located in warehouse 3.

To indicate that either of two conditions must be true to select a record, place the first criterion in the Criteria row for the first column and place the second criterion in the row labeled "or," as shown in Figure 2.11.

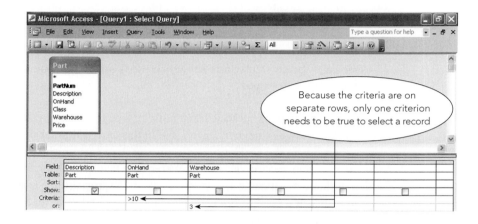

The query results appear in Figure 2.12.

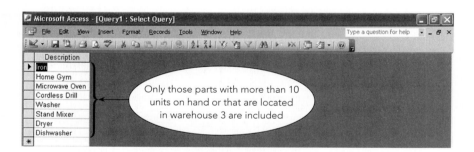

EXAMPLE 6 : List the number, name, and balance for each customer whose balance is between $1,000 and $5,000.

This example requires you to search for a range of values to find all customers with balances between $1,000 and $5,000. When you ask this kind of question, you are really looking for all balances that are greater than $1,000 *and* less than $5,000; the answer to this question requires using a compound criterion, or two criteria in the same field.

To place two criteria in the same field, separate the criteria with the word AND. Figure 2.13 shows the AND condition to select all records with a value of more than 1000 and less than 5000 in the Balance field.

FIGURE 2.13
••••••••••••••••
Query that uses
two conditions for a
single field

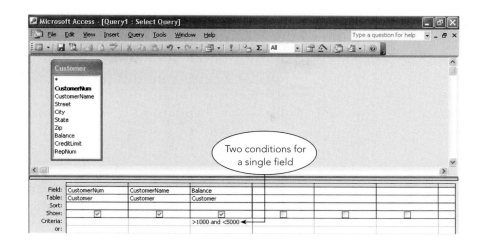

The query results appear in Figure 2.14.

FIGURE 2.14
••••••••••••••••
Query results

Computed Fields

Sometimes you'll need to include calculated fields that are not in the database in queries. A **computed field** or **calculated field** is a field that is the result of a calculation using one or more existing fields. Example 7 illustrates the use of a calculated field.

EXAMPLE 7 : List the number, name, and available credit for all customers.

Available credit is computed by subtracting the balance from the credit limit. Because there is no available credit field in the Customer table, you must calculate it from the existing Balance and CreditLimit fields. To include a computed field in a query, you enter a name for the computed field, followed by a colon, and then followed by an expression in one of the columns in the Field row.

To calculate available credit, you can enter the expression *AvailableCredit:CreditLimit-Balance* directly into the desired Field row in the design grid. When entering an expression in the design grid, the default column size prevents you from being able to see the complete expression. An alternative method is to right-click the column in the Field row to display the shortcut menu, and then click Zoom to open the Zoom dialog box. Then you can type the expression into the Zoom dialog box, as shown in Figure 2.15.

Note: If a field name contains spaces, you must enclose it in square brackets ([ ]). For example, if the field name were Credit Limit instead of CreditLimit, you would need to enter [Credit Limit]. You may also enclose a field name that does not contain spaces in square brackets, but you do not need to do so.

The query results appear in Figure 2.16.

FIGURE 2.16
. . . .
Query results

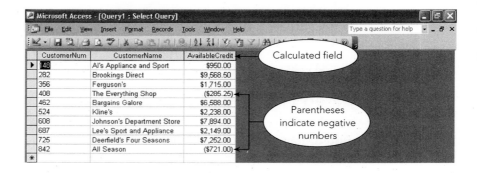

You are not restricted to subtraction in computations. You can use addition (+), multiplication (*), or division (/), as well. You also can include parentheses in your expressions to indicate which computations Access should perform first.

■ Calculating Statistics

All products that support QBE, including Access, support the following built-in **statistics** (called **aggregate functions** in Access): Count, Sum, Avg (average), Max (largest value), Min (smallest value), StDev (standard deviation), Var (variance), First, and Last. To use any of these statistics in a query, you include them in the Total row for the desired column in the design grid. By default, the Total row does not appear automatically in the grid. To include it, you must click the Totals button on the toolbar.

Example 8 illustrates how to use a statistic in a query by counting the number of customers represented by sales rep 35.

EXAMPLE 8 : How many customers are represented by sales rep 35?

To count the number of rows in the Customer table that have the value 35 in the RepNum column, you enter the Count function in the Total row for the CustomerNum column. In the RepNum column, you enter Where in the Total row to indicate that there will also be a criterion. In the Criteria row for the RepNum column, the entry 35 selects only those records for sales rep number 35, as shown in Figure 2.17.

FIGURE 2.17
• • • • • • • • • • • • • • • • •
Query to count
records

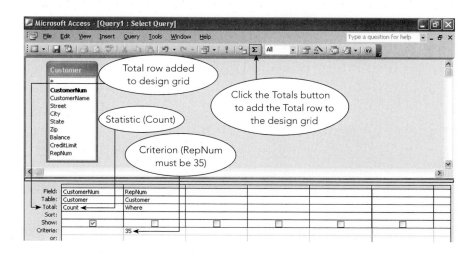

The query results appear in Figure 2.18. Notice that Access titled the new column using the default name, CountOfCustomerNum. You could create your own column name by preceding the field name with the desired column name and a colon in the query design (for example, NumberOfCustomers: CustomerNum changes the default name to "NumberOfCustomers").

FIGURE 2.18
• • • • • • • • • • • • • • • • •
Query results

EXAMPLE 9 : What is the average balance of all customers of sales rep 35?

To calculate the average balance, use the Avg function, as shown in Figure 2.19.

FIGURE 2.19

• • • • • • • • • • • • • • • • • •

Query to calculate
an average

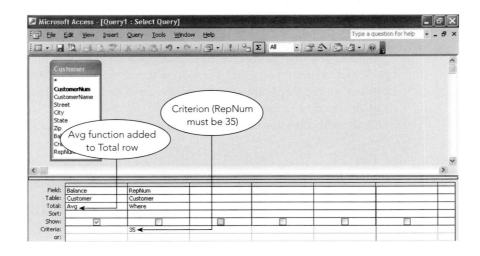

The query results appear in Figure 2.20.

FIGURE 2.20

• • • • • • • • • • • • • • • • •

Query results

Grouping

You can also use functions in combination with grouping, where statistics are calculated for groups of records. You may, for example, need to calculate the average balance for all customers of each sales rep. **Grouping** simply means creating groups of records that share some common characteristic. In grouping by RepNum, for example, the customers of sales rep 20 would form one group, the customers of sales rep 35 would form a second group, and the customers of sales rep 65 would form a third group. The calculations are then made for each group. To group records in Access, select Group By in the Total row for the field to be used for grouping.

EXAMPLE 10   What is the average balance for all customers of each sales rep?

In this example, include the RepNum and Balance fields in the design grid. To group the customer records for each sales rep, select Group By as the entry in the Total row for the RepNum column. To calculate the average balance for each group of customers, enter Avg in the Total row for the Balance column, as shown in Figure 2.21.

FIGURE 2.21
••••••••••••••••••
Query to group
records

The query results appear in Figure 2.22.

FIGURE 2.22
••••••••••••••••••
Query results

■ Sorting

In most queries, the order in which records appear doesn't matter. In other queries, however, the order in which records appear can be very important. You may want to see customers listed alphabetically by customer name, or you may want to see customers listed by rep number. Further, you may want to see customer records listed alphabetically and grouped by sales rep number.

To list the records in a query's results in a particular way, you need to **sort** the records. The field on which records are sorted is called the **sort key**; you can sort records using more than one field when necessary. If you are sorting records by more than one field (such as sorting by customer name within rep number), the more important field (RepNum) is called the **major sort key** (also called the **primary sort key**) and the less important field (CustomerName) is called the **minor sort key** (also called the **secondary sort key**).

To sort in Access, specify the sort order in the Sort row of the design grid for the sort key field.

EXAMPLE 11 : List the customer number, name, balance, and rep number for each customer. Sort the
output alphabetically by customer name.

To sort the records alphabetically using the CustomerName field, select Ascending in the Sort row for the CustomerName column, as shown in Figure 2.23. (To sort the records in reverse alphabetical order, select Descending.)

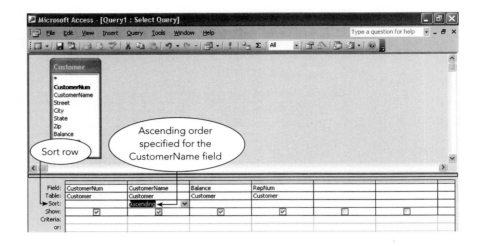

The query results appear in Figure 2.24. Notice that the customer names appear in alphabetical order.

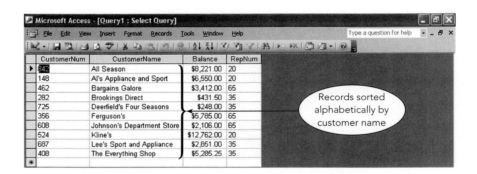

Sorting on Multiple Keys

You can specify more than one sort key in a query; in this case, the sort key on the left in the design grid will be the major (primary) sort key and the sort key on the right will be the minor (secondary) sort key. Example 12 illustrates the process.

EXAMPLE 12 List the customer number, name, balance, and rep number for each customer. Sort the output by sales rep number. Within the customers of each sales rep, sort the output by customer name.

44

To sort records by sales rep number and then by customer name, RepNum is the major sort key and CustomerName is the minor sort key. Figure 2.25 shows an incorrect query design.

FIGURE 2.25
•••••••••••••••
Incorrect query
design to sort on
multiple fields

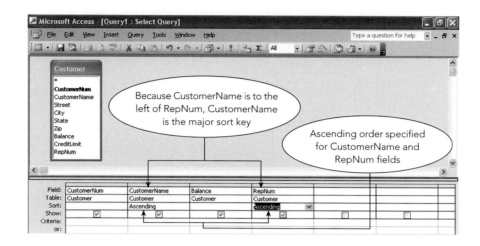

In Figure 2.25, the CustomerName field is to the left of the RepNum field; with this order, CustomerName is the major sort key, so the data will be sorted by customer name, not by sales rep number, as shown in Figure 2.26. If there were two customers with the same name, the data for these customers would be further sorted by sales rep number because RepNum is the minor key.

FIGURE 2.26
•••••••••••••••
Query results

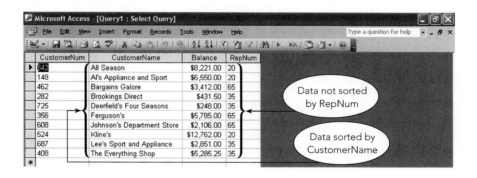

To correct this problem, include the RepNum field in the design grid twice, as shown in Figure 2.27.

FIGURE 2.27
• • • • • • • • • • • • • • • •
Correct query
design to sort on
multiple fields

FIGURE 2.28
• • • • • • • • • • • • • • • •
Query results

Select Ascending as the sort order in the first RepNum column and remove the check mark from the Show check box to omit this RepNum column from the query results. Because this RepNum column is to the left of the other sort key (CustomerName), RepNum is the major sort key. The second RepNum column will be included in the query results in the desired position, as shown in Figure 2.28.

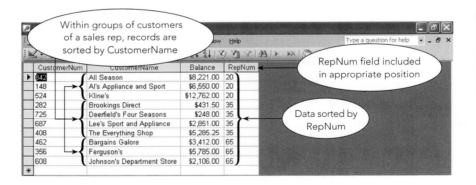

Joining Tables

So far, the queries used in the examples have displayed records from a single table. In many cases, you'll need to create queries to select data from more than one table. To do so, it is necessary to **join** the tables based on matching fields in corresponding columns. To join tables in Access, first you add the field lists for both tables to the upper pane of the Select Query window. Access will draw a line, called a **join line**, between matching fields in the two tables, indicating that the tables are related. (If the corresponding fields have the same field name and at least one of the fields is the primary key of the table that contains it, Access will join the tables automatically.) You then can select fields from either table or from both tables, as you will see in the next example.

46

EXAMPLE 13 List each customer's number and name, along with the number, last name, and first name of each customer's sales rep.

You cannot create this query using a single table—the customer name is in the Customer table and the sales rep name is in the Rep table. The sales rep number can come from either table because it is the matching field. To select the correct data, you need to join the tables by adding the Customer and Rep table field lists to the upper pane and then adding the desired fields from the field lists to the design grid, as shown in Figure 2.29.

FIGURE 2.29
· · · · · · · · · · · · · · · · · ·
Query design to
join two tables

Notice that the Table row indicates the table with which each field is associated. The query results appear in Figure 2.30.

FIGURE 2.30
· · · · · · · · · · · · · · · · · ·
Query results

EXAMPLE 14　For each customer whose credit limit is $10,000, list the customer's number and name, along with the number, last name, and first name of the corresponding sales rep.

The only difference between this query and the one illustrated in Example 13 is that there is an extra restriction—the credit limit must be $10,000. To include this new condition, add the CreditLimit field to the design grid, enter 10000 as the criterion, and remove the check mark from the CreditLimit field's Show box (because the CreditLimit column should not appear in the query results). The query design appears in Figure 2.31.

FIGURE 2.31
. . . .
Query to restrict
records in a join

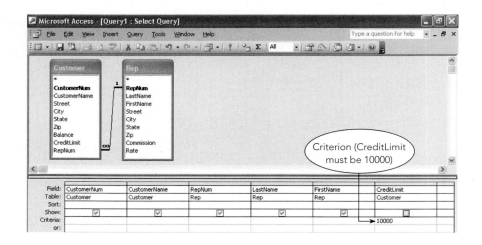

Only the customers with credit limits of $10,000 are included in the query results, as shown in Figure 2.32.

FIGURE 2.32
. . . .
Query results

Joining Multiple Tables

Joining multiple tables is similar to joining two tables. First you add all the tables involved in the join to the upper pane, and then you add the fields to appear in the query results to the design grid in the desired order.

EXAMPLE 15　For each order, list the order number, date, customer number, and customer name. In addition, for each order line within the order, list the part number, description, number of units ordered, and quoted price.

48

This query requires data from four tables: Orders (for basic order data), Customer (for the customer number and name), OrderLine (for the part number, number ordered, and quoted price), and Part (for the description). Figure 2.33 shows the query design. (In Figure 2.33, the field lists have been repositioned to make the relationships clear and the columns in the design grid have been resized so that all the columns are visible.)

FIGURE 2.33
· · · · · · · · · · · · · · · · · · ·
Query to join
multiple tables

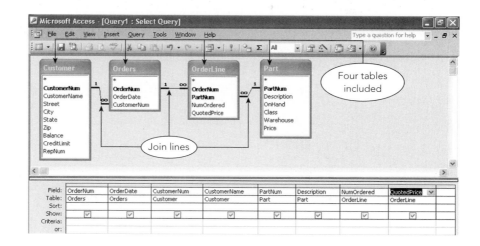

The query results appear in Figure 2.34.

FIGURE 2.34
· · · · · · · · · · · · · · · · · · ·
Query results

Using an Update Query

In addition to retrieving data, you can use a query to update data. A query that changes data is called an **update query**. An update query makes a specified change to all records satisfying the criteria in the query. To change the query type to an update query, click the Query Type button list arrow on the toolbar, and then click Update Query. An extra row, called the Update To row, is added to the design grid when you create an update query. You use this new row to indicate how to update the data selected by the query.

EXAMPLE 16 : The zip code for customers located in the city of Fillmore is incorrect; it should be 33363. Change the zip code for these customers to the correct value.

To change the zip code for only those customers located in Fillmore, include the City column in the design grid and enter a criterion of Fillmore in the Criteria row. To indicate the new value for the zip code, include the Zip column in the design grid and enter the new zip code value in the Update To row of the Zip column, as shown in Figure 2.35. When you click the Run button, Access will indicate how many rows the query will change and give you a chance to cancel the update, if necessary. When you click the Yes button, the query is executed and updates the data specified in the query design.

FIGURE 2.35
· · · · · · · · · · · · · · · · ·
Query design to update data

Using a Delete Query

You can also use queries to delete one or more records at a time. A **delete query** deletes all the records satisfying the criteria entered in the query. For example, you could delete all the order lines associated with a certain order in the OrderLine table with a single delete query.

EXAMPLE 17 : Delete all order lines in which the order number is 21610.

You enter the criteria that will determine the records to be deleted just as you would enter any other criteria. In this example, include the OrderNum field in the design grid and enter the order number 21610 in the Criteria row, as shown in Figure 2.36. To change the query type to a delete query, click the Query Type button list arrow on the toolbar, and then click Delete Query. Notice that an extra row, called the Delete row, is added to the design grid, indicating that this is a delete query. When you click the Run button, Access will indicate how many rows will be deleted and give you a chance to cancel the deletions, if necessary. If you click the Yes button, the query will delete all rows in the OrderLine table in which the order number is 21610.

FIGURE 2.36
••••••••••••••••••
Query design to
delete rows

▪ Using a Make-Table Query

You can use a query to create a new table in either the current database (that is, the database in which you are currently working) or in a separate database. A **make-table query** creates a new table using the query results. The data added to the new table is separate from the original table in which it appears; in other words, you don't move the data to a new table; you create a new table using the data selected by the query.

EXAMPLE 18   Create a new table containing the customer number and customer name, and the number, first name, and last name of the customer's sales rep. Name the new table CustomerRep.

Figure 2.37 shows the query design to select records from the Customer and Rep tables.

FIGURE 2.37
••••••••••••••••••
Make-table query
design

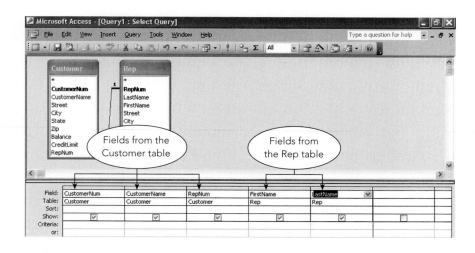

After you have created and tested the query to make sure that it selects the correct records, change the query type to a make-table query by clicking the Query Type button list arrow on the toolbar, and then clicking Make-Table Query. The Make Table dialog box opens. You then can specify the new table's name and where to create it, as shown in Figure 2.38.

FIGURE 2.38
••••••••••••••••
Make Table
dialog box

When you run the make-table query, the records it selects will be added to a new table named CustomerRep in the current database. Figure 2.39 shows the new CustomerRep table created by the make-table query.

FIGURE 2.39
••••••••••••••••
Table created
by the make-
table query

■ Relational Algebra

Relational algebra is a theoretical way of manipulating a relational database. Relational algebra includes operations that act on existing tables to produce new tables, similar to the way the operations of addition and subtraction act on numbers to produce new numbers in the mathematical algebra with which you are familiar.

52

Retrieving data from a relational database through the use of relational algebra involves issuing relational algebra commands to operate on existing tables to form a new table containing the desired information. Sometimes you might need to execute a series of commands to obtain the desired result.

Note: Unlike QBE, relational algebra is not used in current DBMS systems. Its importance is the theoretical base it furnishes to the relational model and the benchmark it provides. Other approaches to querying relational databases are judged by this benchmark.

Note: There is no "standard" method for representing relational algebra commands. This section illustrates one possible approach. What is important is not the particular way the commands are represented, but the results they provide.

As you will notice in the following examples, each command ends with a GIVING clause, followed by a table name. This clause requests the result of the command to be placed in a temporary table with the name you have specified. You can use this temporary table in any way you wish. For instance, you can display it on screen or print a report using the data in the table. After you have finished and are ready to move on to another query, you usually delete the table. (If you wanted to save it for some future use, you could certainly do so. Typically, however, after you have finished working on the query, you won't need to save the table.)

■ Select

The **SELECT** command within relational algebra takes a horizontal subset of a table; that is, it retrieves certain rows from an existing table (based on some user-specified criteria) and saves them as a new table. The SELECT command includes the word WHERE followed by a condition. The rows retrieved will be the rows in which the condition is satisfied.

EXAMPLE 19 : List all information from the Customer table concerning customer 282.

```
SELECT Customer WHERE CustomerNum=282
     GIVING Answer
```

This command creates a new table named Answer that contains only the single row in which the customer number is 282, because that is the only row in which the condition following the word WHERE is true. All the columns from the Customer table will be included in the new Answer table.

EXAMPLE 20 : List all information from the Customer table concerning those customers with credit limits of $7,500.

```
SELECT Customer WHERE CreditLimit=7500
     GIVING Answer
```

This command creates a new table named Answer that contains all the columns from the Customer table, but only those rows in which the credit limit is $7,500.

▓ Project

The **PROJECT** command within relational algebra takes a vertical subset of a table (that is, it causes only certain columns to be included in the new table). The PROJECT command includes the word OVER followed by a list of the columns to be included.

EXAMPLE 21 : List the number and name of all customers.

```
PROJECT Customer OVER (CustomerNum, CustomerName)
    GIVING Answer
```

This command will create a new table named Answer that contains the CustomerNum and CustomerName columns for all the rows in the Customer table.

EXAMPLE 22 : List the number and name of all customers with credit limits of $7,500.

This example requires a two-step process. You first use a SELECT command to create a new table that contains only those customers with credit limits of $7,500. Then you project the new table to restrict the result to only the indicated columns.

```
SELECT Customer WHERE CreditLimit=7500
    GIVING Temp
PROJECT Temp OVER (CustomerNum, CustomerName)
    GIVING Answer
```

The first command creates a new table named Temp that contains all the columns from the Customer table, but only those rows in which the credit limit is $7,500. The second command creates a new table named Answer that contains all the rows from the Temp table (that is, only customers whose credit limit is $7,500), but only the CustomerNum and CustomerName columns.

▓ Join

The join operation is the core operation of relational algebra because it is the command that allows you to extract data from more than one table. In the most common form of the join, two tables are combined based on the values in matching columns, creating a new table containing the columns in both tables. Rows in this new table will be the **concatenation** (combination) of a row from the first table and a row from the second table that match on the common column (often called the **join column**). In other words, two tables are joined *on* the join column.

For example, suppose you want to join the two tables shown in Figure 2.40 on RepNum (the join column), creating a new table named Temp.

54

FIGURE 2.40
••••••••••••••••••
Customer and Rep
tables

Customer

CustomerNum	CustomerName	RepNum
148	Al's Appliance and Sport	20
282	Brookings Direct	35
356	Ferguson's	65
408	The Everything Shop	35
462	Bargains Galore	65
524	Kline's	20
608	Johnson's Department Store	65
687	Lee's Sport and Appliance	35
725	Deerfield's Four Seasons	35
842	All Season	20
701	Peters	05

Rep

RepNum	LastName	FirstName
20	Kaiser	Valerie
35	Hull	Richard
65	Perez	Juan
75	Lewis	Joan

The result of joining the Customer and Rep tables creates the table shown in Figure 2.41. The column that joins the tables (RepNum) appears only once. Other than that, all columns from both tables appear in the result.

FIGURE 2.41
••••••••••••••••••
Table produced by
joining the
Customer and
Rep tables

Temp

CustomerNum	CustomerName	RepNum	LastName	FirstName
148	Al's Appliance and Sport	20	Kaiser	Valerie
282	Brookings Direct	35	Hull	Richard
356	Ferguson's	65	Perez	Juan
408	The Everything Shop	35	Hull	Richard
462	Bargains Galore	65	Perez	Juan
524	Kline's	20	Kaiser	Valerie
608	Johnson's Department Store	65	Perez	Juan
687	Lee's Sport and Appliance	35	Hull	Richard
725	Deerfield's Four Seasons	35	Hull	Richard
842	All Season	20	Kaiser	Valerie

If there is a row in one table that does not match any row in the other table, that row will not appear in the result of the join. Thus, the row for sales rep 75 (Joan Lewis) from the Rep table and the row for customer 701 (Peters) from the Customer table do not appear in the join table because their rows are not common to both tables.

You can restrict the output from the join to include only certain columns by using the PROJECT command, as shown in the following example.

EXAMPLE 23   For each customer, list the customer number, customer name, sales rep number, and sales rep's last name.

```
JOIN Customer Rep
    WHERE Customer.RepNum=Rep.RepNum
    GIVING Temp
PROJECT Temp OVER (CustomerNum, CustomerName, RepNum,
    LastName)
    GIVING Answer
```

In the WHERE clause of the JOIN command, the matching fields are both named RepNum—the field in the Rep table named RepNum is supposed to match the field in the Customer table named RepNum. Because there are two fields named RepNum, you must qualify the field names. Just like in QBE, the RepNum field in the Rep table is written as Rep.RepNum and the RepNum field in the Customer table is written as Customer.RepNum.

In this example, the JOIN command joins the Rep and Customer tables to create a new table named Temp. The PROJECT command creates a new table named Answer that contains all the rows from the Temp table, but only the CustomerNum, CustomerName, RepNum, and LastName columns.

The type of join used in Example 23 is called a **natural join**. Although this type of join is by far the most common kind, there is another possibility. The other type of join, the **outer join**, is similar to the natural join except that it also includes records from each original table that are not common in both tables. In a natural join, these unmatched records do not appear in the new table. In the outer join, unmatched records are included and the values of the fields are vacant, or **null**, for the records that do not have data common in both tables. Performing an outer join for Example 23 produces the table shown in Figure 2.42.

FIGURE 2.42
•••••••••••••••••••
Table produced by
an outer join of the
Customer and
Rep tables

Temp

CustomerNum	CustomerName	RepNum	LastName	FirstName
148	Al's Appliance and Sport	20	Kaiser	Valerie
282	Brookings Direct	35	Hull	Richard
356	Ferguson's	65	Perez	Juan
408	The Everything Shop	35	Hull	Richard
462	Bargains Galore	65	Perez	Juan
524	Kline's	20	Kaiser	Valerie
608	Johnson's Department Store	65	Perez	Juan
687	Lee's Sport and Appliance	35	Hull	Richard
725	Deerfield's Four Seasons	35	Hull	Richard
842	All Season	20	Kaiser	Valerie
701	Peters	05	-	-
-	-	75	Lewis	Joan

■ Normal Set Operations

Relational algebra includes set operations for union, intersection, and difference. The **union** of tables A and B is a table containing all rows that are in either table A or table B or in both table A and table B. The **intersection** of two tables is a table containing all rows that are common in both table A and table B. The **difference** of two tables A and B (referred to as A minus B) is the set of all rows that are in table A but that are not in table B.

Union

There is a restriction on set operations. It does not make sense, for example, to talk about the union of the Rep table and the Customer table because the tables do not contain the same columns. The two tables *must* have the same structure for a union to be appropriate; the formal term is union compatible. Two tables are **union compatible** if they have the same number of columns and if their corresponding columns represent the same type of data. If, for example, the first column in table A contains customer numbers, the first column in table B must also contain customer numbers.

EXAMPLE 24 : List the number and name of those customers that have orders or are represented by sales rep 65, or both.

You can create a table containing the number and name of all customers that have orders by joining the Orders table and the Customer table (Temp1 in the following example) and then projecting the result over CustomerNum and CustomerName (Temp2). You can also create a table containing the number and name of all customers represented by sales rep 65 by selecting from the Customer table (Temp3) and then projecting the result (Temp4). The two tables ultimately created by this process (Temp2 and Temp4) have the same structure. They each have two fields: CustomerNum and CustomerName. Because these two tables are union compatible, it is appropriate to take the union of these two tables. This process is accomplished in relational algebra using the following code:

```
JOIN Orders, Customer
     WHERE Orders.CustomerNum=Customer.CustomerNum
     GIVING Temp1
PROJECT Temp1 OVER CustomerNum, CustomerName
     GIVING Temp2
SELECT Customer WHERE RepNum='65'
     GIVING Temp3
PROJECT Temp3 OVER CustomerNum, CustomerName
     GIVING Temp4
UNION Temp2 WITH Temp4 GIVING Answer
```

Intersection

As you would expect, using the intersection operation is very similar to using the union operation. The only difference is that you replace the UNION command with the **INTERSECT** command, as illustrated in the following example.

EXAMPLE 25   List the number and name of customers that have orders *and* that are represented by sales rep 65.

In this example, you need to intersect the two tables instead of taking their union. The code to accomplish this is as follows:

```
JOIN Orders, Customer
     WHERE Orders.CustomerNum=Customer.CustomerNum
     GIVING Temp1
PROJECT Temp1 OVER CustomerNum, CustomerName
     GIVING Temp2
SELECT Customer WHERE RepNum='65'
     GIVING Temp3
PROJECT Temp3 OVER CustomerNum, CustomerName
     GIVING Temp4
INTERSECT Temp2 WITH Temp4 GIVING Answer
```

Difference

The difference operation is performed by the **SUBTRACT** command in relational algebra.

EXAMPLE 26   List the number and name of those customers that have orders but that are *not* represented by sales rep 65.

This process is virtually identical to the one encountered in the union example, but in this case, you subtract one of the tables from the other instead of taking their union or intersection. This process is accomplished in relational algebra using the following command:

```
JOIN Orders, Customer
     WHERE Orders.CustomerNum=Customer.CustomerNum
     GIVING Temp1
PROJECT Temp1 OVER CustomerNum, CustomerName
     GIVING Temp2
SELECT Customer WHERE RepNum='65'
     GIVING Temp3
PROJECT Temp3 OVER CustomerNum, CustomerName
     GIVING Temp4
SUBTRACT Temp4 FROM Temp2 GIVING Answer
```

The next two sections present the final two important but infrequently used commands in relational algebra: product and division.

■ Product

The **product** of two tables (mathematically called the **Cartesian product**) is the table obtained by concatenating every row in the first table with every row in the second table. Thus, the product of the Orders table and the Part table, which are both shown in Figure 2.43, appears in the figure as the table labeled "Product of Orders and Part."

FIGURE 2.43
. . . .
Product of
two tables

Orders

OrderNum	OrderDate
21608	10/20/2007
21610	10/20/2007
21613	10/21/2007

Part

PartNum	Description
DR93	Gas Range
DW11	Washer

Product of Orders and Part

OrderNum	OrderDate	PartNum	Description
21608	10/20/2007	DR93	Gas Range
21610	10/20/2007	DR93	Gas Range
21613	10/21/2007	DR93	Gas Range
21608	10/20/2007	DW11	Washer
21610	10/20/2007	DW11	Washer
21613	10/21/2007	DW11	Washer

Every row in the Orders table is matched with every row in the Part table. If the Orders table has *m* rows and the Part table has *n* rows, there would be *m times n* rows in the product. If, as is typically the case, the tables have many rows, the number of rows in the product can be so great that it is not practical to form the product. Usually, you would want only those combinations that satisfy certain restrictions; thus, you would almost always use the join operation instead of the product operation.

■ Division

The **division** process is best illustrated by considering the division of a table with two columns by a table with a single column, which is the most common situation in which to use this operation. Consider the first two tables shown in Figure 2.44. The first table contains two columns: OrderNum and PartNum. The second contains only a single column, PartNum.

FIGURE 2.44
....
Dividing one table
by another

OrderLine

OrderNum	PartNum
21608	AT94
21610	DR93
21610	DW11
21613	KL62
21614	KT03
21617	BV06
21617	CD52
21619	DR93
21623	KV29

Part

PartNum
DR93
DW11

Result of dividing OrderLine by Part

OrderNum
21610

The quotient, that is, the result of the division, is a new table with a single column named OrderNum (the column from table A that is *not* in table B). The rows in this new table contain those order numbers from the OrderLine table that "match" with *all* the parts appearing in the Part table. For an order number to appear in the quotient, there must be a row in the OrderLine table with that order number in the OrderNum column and DR93 in the PartNum column. There must also be a row in the OrderLine table with this same order number in the OrderNum column and DW11 in the PartNum column. It doesn't matter if there are other rows in the OrderLine table containing the same order number as long as the rows with DR93 and DW11 are present. With the sample data, only order number 21610 qualifies. Thus, the result is the final table shown in the figure.

■ SUMMARY

- A relation is a two-dimensional table in which the entries are single-valued, each field has a distinct name, all the values in a field are values of the same attribute (the one identified by the field name), the order of fields is immaterial, each row is distinct, and the order of rows is immaterial.

- A relational database is a collection of relations.

- An unnormalized relation is a structure in which entries need not be single-valued but that satisfies all the other properties of a relation.

- A field name is qualified by preceding it with the table name and a period (for example, Rep.RepNum).

- A table's primary key is the field or fields that uniquely identify a given row within the table.

- Query-By-Example BE (QBE) is a visual tool for manipulating relational databases. QBE queries are creating by completing on-screen forms.

- To include a field in an Access query, place the field in the design grid and make sure there is a check mark in the field's Show box.

- To indicate criteria in an Access query, place the criteria in the appropriate columns in the design grid of the Select Query window.

- To indicate AND criteria in an Access query, place both criteria in the same Criteria row of the design grid; to indicate OR criteria, place the criteria on separate Criteria rows of the design grid.

- To create a computed field in Access, enter an appropriate expression in the desired column of the design grid.

- To calculate statistics in Access, include the appropriate aggregate functions in the Total row for the appropriate column of the design grid.

- To sort query results in Access, select Ascending or Descending in the Sort row for the field or fields that are sort keys.

- When sorting query results using more than one field, the leftmost sort key in the design grid is the major sort key (also called the primary sort key) and the sort key to its right is the minor sort key (also called the secondary sort key).

- To join tables in Access, place field lists for both tables in the upper pane of the Select Query window.

- To make the same change to all records that satisfy certain criteria, use an update query.

- To delete all records that satisfy certain criteria, use a delete query.

- To save the results of a query as a table, use a make-table query.

- Relational algebra is a language used to manipulate relational databases.

- The SELECT command in relational algebra selects only certain rows from a table.

- The PROJECT command in relational algebra selects only certain columns from a table.

- The JOIN command in relational algebra combines data from two or more tables based on common columns.

- The UNION command in relational algebra forms the union of two tables. In order for a union operation to make sense, the tables must be union compatible.

- Two tables are union compatible if they have the same number of columns and their corresponding columns represent the same type of data.

- The INTERSECT command in relational algebra forms the intersection of two tables.

- The SUBTRACT command in relational algebra forms the difference of two tables.

- The PRODUCT process in relational algebra is the Cartesian product of two tables.

- The DIVISION process in relational algebra forms the division of two tables.

■ KEY TERMS

aggregate function
AND criterion
attribute
calculated field
Cartesian product
comparison operator
compound condition
compound criteria
computed field
concatenation
criteria
criterion
delete query
design grid
difference
division
field
grouping
INTERSECT
intersection
join
join column
join line
major sort key
make-table query
minor sort key
natural join

null
OR criterion
outer join
primary key
primary sort key
product
PROJECT
qualify
query
Query-By-Example (QBE)
record
relation
relational algebra
relational database
relational operator
repeating group
secondary sort key
SELECT
sort
sort key
statistic
SUBTRACT
tuple
union
union compatible
unnormalized relation
update query

■ REVIEW QUESTIONS

1. What is a relation?

2. What is a relational database?

3. What is an unnormalized relation? Is it a relation according to the definition of the word *relation*?

4. How is the term *attribute* used in the relational model? What is a more common name for it?

5. Describe the shorthand representation of the structure of a relational database. Illustrate this technique by representing the database for Henry Books as shown in Chapter 1.

6. What does it mean to qualify a field name? How would you qualify the Street field in the Customer table?

7. What is a primary key? What is the primary key for each table in the Henry Books database shown in Chapter 1?

8. How do you include a field in an Access query?

9. How do you indicate criteria in an Access query?

10. How do you combine criteria with AND in an Access query? How do you combine them with OR?

11. How do you create a computed field in an Access query?

12. In which row of the Access design grid do you calculate statistics? What statistics can you use in Access queries?

13. How do you sort data in an Access query?

14. When sorting data on more than one field in an Access query, which field is the major sort key? Which field is the minor sort key? What effect do these keys have on the order in which the rows are displayed?

15. How do you join tables in an Access query?

16. When do you use an update query?

17. When do you use a delete query?

18. When do you use a make-table query?

19. What is relational algebra?

20. Describe the purpose of the SELECT command in relational algebra.

21. Describe the purpose of the PROJECT command in relational algebra.

22. Describe the purpose of the JOIN command in relational algebra.

23. Describe the purpose of the UNION command in relational algebra.

24. Are there any restrictions on the tables when using the UNION command?

25. Describe the purpose of the INTERSECT command in relational algebra.

26. Describe the purpose of the SUBTRACT command in relational algebra.

27. Describe the purpose of the PRODUCT process in relational algebra.

28. Describe the results of the DIVISION process in relational algebra.

■ Premiere Products Exercises: QBE

In the following exercises, you will use the data in the Premiere Products database shown in Figure 2.1. (If you use a computer to complete these exercises, use a copy of the Premiere Products database so you will still have the original data when you complete Chapter 3.) In each step, use QBE to obtain the desired results. You can use the query feature in a DBMS to complete the exercises using a computer or you can simply write a description of how you would complete the task. Check with your instructor if you are uncertain of which approach to take.

1. List the number and name of all customers.

2. List the complete Part table.

3. List the number and name of all customers represented by sales rep 35.

4. List the number and name of all customers that are represented by sales rep 35 and have credit limits of $10,000.

5. List the number and name of all customers that are represented by sales rep 35 or have credit limits of $10,000.

6. For each order, list the order number, order date, the number of the customer that placed the order, and the name of the customer that placed the order.

7. List the number and name of all customers represented by Juan Perez.

8. How many customers have a credit limit of $10,000?

9. Find the total of the balances for all customers represented by sales rep 35.

10. Give the part number, description, and on-hand value (OnHand * Price) for each part in item class HW.

11. List all columns and all records in the Part table. Sort the results by part description.

12. List all columns and all records in the Part table. Sort the results by part number within item class.

13. List the item class and the sum of the value of parts on hand. Group the results by item class.

14. Create a new table named SportingGoods to contain the columns PartNum, Description, OnHand, Warehouse, and Price for all rows in which the item class is SG.

15. In the SportingGoods table, change the description of part BV06 to "Fitness Gym."

16. In the SportingGoods table, delete every row in which the price is greater than $1,000.

▒ Premiere Products Exercises: Relational Algebra

In the following exercises, you will use the data in the Premiere Products database shown in Figure 2.1. In each step, indicate how to use relational algebra to obtain the desired results.

1. List the number and name of all sales reps.

2. List all information from the Part table for part FD21.

3. List the order number, order date, customer number, and customer name for each order.

4. List the order number, order date, customer number, and customer name for each order placed by any customer represented by the sales rep whose last name is Kaiser.

5. List the number and date of all orders that were either placed on 10/20/2007 or placed by a customer whose rep number is 20.

6. List the number and date of all orders that were placed on 10/20/2007 by a customer whose rep number is 20.

7. List the number and date of all orders that were placed on 10/20/2007 but not by a customer whose rep number is 20.

■ Henry Books Case

Ray Henry is aware that the ability to easily query a database is one of the most important benefits of using a good DBMS. Now that you have helped him computerize his database, he is anxious to have you help him obtain answers to a variety of questions. In the following exercises, you will use the data in the Henry Books database shown in Figures 1.17 through 1.20 in Chapter 1. (If you use a computer to complete these exercises, use a copy of the Henry Book database so you will still have the original data when you complete Chapter 3.) In each step, use QBE to obtain the desired results. You can use the query feature in a DBMS to complete the exercises using a computer or you can simply write a description of how you would complete the task. Check with your instructor if you are uncertain of which approach to take.

1. List the author number and last name for every author.

2. List the complete Branch table (all rows and all columns).

3. List the name of every publisher located in Boston.

4. List the name of every publisher not located in Boston.

5. List the name of every branch that has at least nine employees.

6. List the book code and title of every book that has the type SFI.

7. List the book code and title of every book that has the type SFI and that is a paperback.

8. List the book code and title of every book that has the type SFI or that has the publisher code PE.

9. List the book code, title, and price for each book with a price that is greater than $5 but less than $10.

10. List the book code and title of every book that has the type FIC and a price of less than $10.

11. Customers who are part of a special program get a 15% discount off regular book prices. To determine the discounted prices, list the book code, title, and discounted price of every book. (Your calculated column should determine 85% of the current price, which is 100% less a 15% discount.)

12. List the book code and title of every book that has the type SFI, HOR, or ART.

13. List the book code, title, and publisher code for all books. Sort the results by title within publisher code.

14. How many books have the type SFI?

15. Calculate the average price for each type of book.

16. For every book, list the book code, book title, publisher code, and publisher name.

17. For every book published by Taunton Press, list the book title and book price.

18. List the book title and book code for every book published by Putnum Publishing Group that has a book price greater than $15.

19. Create a new table named Fiction using the data in the BookCode, Title, PublisherCode, and Price columns in the Book table for those books that have the type FIC.

20. Use an update query to change the price of any book in the Fiction table with a current price of 14.00 to 14.50.

21. Use a delete query to delete all books in the Fiction table that have the publisher code VB.

CHAPTER 3

The Relational Model 2: SQL

OBJECTIVES

- Introduce the structured query language (SQL).

- Use simple and compound conditions in SQL.

- Use computed fields in SQL.

- Use built-in SQL functions.

- Use subqueries in SQL.

- Group records in SQL.

- Join tables using SQL.

- Perform union operations in SQL.

- Use SQL to update database data.

- Use an SQL query to create a table in a database.

Introduction

In this chapter, you will examine the language called **SQL (structured query language)**. Like Access and Query-By-Example (QBE), SQL provides users with the capability of querying a relational database. However, in SQL, you must enter **commands** to obtain the desired results, rather than completing an on-screen form as you do in Access and QBE. SQL uses commands to create tables, update tables, and retrieve data from tables. The commands that are used to retrieve data are usually called queries.

SQL was developed under the name SEQUEL at the IBM San Jose research facilities as the data manipulation language for IBM's prototype relational database management system (DBMS), System R, in the mid-1970s. In 1980, it was renamed SQL [but still pronounced "sequel" although the equally popular pronunciation of "S-Q-L" ("ess-cue-ell") is used in this book] to avoid confusion with an unrelated hardware product called SEQUEL. SQL is used as the data manipulation language for IBM's current production offerings—SQL/DS and DB2—in the relational DBMS arena. Most relational DBMSs use a version of SQL as a data manipulation language. SQL is the standard language for relational database manipulation.

You will begin studying SQL by examining how to use it to create a database. You will examine simple retrieval methods and compound conditions. You will use computed fields in SQL and learn how to sort data. Then you will learn how to use built-in functions, nested queries, and grouping. You will learn how to join tables and use the UNION operator. Finally, you will use SQL to update data in a database. The end of this chapter includes generic versions of all the SQL commands presented in the chapter.

Table Creation

You use the SQL **CREATE TABLE** command to describe the layout of a table. The word TABLE is followed by the name of the table to be created and then by the names and data types of the columns (fields) that comprise the table. The rules for naming tables and columns vary slightly from one version of SQL to another. If you have any doubts about the validity of any of the names you have chosen, you should consult the manual for your SQL version.

Some common restrictions placed on table and column names by DBMSs are:

1. The names cannot exceed 18 characters.

2. The names must start with a letter.

3. The names can contain only letters, numbers, and underscores (_).

4. The names cannot contain spaces.

Note: Unlike some other versions of SQL, Access SQL permits the use of spaces within table and column names. There is a restriction, however, on the way such names are used in SQL commands. When you use a name containing a space in Access SQL, you must enclose it in square brackets. For example, if the name of the CreditLimit column were changed to Credit Limit (with a space between "Credit" and "Limit"), you would write the column as [Credit Limit] because the name includes a space.

Note: In systems that permit the use of uppercase and lowercase letters in table and column names, you can avoid using spaces by capitalizing the first letter of each word in the name and using lowercase letters for the remaining letters in the words. For example, the name of the credit limit column would be CreditLimit. In systems that do not permit the use of spaces or mixed-case letters, some programmers use an underscore character to separate words. For example, the name of the credit limit field would be CREDIT_LIMIT.

For each field in a table, you must specify the type of data that the field can store. Although the actual data types will vary slightly from one implementation of SQL to another, the following list indicates the data types you will often encounter:

1. *INTEGER.* Stores integers, which are numbers without a decimal part. The valid data range is -2147483648 to 2147483647. You can use the contents of INTEGER fields for arithmetic.

2. *SMALLINT.* Stores integers, but uses less space than the INTEGER data type. The valid data range is -32768 to 32767. SMALLINT is a better choice than INTEGER when you are certain that the field will store numbers within the indicated range. You can use the contents of SMALLINT fields for arithmetic.

3. *DECIMAL(p,q).* Stores a decimal number p digits long with q of these digits being decimal places. For example, DECIMAL(5,2) represents a number with three places to the left and two places to the right of the decimal. You can use the contents of DECIMAL fields for arithmetic.

4. *CHAR(n).* Stores a character string n characters long. You use the CHAR type for fields that contain letters and other special characters, and for fields that contain numbers that will not be used for arithmetic. Because neither sales rep numbers nor customer numbers will be used in any arithmetic operations, for example, they both are assigned CHAR as the data type. (Some DBMSs use TEXT rather than CHAR, but they mean the same thing.)

5. *DATE.* Stores dates in the form DD-MON-YYYY or MM/DD/YYYY. For example, May 12, 2007, could be stored as 12-MAY-2007 or 5/12/2007.

EXAMPLE 1 : Describe the layout of the Rep table to the DBMS using SQL.

The general CREATE TABLE command for the Rep table is:

```
CREATE TABLE Rep
(RepNum  CHAR(2),
LastName  CHAR(15),
FirstName  CHAR(15),
Street  CHAR(15),
City  CHAR(15),
State  CHAR(2),
Zip  CHAR(5),
Commission  DECIMAL(7,2),
Rate  DECIMAL(3,2) )
;
```

In this SQL command, which uses the data definition features of SQL, you're describing a table that will be named Rep. It contains nine fields: RepNum, LastName, FirstName, Street, City, State, Zip, Commission, and Rate. RepNum is a character field, two positions in length. LastName is a character field with 15 characters. Commission is a numeric field that stores seven digits, including two decimal places. Similarly, Rate is a numeric field that stores three digits, including two decimal places.

Note: Because many versions of SQL require you to end a command with a semicolon, commands in this text will end with semicolons.

Note: In SQL, commands are free-format. No rule says that a particular word must begin in a particular position on the line. The previous SQL command could have been written as:

```
CREATE TABLE Rep (RepNum CHAR(2),LastName CHAR(15),
FirstName CHAR(15),Street CHAR(15), City CHAR(15),
State CHAR(2),Zip CHAR(5),Commission DECIMAL(7,2),
Rate DECIMAL(3,2) )
;
```

The manner in which the first CREATE TABLE command was written simply makes the command more readable. In general, you should strive for such readability when you write SQL commands.

Note: Access doesn't have a DECIMAL data type. To create numbers with decimals, you must use either the CURRENCY or NUMBER data type. Use the CURRENCY data type for fields that will contain currency values; use the NUMBER data type for all other numeric fields. The corresponding CREATE TABLE command for Access is:

```
CREATE TABLE Rep
(RepNum  CHAR(2),
LastName  CHAR(15),
FirstName  CHAR(15),
Street  CHAR(15),
City  CHAR(15),
State  CHAR(2),
Zip  CHAR(5),
Commission  CURRENCY,
Rate  NUMBER )
;
```

■ Simple Retrieval

The basic form of an SQL retrieval command is simply SELECT-FROM-WHERE. After the word SELECT, you list those fields you wish to display in the query results. This portion of the command is called the **SELECT clause**. The fields will appear in the query results in the order in which they are listed in the SELECT clause. After the word FROM, you list the table or tables that contain the data to display in the query results. This portion of the command is called the **FROM clause**. Finally, after the word WHERE, you list any conditions, such as the credit limit must be $10,000, that you want to apply to the data you want to retrieve. This portion of the command, which is optional, is called the **WHERE clause**.

There are no special formatting rules in SQL. In this text, you place the word FROM on a new line, and then place the word WHERE (when it is used) on the next line to make the command more readable.

Note: This text uses a common style in which words that are part of the SQL language, called **reserved words**, appear in all uppercase letters. All other words in commands appear in a combination of uppercase and lowercase letters.

Note: The SQL version used in the examples is Microsoft Office Access 2003. Although the various versions of SQL are not identical, the differences are relatively minor. After you have mastered one version of SQL, you can apply your skills to learn another version of SQL.

Note: If you plan to work through the examples in this chapter using a computer, you should use a copy of the original Premiere Products database because the version of the database used in subsequent chapters does not include these changes.

EXAMPLE 2 : List the number, name, and balance of all customers.

Because you want to list all customers, you won't need to use the WHERE clause—you don't need to put any restrictions on the data to retrieve. Figure 3.1 shows the query to select the number, name, and balance of all customers.

FIGURE 3.1
· · · · · · · · · · · · · · · ·
SQL query to select customer data

Note: In Access 2003, you can change the font size for queries created in SQL view. The commands and results in this chapter reflect a change in the font size from the default size of 8 points to 10 points.

The results of executing this query appear in Figure 3.2.

FIGURE 3.2
· · · · · · · · · · · · · · · ·
Query results

EXAMPLE 3 : List the complete Part table.

You could use the same approach as shown in Example 2 by listing each field in the Part table in the SELECT clause. However, there is a shortcut. Instead of listing all the field names in the SELECT clause, you can use the * symbol. When used after the word SELECT, the * symbol indicates that you want to include all fields in the query results in the order in which you described them to the system during data definition. If you want to include all the fields in the query results, but in a different order, you could type the names of the fields in the order in which you want them to appear. In this case, assuming the default order is appropriate, the query design appears in Figure 3.3.

FIGURE 3.3
· · · · · · · · · · · · · · · ·
SQL query to list
the complete
Part table

The query results appear in Figure 3.4.

FIGURE 3.4
· · · · · · · · · · · · · · · ·
Query results

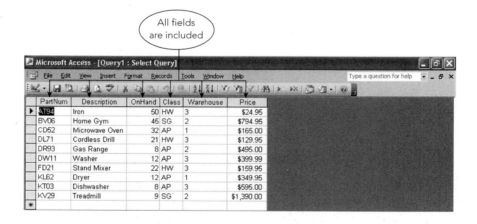

EXAMPLE 4 : List the name of every customer with a $10,000 credit limit.

You include the following condition in the WHERE clause to restrict the query results to only those customers with a credit limit of $10,000:

`WHERE CreditLimit=10000`

Notice that you do not type commas in numbers. The query design appears in Figure 3.5.

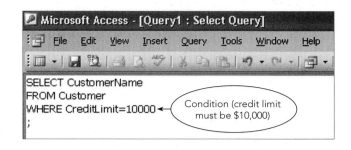

The query results appear in Figure 3.6.

FIGURE 3.6

. . . .

Query results

The condition in the preceding WHERE clause is called a simple condition. A **simple condition** includes the field name, comparison operator, and then either another field name or a value, such as CreditLimit = 10000 or CreditLimit > Balance. Figure 3.7 lists the comparison operators that you can use in SQL commands. Notice that there are two versions of the "not equal to" operator: < > and !=. You must use the correct one for your version of SQL. If you use the wrong one, your system will generate an error, in which case, you'll know to use the other version.

Comparison Operator	Meaning
=	Equal to
<	Less than
>	Greater than
<=	Less than or equal to
>=	Greater than or equal to
< >	Not equal to (used by most implementations of SQL)
!=	Not equal to (used by some implementations of SQL)

In Example 4, the WHERE clause compared a numeric field (CreditLimit) to a number (10000). When a query involves a character field, such as CustomerNum or CustomerName, you must enclose the value to which the field is being compared in single quotation marks, as illustrated in Examples 5 and 6.

EXAMPLE 5 : Find the name of customer 148.

The query design appears in Figure 3.8. Because CustomerNum is a character field, the value 148 is enclosed in single quotation marks.

The query results appear in Figure 3.9. Only a single record appears in the query results because the CustomerNum field is the primary key for the Customer table and there can be only one customer with the number 148.

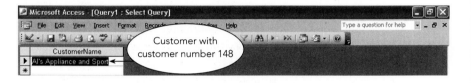

EXAMPLE 6 : Find the customer name for every customer located in the city of Grove.

The query design appears in Figure 3.10.

The query results appear in Figure 3.11. Because more than one customer is located in Grove, there are multiple records in the query results.

74

FIGURE 3.11
....
Query results

You can also use dates in conditions. The way you do so varies slightly from one implementation of SQL to another. In Access, you place number signs around the date (for example, #11/15/2007#). In others, you enter the day of the month, a hyphen, the three-character abbreviation for the month, a hyphen, and the year, all enclosed in quotation marks (for example, '15-NOV-2007').

EXAMPLE 7 ⋮ List the number, name, credit limit, and balance for all customers with credit limits that exceed their balances.

The query design appears in Figure 3.12. Notice that the condition in the WHERE clause compares the contents of two fields.

The query results appear in Figure 3.13.

FIGURE 3.13
....
Query results

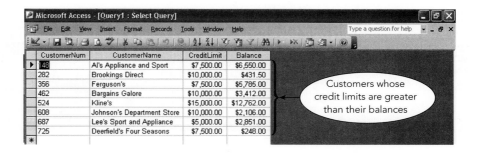

Compound Conditions

The conditions you've seen so far are called simple conditions. The following examples require compound conditions. **Compound conditions** are formed by connecting two or more simple conditions using one or both of the following operators: AND and OR. You can also precede a single condition with the NOT operator to negate a condition.

When you connect simple conditions using the AND operator, all the simple conditions must be true for the compound condition to be true. When you connect simple conditions using the OR operator, the compound condition will be true whenever any of the simple conditions are true. Preceding a condition by the NOT operator reverses the truth or falsity of the original condition. That is, if the original condition is true, the new condition will be false; if the original condition is false, the new one will be true.

EXAMPLE 8   List the descriptions of all parts that are located in warehouse 3 and for which there are more than 20 units on hand.

In this example, you want to list those parts for which *both* the warehouse number is equal to 3 *and* the number of units on hand is greater than 20. Thus, you form a compound condition using the AND operator, as shown in Figure 3.14.

FIGURE 3.14
....
Compound condition that uses the AND operator

The query results appear in Figure 3.15.

FIGURE 3.15
....
Query results

EXAMPLE 9   List the descriptions of all parts that are located in warehouse 3 or for which there are more than 20 units on hand, or both.

As you would expect, you form compound conditions with the OR operator in the same way. The compound condition shown in Figure 3.16 uses the OR operator instead of the AND operator.

FIGURE 3.16
••••••••••••••••••
Compound condi-
tion that uses the
OR operator

The query results appear in Figure 3.17.

FIGURE 3.17
••••••••••••••••••
Query results

EXAMPLE 10 : List the descriptions of all parts that are not in warehouse 3.

For this example, you could use a simple condition and the "not equal to" operator (< >).
As an alternative, you could use the "equals" operator (=) in the condition, but precede the
entire condition with the NOT operator, as shown in Figure 3.18.

FIGURE 3.18
••••••••••••••••••
SQL query with the
NOT operator

The query results appear in Figure 3.19.

FIGURE 3.19
· · · · · · · · · · · · · · · · ·
Query results

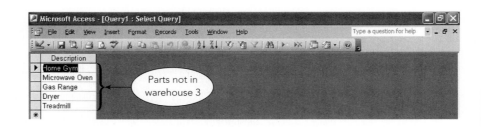

EXAMPLE 11 : List the number, name, and balance of all customers with balances greater than or equal to $1,000 and less than or equal to $5,000.

You could use a WHERE clause and the AND operator (Balance>=1000 AND Balance<=5000). An alternative to this approach uses the BETWEEN operator, as shown in Figure 3.20.

FIGURE 3.20
· · · · · · · · · · · · · · · · ·
SQL query with the
BETWEEN operator

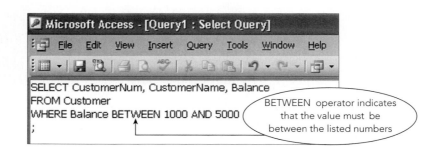

The query results appear in Figure 3.21.

FIGURE 3.21
· · · · · · · · · · · · · · · · ·
Query results

The BETWEEN operator is not an essential feature of SQL; you have just seen that you can obtain the same result without it. Using the BETWEEN operator, however, does make certain SELECT commands easier to construct.

Computed Fields

Similar to QBE, you can include fields in queries that are not in the database but whose values you can compute from existing database fields. A field whose values you derive from existing fields is called a computed or calculated field. Computed fields can involve addition (+), subtraction (-), multiplication (*), or division (/). The query in Example 12, for example, uses subtraction.

EXAMPLE 12 : List the number, name, and available credit for all customers.

There is no field for available credit in the database, but you can compute it using two fields that *are* present in the database: CreditLimit and Balance. The query design shown in Figure 3.22 creates a new field named AvailableCredit, which is computed by subtracting the value in the Balance field from the value in the CreditLimit field (AvailableCredit = CreditLimit - Balance). By using the word AS after the computation followed by AvailableCredit, you can assign a name to the computed field.

FIGURE 3.22
· · · · · · · · · · · · · · · · ·
SQL query with a
computed field

The query results appear in Figure 3.23. The column heading for the computed field is the name that you specified in the SELECT clause.

FIGURE 3.23
· · · · · · · · · · · · · · · · ·
Query results

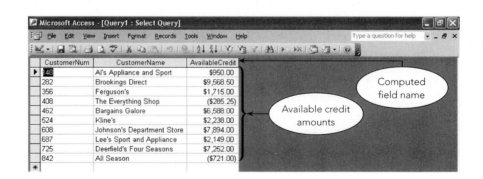

EXAMPLE 13 : List the number, name, and available credit for all customers with credit limits that exceed their balances.

The only difference between Examples 12 and 13 is that Example 13 includes a condition, as shown in Figure 3.24.

FIGURE 3.24
•••••••••••••••••
SQL query with
computed field and
condition

The query results appear in Figure 3.25.

FIGURE 3.25
•••••••••••••••••
Query results

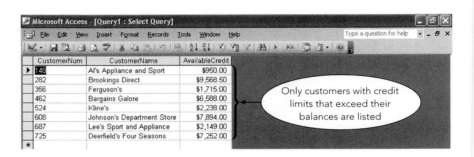

Using Special Operators (LIKE and IN)

In most cases, your conditions will involve exact matches, such as finding all customers whose city is Sheldon. In some cases, however, exact matches will not work. For example, you might know only that the desired value contains a certain collection of characters. In such cases, you use the LIKE operator with a wildcard symbol, as shown in Example 14.

EXAMPLE 14 : List the number, name, and complete address of every customer located on a street that contains the letters "Oxford."

All you know is that the addresses that you want contain a certain collection of characters ("Oxford") somewhere in the Street field, but you don't know where. In Access SQL, the asterisk (*) is used as a wildcard to represent any collection of characters; in some versions of SQL, the percent sign (%) is used as a wildcard. To use a wildcard, use the LIKE operator in the WHERE clause. The query design shown in Figure 3.26 will retrieve information for every customer whose street contains some collection of characters followed by the letters "Oxford," followed potentially by some additional characters.

The query results appear in Figure 3.27.

FIGURE 3.27
. . . .
Query results

Note: Another wildcard symbol in Access SQL is the question mark (?), which represents any individual character. For example, "T?m" represents the letter "T" followed by any single character, followed by the letter "m," and would retrieve records that include the words Tim, Tom, or T3m, for example. Many versions of SQL use the underscore character (_) instead of the question mark to represent any individual character.

Note: In a large database, you should use wildcards only when absolutely necessary. Searches involving wildcards can be extremely slow to process.

Another operator, IN, provides a concise way of phrasing certain conditions, as Example 15 illustrates.

EXAMPLE 15   List the number, name, street, and credit limit for every customer with a credit limit of $7,500, $10,000, or $15,000.

In this query, you use the SQL IN operator to determine whether a credit limit is $7,500, $10,000, or $15,000. You could obtain the same result by using the condition WHERE CreditLimit = 7500 OR CreditLimit = 10000 OR CreditLimit = 15000. The approach shown in Figure 3.28 is simpler, however—the IN clause contains the collection of values 7500, 10000, and 15000. The condition is true for those rows in which the value in the CreditLimit column is in this collection.

FIGURE 3.28

· · · · · · · · · · · · · · · · · ·

SQL query with an
IN operator

The query results appear in Figure 3.29.

FIGURE 3.29

· · · · · · · · · · · · · · · · ·

Query results

Sorting

Recall that the order of rows in a table is considered to be immaterial. From a practical standpoint, this means that when you query a relational database, there are no guarantees concerning the order in which the results will be displayed. The results may appear in the order in which the data was originally entered, but even this is not certain. Thus, if the order in which the data is displayed is important, you should *specifically* request that the results be displayed in a desired order. In SQL, you sort data using the **ORDER BY clause**.

EXAMPLE 16   List the number, name, street, and credit limit of all customers. Order (sort) the customers by name.

The field on which to sort data is called a **sort key**, or simply a **key**. To sort the output, you include the words ORDER BY in the SQL query, followed by the sort key field, as shown in Figure 3.30.

FIGURE 3.30
• • • • • • • • • • • • • • •
SQL query to
sort data

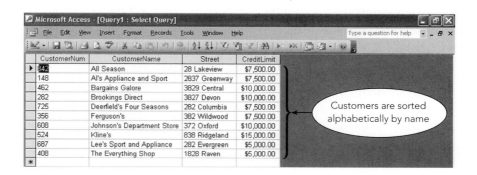

The query results appear in Figure 3.31.

FIGURE 3.31
• • • • • • • • • • • • • • •
Query results

EXAMPLE 17   List the number, name, street, and credit limit of all customers. Order the customers by name within descending credit limit. (In other words, sort the customers by credit limit in descending order. Within each group of customers that have a common credit limit, sort the customers by name.)

When you need to sort data on two fields, the more important key is called the major sort key (also referred to as the primary sort key) and the less important key is called the minor sort key (also referred to as the secondary sort key). In this case, because you need to sort the output by name within credit limit, the CreditLimit field is the major sort key and the CustomerName field is the minor sort key. If there are two sort keys, as in Example 17, the major sort key is listed first. You can specify to sort the output in descending (high-to-low) order by following the sort key with the word DESC, as shown in Figure 3.32.

FIGURE 3.32
••••••••••••••
SQL query to
sort data on
multiple fields

Major (primary) sort key

Descending order

Minor (secondary) sort key

The query results appear in Figure 3.33.

FIGURE 3.33
••••••••••••••••
Query results

Within credit limit, customers are sorted by name

Customers are sorted by credit limit in descending order

■ Built-In Functions

SQL has **built-in functions** (also called **aggregate functions**) to calculate the number of entries, the sum or average of all the entries in a given column, and the largest or smallest values in a given column. In SQL, these functions are called COUNT, SUM, AVG, MAX, and MIN, respectively.

EXAMPLE 18 : How many parts are in item class HW?

In this query, you want to know the number of rows in the query results from selecting only those parts that are in item class HW. You could count the number of part numbers in the query results, or the number of descriptions, or the number of entries in any other field. It doesn't matter which column you choose because all columns will yield the correct answer. Rather than requiring you to pick a column arbitrarily, some versions of SQL allow you to use the * symbol to select any column. In SQL versions that support the * symbol, you could use the query design shown in Figure 3.34.

FIGURE 3.34
•••••••••••••••••
SQL query to count
records

The query results appear in Figure 3.35.

FIGURE 3.35
•••••••••••••••••
Query results

If your implementation of SQL doesn't permit the use of the * symbol, you could write the query as:

```
SELECT COUNT(PartNum)
FROM Part
WHERE Class='HW'
;
```

EXAMPLE 19 : Find the number of customers and the total of their balances.

There are two differences between COUNT and SUM—other than the obvious fact that they are computing different statistics: In the case of SUM, you *must* specify the field for which you want a total, and the field must be numeric. (How could you calculate a sum of names or addresses?) The query design appears in Figure 3.36.

FIGURE 3.36
•••••••••••••••••
SQL query to count
records and
calculate a total

The query results appear in Figure 3.37.

FIGURE 3.37
....
Query results

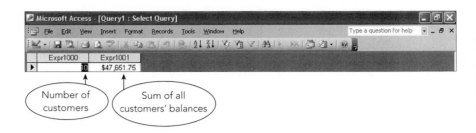

The use of AVG, MAX, and MIN is similar to the use of SUM. The only difference is that different statistics are calculated.

EXAMPLE 20 : Find the total number of customers and the total of their balances. Change the column names for the number of customers and the total of their balances to CustomerCount and BalanceTotal, respectively.

Just as with computed fields, you can use the word AS to assign names to these computations, as shown in Figure 3.38.

FIGURE 3.38
....
SQL query to
perform
calculations and
rename fields

The query results appear in Figure 3.39.

FIGURE 3.39
....
Query results

Nesting Queries

It is possible to place one query inside another to create a nested query. The inner query is called a **subquery**; the subquery is evaluated first. After the subquery has been evaluated, the outer query can be evaluated. Example 21 illustrates the process.

EXAMPLE 21   List the order number for each order that contains an order line for a part located in warehouse 3.

You can find the answer by first using the Part table and creating a list of part numbers for those parts in warehouse 3. Then you can use the OrderLine table to find those order numbers present in any row on which the part number is in the results you created in the inner query. The corresponding query design appears in Figure 3.40.

FIGURE 3.40
• • • • • • • • • • • • • • •
SQL query with subquery

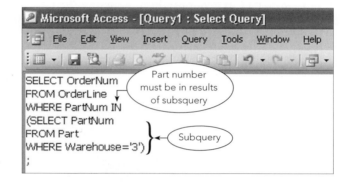

Note: Although it is not required, it is common to enclose subqueries in parentheses for readability.

The query results appear in Figure 3.41.

FIGURE 3.41
• • • • • • • • • • • • • • •
Query results

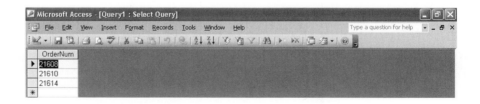

The inner query, which finds all the part numbers in the Part table with a warehouse number of 3, is the subquery. The subquery is evaluated first, producing a list of part numbers. After the subquery has been evaluated, the outer query is evaluated. Order numbers in the results are those that appear in any row in the OrderLine table for which the part number in the row is in the subquery results.

▉ Grouping

Recall from Chapter 2 that grouping means creating groups of records that share some common characteristic. In grouping customers by sales rep number, for example, the customers of sales rep 20 would form one group, the customers of sales rep 35 would form a second group, and the customers of sales rep 65 would form a third group.

In Example 22, you need to group customers by rep number to perform the necessary computations.

EXAMPLE 22 : For each sales rep, list the rep number, the number of customers assigned to the rep, and the average balance of the rep's customers. Group the records by rep number and order the records by rep number.

This type of query requires grouping by rep number to make the correct calculations for each group. To indicate grouping in SQL, you use the **GROUP BY clause**, as shown in Figure 3.42. It is important to note that the GROUP BY clause does not mean that the query results will be sorted. To display the query results in a particular order, you must use the ORDER BY clause. The query design in Figure 3.42 uses the ORDER BY clause to sort the query results by rep number.

FIGURE 3.42

· · · · · · · · · · · · · · · · ·

SQL query to group records

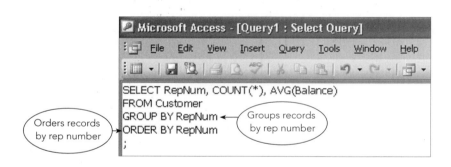

The query results appear in Figure 3.43.

FIGURE 3.43

· · · · · · · · · · · · · · · · ·

Query results

When rows are grouped, one line of output is produced for each group. The only things that may be displayed in the grouped results are statistics calculated for the group or fields whose values are the same for all rows in a group.

QUESTION   **Why is it appropriate to display the rep number?**

ANSWER   Because the output is grouped by rep number, the rep number in one row in a group must be the same as the rep number in any other row in the group.

QUESTION   **Would it be appropriate to display a customer number?**

ANSWER   No, because the customer number will vary from one row in a group to another. (SQL could not determine which customer number to display for the group.)

EXAMPLE 23   For each sales rep with fewer than four customers, list the rep number, the number of customers assigned to the rep, and the average balance of the rep's customers. Rename the count of the number of customers and the average of the balances to NumCustomers and AverageBalance, respectively. Order the groups by rep number.

Examples 22 and 23 are similar, but there are two important differences: You need to rename the fields, and there is a restriction to display the calculations for only those reps having fewer than four customers. In other words, you want to display only those groups for which COUNT(*) is less than four. This restriction does not apply to individual rows, but to *groups*. Because the WHERE clause applies only to rows, it is not the appropriate clause to accomplish the kind of selection you need. Fortunately, the **HAVING clause** is to groups what the WHERE clause is to rows, as shown in Figure 3.44.

FIGURE 3.44

SQL query to restrict the groups that are included

The query results appear in Figure 3.45.

FIGURE 3.45

Query results

In this case, the row created for a group will be displayed only if the count of the number of records in the group is less than four.

You can include both a WHERE clause and a HAVING clause in the same query design, as shown in Figure 3.46.

The query results appear in Figure 3.47.

In this case, the WHERE clause will restrict the rows from the Customer table to those rows in which the credit limit is less than $10,000. These rows will then be grouped by rep number. The HAVING clause then restricts the groups to be displayed to those for which the count of the rows in the group is less than three.

Joining Tables

Many queries require data from more than one table. Just as with QBE and relational algebra, it is necessary to be able to join tables; that is, you need to be able to find rows in two or more tables that have identical values in matching fields. In SQL, this is accomplished by entering the appropriate conditions in the WHERE clause.

EXAMPLE 24   List the number and name of each customer together with the number, last name, and first name of the sales rep who represents the customer. Order the records by customer number.

Because the numbers and names of customers are in the Customer table and the numbers and names of sales reps are in the Rep table, you need to include both tables in your SQL query. To join the tables, you'll construct the SQL command as follows:

1. In the SELECT clause, list all fields you want to display.

2. In the FROM clause, list all tables involved in the query.

3. In the WHERE clause, give the condition that will restrict the data to be retrieved to only those rows from the two tables that match; that is, you'll restrict it to the rows that have common values in matching fields.

Just as in relational algebra, it is often necessary to qualify a field, that is, to specify the particular field you are referencing. To qualify a field name, precede the name of the field with the name of the table, followed by a period. The RepNum field in the Rep table is written as Rep.RepNum, and the RepNum field in the Customer table is written as Customer.RepNum. The query design appears in Figure 3.48.

FIGURE 3.48
•••••••••••••••••
SQL query to
join tables

The query results appear in Figure 3.49.

FIGURE 3.49
•••••••••••••••••
Query results

When there is potential ambiguity when listing field names, you *must* qualify the fields involved. It is permissible to qualify other fields as well, even if there is no possible confusion. Some people prefer to qualify all fields, and this is certainly not a bad approach. In this text, however, you will qualify fields only when it is necessary to do so.

EXAMPLE 25

List the number and name of each customer whose credit limit is $10,000, together with the number, last name, and first name of the sales rep who represents the customer. Order the records by customer number.

In Example 24, the condition in the WHERE clause served only to relate a customer to a sales rep. Although relating a customer to a sales rep is essential in this example as well, you also need to restrict the output to only those customers whose credit limit is $10,000. You can accomplish this goal by using the AND operator to create a compound condition, as shown in Figure 3.50.

FIGURE 3.50
• • • • • • • • • • • • • • •
SQL query to restrict the records in a join

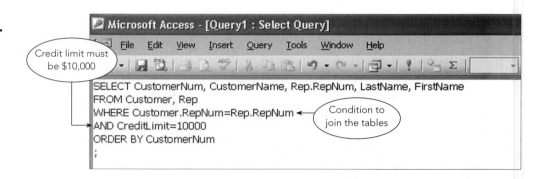

The query results appear in Figure 3.51.

FIGURE 3.51
• • • • • • • • • • • • • • • •
Query results

It is possible to join more than two tables, as illustrated in Example 26. For each pair of tables to join, you must include a condition indicating how the tables are related.

EXAMPLE 26

For every order, list the order number, order date, customer number, and customer name. In addition, for each order line within the order, list the part number, description, number ordered, and quoted price. Order the records by order number.

The order number and date are stored in the Orders table. The customer number and name are stored in the Customer table. The part number and description are stored in the Part table. The number ordered and quoted price are stored in the OrderLine table. Thus, you need to join *four* tables: Orders, Customer, OrderLine, and Part. The procedure for joining more than two tables is essentially the same as the one for joining two tables. The difference is that the condition in the WHERE clause will be a compound condition, as shown in Figure 3.52. The first condition relates an order to a customer using the common CustomerNum columns. The second condition relates the order to an order line using the common OrderNum columns. The final condition relates the order line to a part using the common PartNum columns.

FIGURE 3.52
....
SQL query to join multiple tables

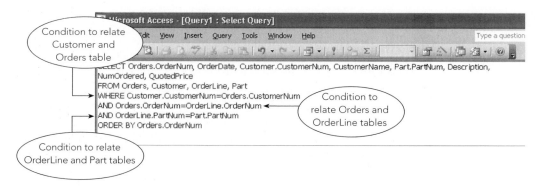

The query results appear in Figure 3.53.

FIGURE 3.53
....
Query results

The query shown in Figure 3.52 is more complex than many of the previous ones. You might think that SQL is not such an easy language to use after all. If you take it one step at a time, however, the query in Example 26 really isn't that difficult. To construct a detailed query in a step-by-step fashion, do the following:

1. List in the SELECT clause all the columns that you want to display. If the name of a column appears in more than one table, precede the column name with the table name (that is, qualify the column name).

2. List in the FROM clause all the tables involved in the query. Usually you include the tables that contain the columns listed in the SELECT clause. Occasionally, however, there might be a table that does not contain any columns used in the SELECT clause but that does contain columns used in the WHERE clause. In this case, you must also

list the table. For example, if you do not need to list a customer number or name, but you do need to list the sales rep name, you wouldn't include any columns from the Customer table in the SELECT clause. The Customer table is still required, however, because you must include columns from it in the WHERE clause.

3. Take one pair of related tables at a time, and indicate in the WHERE clause the condition that relates the tables. Join these conditions with the AND operator. If there are any other conditions, include them in the WHERE clause and connect them to the other conditions with the AND operator.

■ Union

Recall from Chapter 2 that the union of two tables is a table containing all rows that are in either the first table, the second table, or both tables. The two tables involved in a union *must* have the same structure, or be union compatible; in other words, they must have the same number of fields and their corresponding fields must have the same data types. If, for example, the first field in one table contains customer numbers, the first field in the other table also must contain customer numbers.

EXAMPLE 27   List the number and name of all customers that are either represented by sales rep 35 or that currently have orders on file, or both.

You can create a table containing the number and name of all customers that are represented by sales rep 35 by selecting customer numbers and names from the Customer table in which the sales rep number is 35. Then you can create another table containing the number and name of every customer that currently has orders on file by joining the Customer and Orders tables. The two tables created by this process have the same structure—fields named CustomerNum and CustomerName. Because the tables are union compatible, it is possible to take the union of these two tables, which is the appropriate operation for this example, as shown in Figure 3.54.

FIGURE 3.54
....
SQL query to
perform a union

The query results appear in Figure 3.55.

FIGURE 3.55
••••••••••••••••
Query results

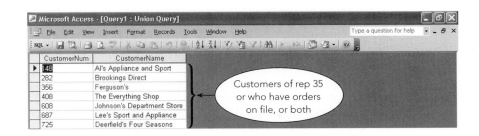

If a SQL implementation truly supports the union operation, it will remove any duplicate rows (for instance, any customers that are represented by sales rep 35 *and* that currently have orders on file will not appear twice). Some SQL implementations have a union operation but will not remove duplicate values.

Updating Tables

There are more uses to SQL than simply retrieving data from a database and creating tables. SQL has several other capabilities, including the ability to update a database, as demonstrated in the following examples.

Note: If you plan to work through the examples in this section using Access, you should use a copy of the original Premiere Products database because the version of the database used in subsequent chapters does not include these changes. As an alternative, if you are using a DBMS (such as Oracle) that supports the ROLLBACK command, which reverses changes to a database, you can ensure your changes are undone by typing the word ROLLBACK before exiting the DBMS. If you have any questions concerning which of these (or other) approaches is appropriate for you, check with your instructor.

EXAMPLE 28 : Change the street address of customer 524 to 1445 Rivard.

You can use the SQL **UPDATE** command to make changes to existing data. After the word UPDATE, you indicate the table to be updated. After the word SET, you indicate the field to be changed, followed by an equals sign and the new value. Finally, you can include a condition in the WHERE clause, in which case, only the records that satisfy the condition will be changed. The SQL command for this example appears in Figure 3.56. When you run this query in Access, a dialog box opens and indicates the number of records the UPDATE command will affect. In this case, you would update only one record because the WHERE clause selects customer 524.

FIGURE 3.56
••••••••••••••••••
SQL query to
update data

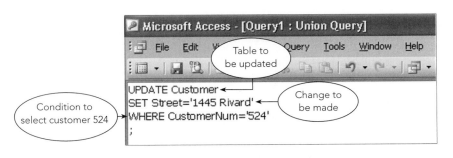

EXAMPLE 29   Add a new sales rep to the Rep table. Her number is 16, her name is Sharon Rands, and her address is 826 Raymond, Altonville, FL 32543. She has not yet earned any commission, but her commission rate is 5% (0.05).

To add new data to a table, you use the **INSERT** command. After the words INSERT INTO, you list the name of the table, followed by the word VALUES. Then you list the values for each of the columns in parentheses, as shown in Figure 3.57. Character values must be enclosed within single quotation marks. When you run this query in Access, a dialog box opens and indicates the number of records the INSERT command will append to the table. In this case, you would add one record to the Rep table.

FIGURE 3.57
· · · · · · · · · · · · · · · · ·
SQL query to
insert a row

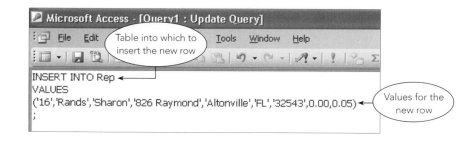

EXAMPLE 30   Delete any row in the OrderLine table in which the part number is BV06.

To delete data from the database, use the **DELETE** command, which consists of the words DELETE followed by a FROM clause identifying the table. Use a WHERE clause to specify a condition. Any records satisfying the condition will be deleted. The DELETE command for this example is shown in Figure 3.58. When you run this query in Access, a dialog box opens and indicates the number of records the DELETE command will delete. In this case, you would delete only one record because the WHERE clause selects part number BV06.

FIGURE 3.58
· · · · · · · · · · · · · · · · ·
SQL query to
delete rows

Creating a Table from a Query

You can save the results of a query as a table by including the **INTO clause** in the query, as illustrated in Example 31.

EXAMPLE 31   Create a new table named SmallCust, consisting of all fields from the Customer table and those rows in which the credit limit is less than or equal to $7,500.

To create the SmallCust table, create a query to select all fields from the Customer table, include a WHERE clause to restrict the rows to those in which CreditLimit <= 7500, and include an INTO clause. The INTO clause precedes the FROM clause and consists of the word INTO followed by the name of the table to be created. The query design appears in Figure 3.59. When you run this query in Access, a dialog box opens and indicates the number of records the INTO clause will paste into the new table. In this case, you would add six rows to the SmallCust table.

FIGURE 3.59
· · · · · · · · · · · · · · · · ·
Query to create a
new table

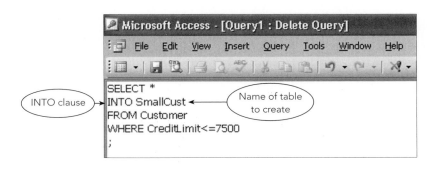

After you execute this query, you can use the SmallCust table shown in Figure 3.60, which is just like any other table you might have created using the CREATE TABLE command.

FIGURE 3.60
· · · · · · · · · · · · · · · · ·
SmallCust table
created by query

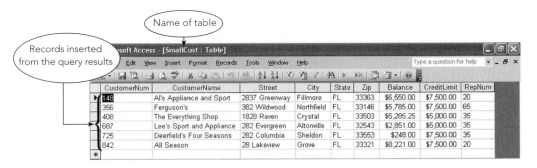

■ Summary of SQL Commands

This section contains generic versions of SQL commands for every example presented in this chapter (the example numbers match the ones used in the chapter, making it easy to return to the page in the chapter on which the example is described). In most cases, commands in Access are identical to the generic versions. For those commands that differ in other SQL implementations, both the generic version and the Access version are included.

EXAMPLE 1 : Describe the layout of the Rep table to the DBMS using SQL.

```
CREATE TABLE Rep
(RepNum  CHAR(2),
LastName  CHAR(15),
FirstName  CHAR(15),
Street  CHAR(15),
City  CHAR(15),
State  CHAR(2),
Zip  CHAR(5),
Commission  DECIMAL(7,2),
Rate  DECIMAL(3,2) )
;
```

Access:

```
CREATE TABLE Rep
(RepNum  CHAR(2),
LastName  CHAR(15),
FirstName  CHAR(15),
Street  CHAR(15),
City  CHAR(15),
State  CHAR(2),
Zip  CHAR(5),
Commission  CURRENCY,
Rate  NUMBER )
;
```

EXAMPLE 2 : List the number, name, and balance of all customers.

```
SELECT CustomerNum, CustomerName, Balance
FROM Customer
;
```

EXAMPLE 3 : List the complete Part table.

```
SELECT *
FROM Part
;
```

EXAMPLE 4 : List the name of every customer with a $10,000 credit limit.

```
SELECT CustomerName
FROM Customer
WHERE CreditLimit=10000
;
```

EXAMPLE 5 : Find the name of customer 148.

```
SELECT CustomerName
FROM Customer
WHERE CustomerNum='148'
;
```

EXAMPLE 6 : Find the customer name for every customer located in the city of Grove.

```
SELECT CustomerName
FROM Customer
WHERE City='Grove'
;
```

EXAMPLE 7 : List the number, name, credit limit, and balance for all customers with credit limits that exceed their balances.

```
SELECT CustomerNum, CustomerName, CreditLimit, Balance
FROM Customer
WHERE CreditLimit>Balance
;
```

EXAMPLE 8 : List the descriptions of all parts that are located in warehouse 3 and for which there are more than 20 units on hand.

```
SELECT Description
FROM Part
WHERE Warehouse='3'
AND OnHand>20
;
```

EXAMPLE 9 : List the descriptions of all parts that are located in warehouse 3 or for which there are more than 20 units on hand, or both.

```
SELECT Description
FROM Part
WHERE Warehouse='3'
OR OnHand>20
;
```

EXAMPLE 10 : List the descriptions of all parts that are not in warehouse 3.

```
SELECT Description
FROM Part
WHERE NOT Warehouse='3'
;
```

EXAMPLE 11 : List the number, name, and balance of all customers with balances greater than or equal to $1,000 and less than or equal to $5,000.

```
SELECT CustomerNum, CustomerName, Balance
FROM Customer
WHERE Balance BETWEEN 1000 AND 5000
;
```

E XAMPLE 12

List the number, name, and available credit for all customers.

```
SELECT CustomerNum, CustomerName, CreditLimit-Balance
AS AvailableCredit
FROM Customer
;
```

E XAMPLE 13

List the number, name, and available credit for all customers with credit limits that exceed their balances.

```
SELECT CustomerNum, CustomerName, CreditLimit-Balance
AS AvailableCredit
FROM Customer
WHERE CreditLimit>Balance
;
```

E XAMPLE 14

List the number, name, and complete address of every customer located on a street that contains the letters "Oxford."

```
SELECT CustomerNum, CustomerName, Street, City, State, Zip
FROM Customer
WHERE Street LIKE '%Oxford%'
;
```

Access:

```
SELECT CustomerNum, CustomerName, Street, City, State, Zip
FROM Customer
WHERE Street LIKE '*Oxford*'
;
```

E XAMPLE 15

List the number, name, street, and credit limit for every customer with a credit limit of $7,500, $10,000, or $15,000.

```
SELECT CustomerNum, CustomerName, Street, CreditLimit
FROM Customer
WHERE CreditLimit IN (7500, 10000, 15000)
;
```

E XAMPLE 16

List the number, name, street, and credit limit of all customers. Order (sort) the customers by name.

```
SELECT CustomerNum, CustomerName, Street, CreditLimit
FROM Customer
ORDER BY CustomerName
;
```

100

EXAMPLE 17 List the number, name, street, and credit limit of all customers. Order the customers by name within descending credit limit.

```
SELECT CustomerNum, CustomerName, Street, CreditLimit
FROM Customer
ORDER BY CreditLimit DESC, CustomerName
;
```

EXAMPLE 18 How many parts are in item class HW?

```
SELECT COUNT(*)
FROM Part
WHERE Class='HW'
;
```

EXAMPLE 19 Find the number of customers and the total of their balances.

```
SELECT COUNT(*), SUM(Balance)
FROM Customer
;
```

EXAMPLE 20 Find the total number of customers and the total of their balances. Change the column names for the number of customers and the total of their balances to CustomerCount and BalanceTotal, respectively.

```
SELECT COUNT(*) AS CustomerCount, SUM(Balance) AS BalanceTotal
From Customer
;
```

EXAMPLE 21 List the order number for each order that contains an order line for a part located in warehouse 3.

```
SELECT OrderNum
FROM OrderLine
WHERE PartNum IN
(SELECT PartNum
FROM Part
WHERE Warehouse='3')
;
```

EXAMPLE 22 For each sales rep, list the rep number, the number of customers assigned to the rep, and the average balance of the rep's customers. Group the records by rep number and order the records by rep number.

```
SELECT RepNum, COUNT(*), AVG(Balance)
FROM Customer
GROUP BY RepNum
ORDER BY RepNum
;
```

EXAMPLE 23 : For each sales rep with fewer than four customers, list the rep number, the number of customers assigned to the rep, and the average balance of the rep's customers. Rename the count of the number of customers and the average of the balances to NumCustomers and AverageBalance, respectively. Order the groups by rep number.

```
SELECT RepNum, COUNT(*) AS NumCustomers, AVG(Balance)
AS AverageBalance
FROM Customer
GROUP BY RepNum
HAVING COUNT(*)<4
ORDER BY RepNum
;
```

EXAMPLE 24 : List the number and name of each customer together with the number, last name, and first name of the sales rep who represents the customer. Order the records by customer number.

```
SELECT CustomerNum, CustomerName, Rep.RepNum, LastName, FirstName
FROM Customer, Rep
WHERE Customer.RepNum=Rep.RepNum
ORDER BY CustomerNum
;
```

EXAMPLE 25 : List the number and name of each customer whose credit limit is $10,000, together with the number, last name, and first name of the sales rep who represents the customer. Order the records by customer number.

```
SELECT CustomerNum, CustomerName, Rep.RepNum, LastName, FirstName
FROM Customer, Rep
WHERE Customer.RepNum=Rep.RepNum
AND CreditLimit=10000
ORDER BY CustomerNum
;
```

EXAMPLE 26 : For every order, list the order number, order date, customer number, and customer name. In addition, for each order line within the order, list the part number, description, number ordered, and quoted price. Order the records by order number.

```
SELECT Orders.OrderNum, OrderDate, Customer.CustomerNum,
CustomerName, Part.PartNum, Description, NumOrdered, QuotedPrice
FROM Orders, Customer, OrderLine, Part
WHERE Customer.CustomerNum=Orders.CustomerNum
AND Orders.OrderNum=OrderLine.OrderNum
AND OrderLine.PartNum=Part.PartNum
ORDER BY Orders.OrderNum
;
```

EXAMPLE 27 : List the number and name of all customers that are either represented by sales rep 35 or that currently have orders on file, or both.

```
SELECT CustomerNum, CustomerName
FROM Customer
WHERE RepNum='35'
UNION
SELECT Customer.CustomerNum, CustomerName
FROM Customer, Orders
WHERE Customer.CustomerNum=Orders.CustomerNum
;
```

EXAMPLE 28 : Change the street address of customer 524 to 1445 Rivard.

```
UPDATE Customer
SET Street='1445 Rivard'
WHERE CustomerNum='524'
;
```

EXAMPLE 29 : Add a new sales rep to the Rep table. Her number is 16, her name is Sharon Rands, and her address is 826 Raymond, Altonville, FL 32543. She has not yet earned any commission, but her commission rate is 5% (0.05).

```
INSERT INTO Rep
VALUES
('16','Rands','Sharon','826 Raymond','Altonville','FL','32543',0.00,0.05)
;
```

EXAMPLE 30 : Delete any row in the OrderLine table in which the part number is BV06.

```
DELETE
FROM OrderLine
WHERE PartNum='BV06'
;
```

EXAMPLE 31 : Create a new table named SmallCust, consisting of all fields from the Customer table and those rows in which the credit limit is less than or equal to $7,500.

```
SELECT *
INTO SmallCust
FROM Customer
WHERE CreditLimit<=7500
;
```

Note: If your version of SQL does not support the INTO clause, you can create the SmallCust table using a CREATE TABLE command (see Example 1). Then you can use the following INSERT command:

```
INSERT INTO SmallCust
SELECT *
FROM Customer
WHERE CreditLimit<=7500
;
```

■ SUMMARY

- Structured query language (SQL) is a language that is used to manipulate relational databases.

- The basic form of an SQL query is SELECT-FROM-WHERE.

- The CREATE TABLE command describes the layout of a table.

- In SQL retrieval commands, fields are listed after SELECT, tables are listed after FROM, and conditions are listed after WHERE.

- In conditions, character values must be enclosed in single quotation marks.

- Compound conditions are formed by combining simple conditions using either or both of the following operators: AND and OR.

- Sorting is accomplished using the ORDER BY clause. The field on which the records are sorted is called the sort key. If the data is sorted in more than one field, the more important field is called the major sort key, or primary sort key. The less important field is called the minor sort key, or secondary sort key.

- Joining tables is accomplished in SQL by using a condition that relates matching rows in the tables to be joined.

- SQL has the built-in (also called aggregate) functions COUNT, SUM, AVG, MAX, and MIN.

- One SQL query may be placed inside another. The inner query is called a subquery and is evaluated first.

- The union of the results of two queries is specified by placing the UNION operator between the two queries.

- Calculated fields are specified in SQL queries by including the calculation, followed by the word AS, and then followed by the name of the calculated field.

- The INSERT command is used to add a new row to a table.

- The UPDATE command is used to change existing data.

- The DELETE command is used to delete records.

- The INTO clause is used in a SELECT command to create a table containing the results of the query.

■ KEY TERMS

aggregate function
built-in function
CHAR(n)
command
compound condition
CREATE TABLE
DATE
DECIMAL(p,q)
DELETE
FROM clause
GROUP BY clause
HAVING clause
INSERT

INTEGER
INTO clause
key
ORDER BY clause
reserved word
SELECT clause
simple condition
SMALLINT
sort key
SQL (structured query language)
subquery
UPDATE
WHERE clause

■ REVIEW QUESTIONS

1. Describe the process of creating a table in SQL and the different data types you can use for fields.

2. What is the purpose of the WHERE clause in SQL? Which comparison operators can you use in a WHERE clause?

3. How do you write a compound condition in an SQL query? When is it true?

4. What is a computed field? How can you use one in an SQL query? How do you assign a name to a computed field?

5. How do you use the LIKE and IN operators in an SQL query?

6. How do you sort data in SQL? If there is more than one sort key, how do you indicate which one is the major sort key? How do you sort data in descending order?

7. What are the SQL built-in functions? How do you use them in an SQL query?

8. What is a subquery? When is a subquery executed?

9. How do you group data in SQL? When you group data in SQL, are there any restrictions on the items that you can include in the SELECT clause?

10. How do you join tables in SQL?

11. How do you qualify the name of a field in an SQL query? When is it necessary to do so?

12. How do you take the union of two tables in SQL? What criteria must the tables meet to make a union possible?

13. Describe the three update commands in SQL.

14. How do you save the results of an SQL query as a table?

■ Premiere Products Exercises

In the following exercises, you will use the data in the Premiere Products database shown in Figure 2.1 in Chapter 2. (If you use a computer to complete these exercises, use a copy of the original Premiere Products database so you will still have the original data when you complete Chapter 4.) In each step, use SQL to obtain the desired results. You can use a DBMS to complete the exercises using a computer, or you can simply write the SQL command to complete each step. Check with your instructor if you are uncertain of which approach to take.

1. List the number and name of all customers.

2. List the complete Part table.

3. List the number and name of every customer represented by sales rep 35.

4. List the number and name of all customers that are represented by sales rep 35 and have credit limits of $10,000.

5. List the number and name of all customers that are represented by sales rep 35 or have credit limits of $10,000.

6. For each order, list the order number, order date, the number of the customer that placed the order, and the name of the customer that placed the order.

7. List the number and name of all customers represented by Juan Perez.

8. How many orders were placed on 10/20/2007?

9. Find the total of the balances for all customers represented by sales rep 35.

10. Give the part number, description, and on-hand value (OnHand * Price) for each part in item class HW.

11. List all columns and all rows in the Part table. Sort the results by part description.

12. List all columns and all rows in the Part table. Sort the results by part number within item class.

13. List the item class and the sum of the value of parts on hand. Group the results by item class.

14. Create a new table named SportingGoods to contain the columns PartNum, Description, OnHand, Warehouse, and Price for all rows in which the item class is SG.

15. In the SportingGoods table, change the description of part BV06 to "Fitness Gym."

16. In the SportingGoods table, delete every row in which the price is greater than $1,000.

■ Henry Books Case

Ray Henry is very aware of the importance of the SQL language in database management. He realizes that he can use SQL to perform the same functions that you performed with queries in Chapter 2. In each of the following steps, use SQL to obtain the desired results using the data shown in Figures 1.17 through 1.20 in Chapter 1. (If you use a computer to complete these exercises, use a copy of the original Henry Books database so you will still have the original data when you complete Chapter 4.) You can use a DBMS to complete the exercises using a computer, or you can simply write the SQL command to complete each step. Check with your instructor if you are uncertain of which approach to take.

1. List the author number and last name for every author.

2. List the complete Branch table (all rows and all columns).

3. List the name of every publisher located in Boston.

4. List the name of every publisher not located in Boston.

5. List the name of every branch that has at least 9 employees.

6. List the book code and title of every book that has the type SFI.

7. List the book code and title of every book that has the type SFI and that is a paper-back.

8. List the book code and title of every book that has the type SFI or that has the publisher code PE.

9. List the book code, title, and price for each book with a price that is greater than $5 but less than $10.

10. List the book code and title of every book that has the type FIC and a price of less than $10.

11. Customers that are part of a special program get a 15% discount off regular book prices. To determine the discounted prices, list the book code, title, and discounted price of every book. (Your calculated column should calculate 85% of the current price, which is 100% less a 15% discount.)

12. List the book code and title of every book that has the type SFI, HOR, or ART.

13. List the book code, title, and publisher code for all books. Sort the results by title within publisher code.

14. How many books have the type SFI?

15. Calculate the average price for each type of book.

16. For every book, list the book code, book title, publisher code, and publisher name.

17. For every book published by Taunton Press, list the book title and book price.

18. List the book title and book code for every book published by Putnam Publishing Group that has a book price greater than $15.

19. Create a new table named Fiction using the data in the BookCode, Title, PublisherCode, and Price columns in the Book table for those books that have the type FIC.

20. Use an update query to change the price of any book in the Fiction table with a current price of 14.00 to 14.50.

21. Use a delete query to delete all books in the Fiction table that have the publisher code VB.

CHAPTER 4

The Relational Model 3: Advanced Topics

OBJECTIVES

- Define, describe, and use views.

- Use indexes to improve database performance.

- Examine the security features of a database management system (DBMS).

- Discuss entity, referential, and legal-values integrity.

- Make changes to the structure of a relational database.

- Define and use the system catalog.

Introduction

In Chapter 3, you used structured query language (SQL) to define table data and manipulate it. In this chapter, you will investigate some other aspects of the relational model. You will learn about views, which represent a way of giving each user his or her own view of the data in a database. You will examine indexes and use them to improve database performance. You also will investigate the features of a database management system (DBMS) that provide security. Then you will learn about important integrity rules and examine ways to change the structure of a database. Finally, you will use the system catalog found in many relational DBMSs to provide users with information about the structure of a database.

Note: In this chapter, concepts are introduced using SQL and followed by the method that you would use to accomplish the same task in Microsoft Access.

Note: If you plan to work through the examples in this chapter using a computer, you should use a copy of the original Premiere Products database because the version of the database used in this chapter does not include the changes made in Chapter 3.

Views

Most DBMSs support the creation of views. A **view** is an application program's or an individual user's picture of the database. An individual can use a view to create reports, charts, and other objects that show database data. In many cases, a user can use a view to examine table data as well. Because a view is usually much less involved than the full database, its use can represent a great simplification. Views also provide a measure of security because omitting sensitive tables or fields from a view will render them unavailable to anyone who is accessing the database via that view.

To illustrate the idea of a view, let's suppose that Juan is interested in the part number, part description, units on hand, and unit price for those Premiere Products parts that are in class HW. He is not interested in any of the other fields in the Part table, nor is he interested in any of the rows that correspond to parts in other item classes. Viewing this data would be simpler for Juan if the other rows and fields were not even present.

Although you cannot change the structure of the Part table and omit some of its rows just for Juan, you can do the next best thing. You can provide him a view that consists of precisely the rows and fields he needs to access. Using SQL, the following CREATE VIEW command creates the view that Juan can use to view the data he needs:

```
CREATE VIEW Housewares AS
SELECT PartNum, Description, OnHand, Price
FROM Part
WHERE Class='HW'
;
```

The SELECT command, which is called the **defining query**, indicates precisely what to include in the view. Conceptually, given the current data in the Premiere Products database, this view will contain the data shown in Figure 4.1. The data does not really exist in this form, however, nor will it *ever* exist in this form. It is tempting to think that when this view is used, the query is executed and will produce some sort of temporary table, named Housewares, which Juan then could access, but this is *not* what actually happens.

110

FIGURE 4.1
••••••••••••••••••
Housewares view

Housewares

PartNum	Description	OnHand	Price
AT94	Iron	50	$24.95
DL71	Cordless Drill	21	$129.95
FD21	Stand Mixer	22	$159.95

Instead, the query acts as a sort of window into the database, as shown in Figure 4.2. As far as Juan is concerned, the entire database is just the darker shaded portion of the Part table. Juan can see any change that affects the darker portion of the Part table, but he is totally unaware of any other changes that are made in the database.

FIGURE 4.2
••••••••••••••••••
Housewares view of
the Premiere
Products database

Rep

RepNum	LastName	FirstName	Street	City	State	Zip	Commission	Rate
20	Kaiser	Valerie	624 Randall	Grove	FL	33321	$20,542.50	0.05
35	Hull	Richard	532 Jackson	Sheldon	FL	33553	$39,216.00	0.07
65	Perez	Juan	1626 Taylor	Fillmore	FL	33336	$23,487.00	0.05

Customer

CustomerNum	CustomerName	Street	City	State	Zip	Balance	CreditLimit	RepNum
148	Al's Appliance and Sport	2837 Greenway	Fillmore	FL	33336	$6,550.00	$7,500.00	20
282	Brookings Direct	3827 Devon	Grove	FL	33321	$431.50	$10,000.00	35
356	Ferguson's	382 Wildwood	Northfield	FL	33146	$5,785.00	$7,500.00	65
408	The Everything Shop	1828 Raven	Crystal	FL	33503	$5,285.25	$5,000.00	35
462	Bargains Galore	3829 Central	Grove	FL	33321	$3,412.00	$10,000.00	65
524	Kline's	838 Ridgeland	Fillmore	FL	33336	$12,762.00	$15,000.00	20
608	Johnson's Department Store	372 Oxford	Sheldon	FL	33553	$2,106.00	$10,000.00	65
687	Lee's Sport and Appliance	282 Evergreen	Altonville	FL	32543	$2,851.00	$5,000.00	35
725	Deerfield's Four Seasons	282 Columbia	Sheldon	FL	33553	$248.00	$7,500.00	35
842	All Season	28 Lakeview	Grove	FL	33321	$8,221.00	$7,500.00	20

(continued)

Orders

OrderNum	OrderDate	CustomerNum
21608	10/20/2007	148
21610	10/20/2007	356
21613	10/21/2007	408
21614	10/21/2007	282
21617	10/23/2007	608
21619	10/23/2007	148
21623	10/23/2007	608

OrderLine

OrderNum	PartNum	NumOrdered	QuotedPrice
21608	AT94	11	$21.95
21610	DR93	1	$495.00
21610	DW11	1	$399.99
21613	KL62	4	$329.95
21614	KT03	2	$595.00
21617	BV06	2	$794.95
21617	CD52	4	$150.00
21619	DR93	1	$495.00
21623	KV29	2	$1,290.00

Part

PartNum	Description	OnHand	Class	Warehouse	Price
AT94	Iron	50	HW	3	$24.95
BV06	Home Gym	45	SG	2	$794.95
CD52	Microwave Oven	32	AP	1	$165.00
DL71	Cordless Drill	21	HW	3	$129.95
DR93	Gas Range	8	AP	2	$495.00
DW11	Washer	12	AP	3	$399.99
FD21	Stand Mixer	22	HW	3	$159.95
KL62	Dryer	12	AP	1	$349.95
KT03	Dishwasher	8	AP	3	$595.00
KV29	Treadmill	9	SG	2	$1,390.00

When you create a query that involves a view, the DBMS changes the query to one that selects data from the table(s) in the database that created the view. Suppose, for example, that Juan creates the following query:

```
SELECT *
FROM Housewares
WHERE OnHand<25
;
```

The DBMS does *not* execute the query in this form. Instead, it merges the query Juan entered with the query that defines the view to form the query that is actually executed. When the DBMS merges the query that creates the view with the query to select rows where the OnHand value is less than 25, the query that the DBMS actually executes is:

```
SELECT PartNum, Description, OnHand, Price
FROM Part
WHERE Class='HW'
AND OnHand<25
;
```

In the query that the DBMS executes, the FROM clause lists the Part table rather than the Housewares view, the SELECT clause lists fields from the Part table instead of * to select all fields from the Housewares view, and the WHERE clause contains a compound condition to select only those parts in the HW class (as Juan sees in the Housewares view) and only those parts with OnHand values of less than 25.

Juan, however, is unaware that this kind of activity is taking place. To Juan, it seems that he really is using a table named Housewares. One advantage of this approach is that because the Housewares view never exists in its own right, any update to the Part table is *immediately* available in Juan's Housewares view. If the Housewares view were really a table, this would not be the case.

To create a view in Access, you simply create and save a query. For example, to create the Housewares view, you would include the PartNum, Description, OnHand, and Price fields from the Part table. You would also include the Class field in the design grid and enter HW as the criterion. Because the Class field isn't included in the view, you would remove the check mark from the Class field's Show check box. Finally, you would save the query using the name Housewares, as shown in Figure 4.3.

FIGURE 4.3
• • • • • • • • • • • • • •
Access query design of the Housewares view

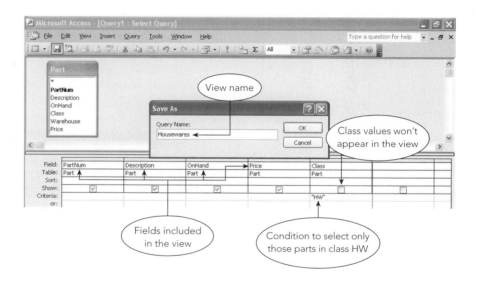

After creating the view, you can use it right away. Figure 4.4 shows the data in the Housewares view. You can create a form for the view, base a report on the view, and treat the view just like it were a table.

FIGURE 4.4
• • • • • • • • • • • • • •
Housewares view datasheet

What if Juan wanted different names for the fields? You can use SQL to change the field names in a view by including the new field names in the CREATE VIEW statement. For example, if Juan wanted the names of the PartNum, Description, OnHand, and Price fields to be PNum, PDesc, OnHd, and Price, respectively, the CREATE VIEW statement would be:

```
CREATE VIEW Housewares (PNum, PDesc, OnHd, Price) AS
SELECT PartNum, Description, OnHand, Price
FROM Part
WHERE Class='HW'
;
```

Now when Juan accesses the Housewares view, he would use the field names PNum, PDesc, OnHd, and Price rather than PartNum, Description, OnHand, and Price.

In Access, you can change the field names by preceding the name of the field with the name you prefer followed by a colon, as shown in Figure 4.5. For example, the desired name, PNum, and a colon precede the field name PartNum.

FIGURE 4.5

Access query design of the Housewares view with changed field names

In the query results, the column headings are PNum, PDesc, OnHd, and Price, as shown in Figure 4.6.

FIGURE 4.6

Datasheet for the Housewares view with changed field names

The Housewares view is an example of a **row-and-column subset view**; that is, it consists of a subset of the rows and columns in some individual table, which, in this case, is the Part table. Because the query can be any SQL query, a view could also join two or more tables.

Suppose, for example, that Francesca needs to know the number and name of each sales rep, along with the number and name of the customers represented by each sales rep. It would be much simpler for her if this information were stored in a single table instead of in two tables that she would have to join together. She would really like a single table that contains the sales rep number, sales rep name, customer number, and customer name. Suppose she would also like these fields to be named SNum, SLast, SFirst, CNum, and CName, respectively. She could use a join in the CREATE VIEW statement, as follows:

```
CREATE VIEW SalesCust (SNum, SLast, SFirst, CNum, CName) AS
SELECT Rep.RepNum, LastName, FirstName, CustomerNum, CustomerName
FROM Rep, Customer
WHERE Rep.RepNum=Customer.RepNum
;
```

Given the current data in the Premiere Products database, conceptually this view is the table shown in Figure 4.7.

FIGURE 4.7
SalesCust view

SalesCust

SNum	SLast	SFirst	CNum	CName
20	Kaiser	Valerie	148	Al's Appliance and Sport
20	Kaiser	Valerie	524	Kline's
20	Kaiser	Valerie	842	All Season
35	Hull	Richard	282	Brookings Direct
35	Hull	Richard	408	The Everything Shop
35	Hull	Richard	687	Lee's Sport and Appliance
35	Hull	Richard	725	Deerfield's Four Seasons
65	Perez	Juan	356	Ferguson's
65	Perez	Juan	462	Bargains Galore
65	Perez	Juan	608	Johnson's Department Store

To Francesca, the SalesCust view is a real table; she does not need to know what goes on behind the scenes in order to use it. She could find the number and name of the sales rep who represents customer 282, for example, by using the following query:

```
SELECT SNum, SLast, SFirst
FROM SalesCust
WHERE CNum='282'
;
```

Francesca is completely unaware that, behind the scenes, the DBMS converts her query to:

```
SELECT Rep.RepNum AS SNum, LastName AS SLast, FirstName AS SFirst
FROM Rep, Customer
WHERE Rep.RepNum=Customer.RepNum
AND CustomerNum='282'
;
```

In Access, the query for the SalesCust view appears in Figure 4.8.

FIGURE 4.8
•••••••••••••••••••
Access query
design of the
SalesCust view

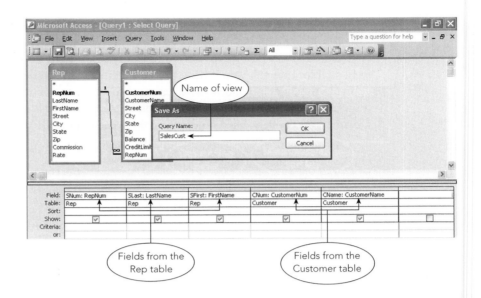

The datasheet for the SalesCust view appears in Figure 4.9.

FIGURE 4.9
•••••••••••••••••••
Datasheet for the
SalesCust view

The use of views provides several advantages:

- Views provide data independence. If the database structure changes (by adding fields or changing relationships between tables, for example) in such a way that the view can still be derived from existing data, the user can still access and use the same view. If adding extra fields to tables in the database is the only change, and these fields are not required by the view's user, the defining query may not even need to be changed in order for the user to continue using the view. If relationships are changed, the defining query may be different, but because users need not be aware of the defining query, this difference is unknown to them. They continue accessing the database through the same view, as though nothing has changed.

- Because each user has his or her own view, different users can view the same data in different ways.

- A view should contain only those fields required by a given user. This practice has two advantages. First, because the view will, in all probability, contain fewer fields than the overall database and because the view is conceptually a single table, rather than a collection of tables, it greatly simplifies the user's perception of the database. Second, views provide a measure of security. Fields that are not included in the view are not accessible to the view's user. For example, omitting the Balance field from a view will ensure that a user of the view cannot access any customer's balance. Likewise, rows that are not included in the view are not accessible. A user of the Housewares view, for example, cannot obtain any information about parts in the AP or SG item classes.

■ Indexes

If you want to find a discussion of a given topic in a book, you could scan the entire book from start to finish, looking for references to the topic you had in mind. More than likely, however, you wouldn't to resort to this technique. If the book had a good index, you could use it to quickly identify the pages on which your topic is discussed.

Within relational model systems on both mainframes and personal computers, the main mechanism for increasing the efficiency with which data is retrieved from the database is the **index**. Conceptually, these indexes are very much like the index in a book. Consider Figure 4.10, for example, which shows the Customer table for Premiere Products together with one extra field, RecordNum. This extra field gives the location of the record in the file (customer 148 is the first record in the table and is on record 1; customer 282 is on record 2; and so on). These record numbers are automatically assigned and used by the DBMS, not by the users, and that is why you do not normally see them. For illustrative purposes, Figure 4.10 includes a RecordNum column to show how an index works.

FIGURE 4.10
· · · · · · · · · · · · · · · · · · ·
Customer table with
record numbers

Customer

RecordNum	CustomerNum	CustomerName	...	Balance	CreditLimit	RepNum
1	148	Al's Appliance and Sport	...	$6,550.00	$7,500.00	20
2	282	Brookings Direct	...	$431.50	$10,000.00	35
3	356	Ferguson's	...	$5,785.00	$7,500.00	65
4	408	The Everything Shop	...	$5,285.25	$5,000.00	35
5	462	Bargains Galore	...	$3,412.00	$10,000.00	65
6	524	Kline's	...	$12,762.00	$15,000.00	20
7	608	Johnson's Department Store	...	$2,106.00	$10,000.00	65
8	687	Lee's Sport and Appliance	...	$2,851.00	$5,000.00	35
9	725	Deerfield's Four Seasons	...	$248.00	$7,500.00	35
10	842	All Season	...	$8,221.00	$7,500.00	20

In order to rapidly access a customer's record on the basis of his or her record number, you might choose to create and use an index, as shown in Figure 4.11.

FIGURE 4.11
·················
Index for the
Customer table on
the CustomerNum
field

CustomerNum Index

CustomerNum	RecordNum
148	1
282	2
356	3
408	4
462	5
524	6
608	7
687	8
725	9
842	10

The index has two fields. The first field contains a customer number and the second field contains the number of the record on which the customer number is found. Because customer numbers are unique, there is only a single corresponding record number in this index. This is not always the case, however. Suppose, for example, that you wanted to quickly access all customers with a specific credit limit or all customers that are represented by a specific sales rep. You might choose to create and use an index on credit limit as well as an index on sales rep number. These two indexes are shown in Figure 4.12.

FIGURE 4.12
·················
Indexes for the
Customer table on
the CreditLimit and
RepNum fields

CreditLimit Index

CreditLimit	RecordNum
$5,000	4, 8
$7,500	1, 3, 9, 10
$10,000	2, 5, 7
$15,000	6

RepNum Index

RepNum	RecordNum
20	1, 6, 10
35	2, 4, 8, 9
65	3, 5, 7

By examining the CreditLimit index in Figure 4.12, you can see that each credit limit occurs in the index along with the numbers of the records on which that credit limit occurs. Credit limit $7,500, for example, occurs on records 1, 3, 9, and 10. Further, the credit limits appear in the index in numerical order. If the DBMS were to use this index to find those records on which the credit limit is $10,000, for example, it could scan the credit limits in the index to find $10,000. After it did, it would determine the corresponding record numbers (2, 5, and 7) and then immediately go to those records in the Customer table, thus finding these customers much more quickly than if it had to scan the entire Customer table one record at a time. Thus, indexes can make the process of retrieving records fast and efficient.

Note: With relatively small tables, the increased efficiency associated with indexes will not be readily apparent. In practice, it is common to encounter tables with thousands, tens of thousands, or even hundreds of thousands of records. In such cases, the increase in efficiency is dramatic. In fact, without indexes, many operations in such databases would simply not be practical—they would take too long to complete.

The field or combination of fields on which the index is built is called the **index key**. Thus, in the first index, the index key is CustomerNum; in the second index, the index key is CreditLimit; and in the third index, the RepNum field is the index key. Typically, the index key for an index can be any field or combination of fields in any table.

After creating an index, you can use it to facilitate data retrieval. In powerful mainframe relational systems, the decision concerning which index or indexes to use (if any) during a particular type of retrieval is a function of the DBMS.

As you would expect, the use of any index is not purely advantageous or disadvantageous. An advantage was already mentioned: An index makes certain types of retrieval more efficient. There are two disadvantages. First, the index occupies space on disk. Using this space for an index, however, is technically unnecessary because any retrieval that can be made using the index can also be made without the index, although less efficiently. The other disadvantage is that the DBMS must update the index whenever corresponding data in the database is updated. Without the index, the DBMS would not need to make these updates. The main question that you must ask when considering whether to create a given index is this: Do the benefits derived during retrieval outweigh the additional storage required and the extra processing involved in update operations?

You can add and delete indexes as necessary. You can create an index after the database is built—it doesn't need to be created at the same time as the database. Likewise, if it appears that an existing index is unnecessary, you can delete it.

The exact process for creating an index varies from one DBMS to another. A common SQL command to create an index is:

```
CREATE INDEX CustomerName
ON Customer (CustomerName)
;
```

This command creates an index named CustomerName. The index is for the Customer table and the index key is the CustomerName field. In this example, the index name is the same as the index key. This format is not a requirement, but it is a good general practice.

Figure 4.13 shows the creation of an index on the CustomerName field in the Customer table using Access. As illustrated in the figure, there are three choices for index options: No, Yes (Duplicates OK), and Yes (No Duplicates).

FIGURE 4.13
· · · · · · · · · · · · · · · · · ·
Creating an index
on a single field in
Access

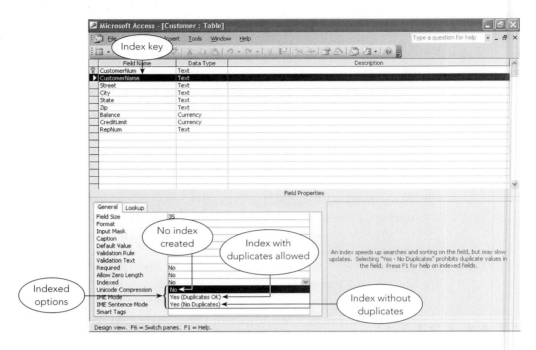

You would select No if you did not want to create an index on the CustomerName field or if you needed to remove a previously created index; No is the default option. You would select Yes (Duplicates OK) to create an index and allow duplicate values. In that case, Access would allow more than one customer with the same name. If you select Yes (No Duplicates), Access would create the index but you would not be able to add a customer with the same name as another customer already in the database.

When you create an index whose key is a single field, you have created a **single-field index** (also called a **single-column index**). A **multiple-field index** (also called a **multiple-column index**) is an index with more than one key field. When creating a multiple-field index, you list the more important key first. In addition, if data for either key appears in descending order, you must follow the field name with the letters DESC.

To create an index named RepBal with the keys RepNum and Balance and with the balances listed in descending order, you could use the following SQL command:

```
CREATE INDEX RepBal
ON Customer (RepNum, Balance DESC)
;
```

Creating multiple-field indexes in Access involves a slightly different process from creating single-field indexes. To create multiple-field indexes, click the Indexes button on the toolbar, enter a name for the index, and then select the fields that make up the index key. If data for any of the fields is to appear in descending order, change the corresponding entry in the Sort Order column to Descending, as shown in Figure 4.14.

FIGURE 4.14
••••••••••••••••••
Creating a multiple-
field index in
Access

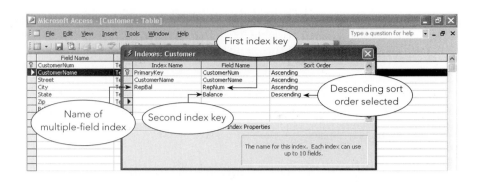

The SQL command used to drop (delete) an index that is no longer necessary is DROP INDEX, which consists of the words DROP INDEX followed by the name of the index to drop. To drop the RepBal index, for example, the command would be:

```
DROP INDEX RepBal
;
```

To delete an index in Access, select the index in the Indexes window (see Figure 4.14), right-click it, and then click Delete Rows on the shortcut menu.

■ Security

Security is the prevention of unauthorized access to the database. Within an organization, the database administrator determines the types of access various users can have to the database. Some users may be able to retrieve and update anything in the database. Other users may be able to retrieve any data from the database but not make any changes to it. Still other users may be able to access only a portion of the database. For example, Bill Kaiser may be able to retrieve and update sales rep and customer data, but not be permitted to retrieve data about parts and orders. Mary Smith may be able to retrieve part data and nothing else. Kyung Park may be able to retrieve and update data on parts in the HW class, but not in other classes.

After the database administrator has determined the access different users of the database will have, it is up to the DBMS to enforce it. In particular, it is up to whatever security mechanism the DBMS provides. In SQL systems, there are two security mechanisms. You have already seen that views furnish a certain amount of security. (If users are accessing the database through a view, they cannot access any data that is not included in the view.) The main mechanism for providing access to a database, however, is the GRANT statement.

The basic idea is that different types of privileges can be granted to users and, if necessary, later revoked. These privileges include such things as the right to select, insert, update, and delete table data. You can grant and revoke user privileges using the GRANT and REVOKE statements. Following are examples of these two statements.

The following command will enable Jones to retrieve data from the Customer table, but not to take any other action.

```
GRANT SELECT ON Customer TO Jones
;
```

The following command will enable users Smith and Park to be able to add new records to the Part table.

```
GRANT INSERT ON Part TO Smith, Park
;
```

The following command will revoke the ability to retrieve Customer records from Jones; that is, Jones will no longer have the privilege granted earlier.

```
REVOKE SELECT ON Customer FROM Jones
;
```

Access includes security features on the Security submenu, which is accessed by selecting the Security command on the Tools menu. Included among these features is the User-Level Security Wizard, which provides assistance in assigning appropriate permissions to various groups of users.

■ Integrity Rules

A relational DBMS must enforce two important integrity rules that were defined by Dr. E. F. Codd (Codd, E.F. "Extending the Relational Database Model to Capture More Meaning." In *ACM TODS 4*, no. 4 (December 1979). Both rules are related to two special types of keys: primary keys and foreign keys. The two integrity rules are called entity integrity and referential integrity.

Entity Integrity

In some DBMSs, when you describe a database, you can indicate that certain fields can accept a special value, called **null**. Essentially, setting the value in a given field to null is similar to not entering a value in it at all. Nulls are used when a value is unknown or inapplicable. It is *not* the same as a blank or zero value, both of which are actual values. For example, a value of zero in the Balance field for a particular customer indicates that the customer has a zero balance. A value of null in a customer's Balance field, on the other hand, indicates that, for whatever reason, the customer's balance is unknown.

If you indicate that the Balance field can be null, you are saying that this situation (a customer with an unknown balance) is something you want to allow. If you don't want to allow unknown values, you indicate that Balance cannot be null.

The decision as to whether to allow nulls is generally made on a field-by-field basis. There is one type of field for which you should *never* allow nulls, however, and that is the primary key. After all, the primary key is supposed to uniquely identify a given row, and this could not happen if nulls were allowed. How, for example, could you tell two customers apart if both had null customer numbers? The restriction that the primary key cannot allow null values is called entity integrity.

Definition: **Entity integrity** is the rule that no field that is part of the primary key may accept null values.

Entity integrity guarantees that each record will indeed have its own identity. In other words, preventing the primary key from accepting null values ensures that you can distinguish one record from another. Typically, the DBMS handles this distinction automatically. All you need to do is specify which field or fields comprise the primary key.

In SQL, you can specify the primary key by entering a PRIMARY KEY clause in either an ALTER TABLE (covered later in this chapter) or a CREATE TABLE command. For example, to use the PRIMARY KEY clause to indicate that CustomerNum is the primary key for the Customer table, the clause would be:

```
PRIMARY KEY (CustomerNum)
```

In general, the PRIMARY KEY clause has the form PRIMARY KEY followed in parentheses by the field (or fields) that make up the primary key. If more than one field is included, the fields are separated by commas. Thus, the PRIMARY KEY clause for the OrderLine table would be:

```
PRIMARY KEY (OrderNum, PartNum)
```

In Access, you designate the primary key by selecting the primary key field in Table Design view and clicking the Primary Key button on the toolbar. A key symbol will appear in the field's row selector to indicate that it is the primary key, as shown in Figure 4.15.

FIGURE 4.15

Specifying a primary key in Access

If the primary key consists of more than one field, select the first field, press and hold down the Shift key, and then click the other field(s) that make up the primary key. Clicking the Primary Key button adds the key symbol to the row selectors of the primary key fields, as shown in Figure 4.16.

FIGURE 4.16

Specifying a primary key consisting of more than one field in Access

Referential Integrity

In the relational model you've examined thus far, you have created the relationships between tables by having common fields in two or more tables. The relationship between sales reps and customers, for example, is accomplished by including the primary key of the Rep table (RepNum) as a field in the Customer table.

This approach has its drawbacks. First of all, relationships are not very obvious. If you were not already familiar with the relationships within the Premiere Products database, you would have to find the matching fields in separate tables in order to locate the relationship. Even then, you couldn't be sure that the matching field names indicate a relationship. Two fields having the same name could be just a coincidence—the fields might have nothing to do with each other. Second, what if the primary key in the Rep table is RepNum but the corresponding field in the Customer table were named SlsrNo? Unless you were aware that these two fields were identical, the relationship between customers and sales reps would not be clear. In a database having as few tables and fields as the Premiere Products database, these problems might be manageable. But picture a database that has 20 tables, each containing an average of 30 fields. As the number of tables and fields increases, so do the potential problems.

There is also another issue with the relational model. Nothing about the model itself would prevent a user from storing data about a customer whose sales rep number did not correspond to any sales rep already in the database. Clearly this is not a desirable situation.

Fortunately, a solution exists for both issues. It involves using foreign keys.

Definition: A **foreign key** is a field (or collection of fields) in a table whose value is required to match the value of the primary key for a second table.

The RepNum field in the Customer table is a foreign key that must match the primary key of the Rep table. In practice, this means that the sales rep number for any customer must be the same as the number of some sales rep who is already in the database.

There is one possible exception to this rule. Perhaps Premiere Products does not require a customer to have a sales rep—it is strictly optional. This situation could be indicated in the Customer table by setting such a customer's sales rep number to null. Technically, however, a null sales rep number would violate the restrictions that you have indicated for a foreign key. Thus, if you were to use a null sales rep number, you would have to modify the definition of a foreign key to include the possibility of nulls. You would insist, though, that if the foreign key contained a value *other than null*, it would have to match the value of the primary key in some row in the other table. (In the example, for instance, a customer's sales rep number could be null, but if it were not null, it would have to be the number of an actual sales rep.) This general property is called referential integrity.

Definition: **Referential integrity** is the rule that if table A contains a foreign key that matches the primary key of table B, the values of this foreign key must either match the value of the primary key for some row in table B or be null.

Note: Usually a foreign key is in a different table from the primary key it is required to match. In the Premiere Products database, for example, to be able to determine the rep for any customer, you include the rep number as a foreign key in the Customer table that must match the primary key in the Rep table. It is possible for the foreign key and the matching primary key to be in the same table, however. As an example of this situation, suppose that one of the requirements in a particular database is that, given an employee, you must be able to determine the manager of that employee. You might have an Employee table with a primary key of EmployeeNum (the employee number). In order to determine the employee's manager, you would include the manager's employee number as a foreign key in the Employee table. Because the manager is also an employee, however, the manager will be in the same Employee table. Thus, this foreign key in the Employee table would need to match the primary key in the same Employee table. The only restriction is that the foreign key *must* have a name that is different from the primary key because the fields are in the same table. For example, you could name the foreign key ManagerEmployeeNum.

Using foreign keys solves the previously mentioned problems. Indicating that the RepNum field in the Customer table is a foreign key that must match the RepNum field in the Rep table explicitly specifies the relationship between customers and sales reps—you do not need to look for common fields in several tables. Further, with foreign keys, matching fields that have different names no longer pose a problem. For example, it would not matter if the name of the foreign key in the Customer table were SlsrNo and the primary key in the Rep table were RepNum; the only thing that *would* matter is that this field was a foreign key that matched the Rep table. Finally, through referential integrity, it is possible for a customer not to have a sales rep number, but it is not possible for a customer to have an *invalid* sales rep number; that is, a customer's sales rep number *must* either be null or the number of a sales rep who is already in the database.

In SQL, you specify referential integrity using FOREIGN KEY clauses in either the CREATE TABLE or ALTER TABLE commands. To specify a foreign key, you need to specify both the field that is a foreign key and the table whose primary key it is to match. In the Customer table, for example, the RepNum field is a foreign key that must match the primary key in the Rep table, as follows:

```
FOREIGN KEY (RepNum) REFERENCES Rep
```

The general form is FOREIGN KEY, followed by the field or combination of fields that make up the foreign key, followed by the word REFERENCES, and then by the name of the table containing the primary key that the foreign key is supposed to match.

In Access, referential integrity is specified as part of the process of defining relationships, as shown in Figure 4.17.

FIGURE 4.17
. . . .
Using the Relationships window to relate tables in Access

You use the mouse to drag the primary key of the Rep table (RepNum) to the foreign key of the Customer table (RepNum). After releasing the mouse button, you can request that Access enforce referential integrity, as shown in Figure 4.18. You also can specify whether update or delete operations will "cascade." Selecting the **cascade delete** option ensures that the deletion of a sales rep record also deletes all customer records related to that sales rep; selecting the **cascade update** option ensures that changes made to the primary key of a sales rep record are also made in the related customer record. In Figure 4.18, the cascade delete and cascade update options are not selected.

With referential integrity enforced, users are not allowed to enter a customer record whose sales rep number does not match any sales rep currently in the Rep table. An error message, such as the one shown in Figure 4.19, appears when an attempt is made to enter an invalid sales rep number.

FIGURE 4.19
∙∙∙∙∙∙∙∙∙∙∙∙∙∙∙∙∙∙∙∙
Attempted violation
of referential
integrity when
adding a record

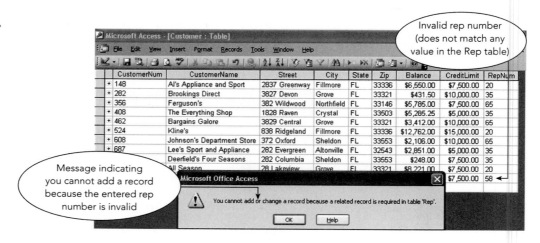

Deleting a sales rep who currently has customers on file would also violate referential integrity because the sales rep's customers would no longer match any sales rep in the Rep table. The DBMS must refuse to carry out this type of deletion and then produce an error message, such as the one shown in Figure 4.20. If sales rep 20 leaves Premiere Products, all of her customers would need to be assigned to other sales reps before deleting her record from the Rep table.

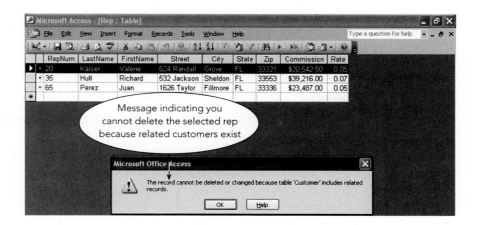

Legal-Values Integrity

In addition to the two integrity rules defined by Codd, there is a third type of integrity, called **legal-values integrity**. Often there is a particular set of values, called the legal values, that are allowable in a field. Legal-values integrity is the property that states that no record can exist in the database with a value in the field other than one of the legal values. For example, the only legal values for the CreditLimit field are $5,000, $7,500, $10,000, or $15,000. A record on which the credit limit is $12,500 would not be permitted.

In SQL, you use the CHECK clause to enforce legal-values integrity. For example, to ensure that the only legal values for credit limits are $5,000, $7,500, $10,000, or $15,000, the CHECK clause that you would include in a CREATE TABLE or ALTER TABLE command is:

```
CHECK (CreditLimit IN (5000, 7500, 10000, 15000))
```

The general form of the CHECK clause is the word CHECK followed by a condition. In the previous CHECK clause, the credit limit must be in the set consisting of 5000, 7500, 10000, or 15000. If any update to the database would result in the condition being violated, the DBMS would automatically reject the update.

In Access, you can restrict the legal values accepted by a field by entering an appropriate **validation rule** that data entered in the field must follow. Figure 4.21 shows the validation rule that restricts entries in the CreditLimit field to 5000, 7500, 10000, and 15000. Along with the validation rule, you usually will enter **validation text** to inform the user of the reason for the rejection when the user attempts to enter data that violates the rule.

FIGURE 4.21
· · · · · · · · · · · · · · · · · ·
Specifying a valida-
tion rule in Access

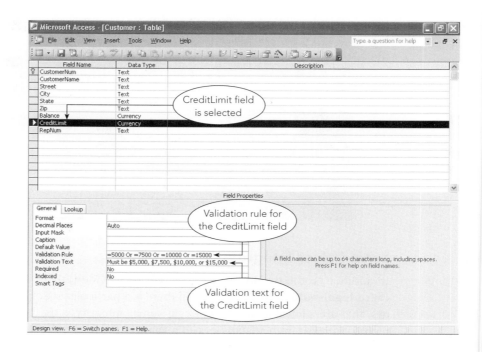

Structure Changes

An important feature of relational DBMSs is the ease with which you can change the database structure by adding and removing tables and fields, by changing the characteristics of
existing fields, or by creating and dropping indexes. Although the exact manner in which
these changes are accomplished varies from one system to another, most systems allow all
of these changes to be made quickly and easily.

Changes to a table's structure are made using the SQL ALTER TABLE command.
Virtually every implementation of SQL allows the creation of new fields in existing tables.
For example, suppose that you need to maintain a customer type for each customer in the
Premiere Products database. You can decide to assign regular customers type R, distribu-
tors type D, and special customers type S. To implement this change, you would need to
add a new field to the Customer table as follows:

```
ALTER TABLE Customer
ADD CustType CHAR(1)
;
```

In Access, you can add a field in Table Design view at any time. Figure 4.22 shows the
Customer table with the addition of the CustType field.

FIGURE 4.22
• • • • • • • • • • • • • • • • • • •
Adding a field in
Access

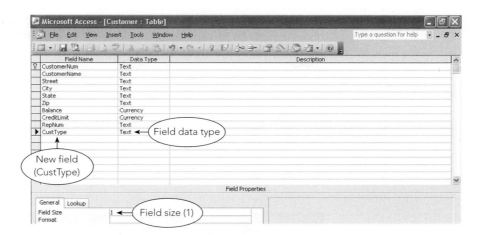

At this point, the Customer table contains an extra field, CustType. For rows (customers) added from this point on, the value of CustType is entered just like any other field. For existing rows, CustType is typically assigned a null value by the DBMS automatically. The user then can change these values, if desired.

Some systems allow changes to the properties of existing fields (but Oracle is one that does not). A typical use of such a provision would be to increase the length of a character field that was found to be inadequate. Assuming that the CustomerName field in the Customer table needed to be increased from 35 to 40 characters, the SQL ALTER TABLE statement would be:

```
ALTER TABLE Customer
CHANGE COLUMN CustomerName TO CHAR(40)
;
```

In Access, you can change field characteristics in Table Design view, as shown in Figure 4.23, in which the size of the CustomerName field, which was 35, has been changed to 40.

FIGURE 4.23
• • • • • • • • • • • • • • • • • • •
Changing a field
characteristic in
Access

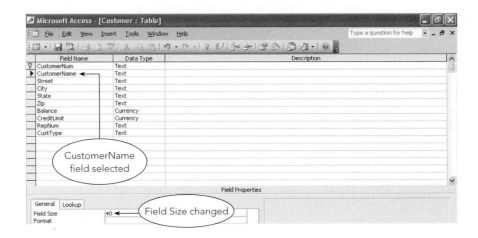

Some systems, but certainly not all, allow existing fields to be deleted. (Oracle is one system that does not allow existing fields to be deleted.) The SQL command for deleting the Warehouse field from the Part table would typically be:

```
ALTER TABLE Part
DELETE Warehouse
;
```

In Access, you can delete a field in Table Design view by selecting the field and pressing the Delete key. Access then will ask you to confirm the deletion of the field, as shown in Figure 4.24. Clicking the Yes button permanently deletes the field.

FIGURE 4.24
· · · · · · · · · · · · · · · · · ·
Deleting a field in
Access

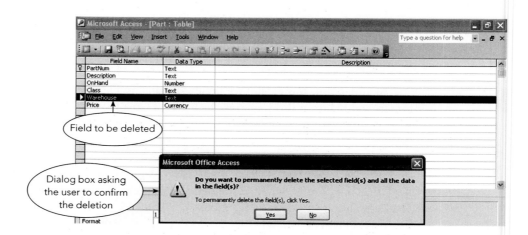

With the SQL DROP TABLE command, you can delete a table that is no longer needed. If the SmallCust table as created in Chapter 3 was no longer needed in the Premiere Products database, you could remove it by entering the following command:

```
DROP TABLE SmallCust
;
```

The table and all indexes and views defined on the table would be deleted. The DROP TABLE command deletes the table structure as well as its data.

In Access, you can drop (delete) a table by right-clicking the table in the Database window and then clicking Delete on the shortcut menu, as shown in Figure 4.25.

FIGURE 4.25
· · · · · · · · · · · · · · · · · · ·
Deleting a table in
Access

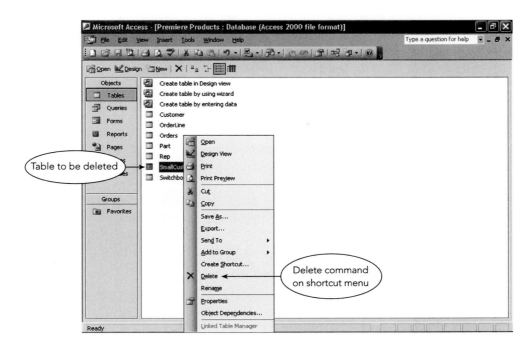

Making Complex Changes

In some cases, you might need to change a table's structure in ways that are beyond the capabilities of your DBMS. Perhaps you need to eliminate a field, change the field order, or combine data from two tables into one, but your system does not allow these types of changes. For example, some systems, including Oracle, do not allow you to reduce the size of a field or to change its data type. In these situations, you can use the CREATE TABLE command to describe the new table, and then insert values into it using the INSERT command combined with an appropriate SELECT clause, as you saw in Chapter 3. If you are using a version of SQL that supports the SELECT INTO command, as Access does, you can use it to create the new table in a single operation.

█ System Catalog

Information about tables in the database is kept in the **system catalog** (or the **catalog**). The catalog is maintained automatically by the DBMS. Whenever a new table is added, the structure of an existing table is changed, or a table is deleted, the DBMS updates the catalog to reflect these changes.

This section describes the types of items kept in the catalog and the way in which you can query it to determine information about the database structure. (This description represents the way catalogs are used in DB2, IBM's mainframe relational DBMS.) Although catalogs in individual relational DBMSs will vary from the examples shown here, the general ideas apply to most relational systems.

The catalog you'll consider contains two tables: **Systables** (information about the tables known to SQL) and **Syscolumns** (information about the columns or fields within these tables). An actual catalog contains other tables as well, such as **Sysindexes** (information about the indexes that are defined on these tables) and **Sysviews** (information about the views that have been created). Although these tables would have many fields, only a few are of concern here.

As shown in Figure 4.26, the Systables table contains the Name, Creator, and Colcount fields. The Name field identifies the name of a table, the Creator field identifies the person or group that created the table, and the Colcount field contains the number of fields within the table that is being described. If, for example, the user named Brown created the Rep table and the Rep table has nine fields, there would be a row in the Systables table in which the Name is Rep, the Creator is Brown, and the Colcount is 9. Similar rows would exist for all tables known to the system.

FIGURE 4.26
· · · · · · · · · · · · · · · · ·
Systables table

Systables

Name	Creator	Colcount
Customer	Brown	10
Part	Brown	6
Orders	Brown	3
Order Line	Brown	4
Rep	Brown	9

The Syscolumns table contains the Colname, Tbname, and Coltype fields, as shown in Figure 4.27. The Colname field identifies the name of a field in one of the tables. The table in which the field is found is stored in the Tbname field, and the data type for the field is found in the Coltype field. There is a row in the Syscolumns table for each field in the Rep table, for example. On each of these rows, Tbname is Rep. On one of these rows, Colname is RepNum and Coltype is CHAR(2). On another row, Colname is LastName and Coltype is CHAR(15).

FIGURE 4.27
....
Syscolumns table

Syscolumns

Colname	Tbname	Coltype
Balance	Customer	DECIMAL(8,2)
City	Customer	CHAR(15)
City	Rep	CHAR(15)
Class	Part	CHAR(2)
Commission	Rep	DECIMAL(7,2)
CreditLimit	Customer	DECIMAL(8,2)
CustomerName	Customer	CHAR(35)
CustomerNum	Customer	CHAR(3)
CustomerNum	Orders	CHAR(3)
Description	Part	CHAR(15)
FirstName	Rep	CHAR(15)
LastName	Rep	CHAR(15)
NumOrdered	OrderLine	DECIMAL(3,0)
OnHand	Part	DECIMAL(4,0)
OrderDate	Orders	DATE
OrderNum	OrderLine	CHAR(5)
OrderNum	Orders	CHAR(5)
PartNum	OrderLine	CHAR(4)
PartNum	Part	CHAR(4)
Price	Part	DECIMAL(6,2)
QuotedPrice	OrderLine	DECIMAL(6,2)
Rate	Rep	DECIMAL (3,2)
RepNum	Customer	CHAR(2)
RepNum	Rep	CHAR(2)
State	Customer	CHAR(2)
State	Rep	CHAR(2)
Street	Customer	CHAR(15)
Street	Rep	CHAR(15)
Warehouse	Part	CHAR(2)
Zip	Customer	CHAR(5)
Zip	Rep	CHAR(5)

A DBMS furnishes ways of using the catalog to determine information about the structure of the database. In some cases, this simply involves using SQL to query the tables in the catalog. For example, to list the name and type of all fields (columns) in the Part table, you could use the following SQL command:

```
SELECT Colname, Coltype
FROM Syscolumns
WHERE Tbname='Part'
;
```

In other cases, there are special tools to provide the desired documentation. For example, Access has a tool called the **Documenter**, which allows you to easily print detailed documentation concerning any table, query, report, form, or other object in the database. To document an object in an Access database, click Tools on the menu bar, point to Analyze, and then click Documenter.

■ SUMMARY

- Views are used to give each user his or her own view of the data in a database. A view is defined in structured query language (SQL) by using a defining query. When you enter a query that references a view, it is merged with the defining query to produce the query that is actually executed. In Access, views are created by saving queries.

- Indexes are often used to facilitate data retrieval from the database. You can create an index on any field or combination of fields.

- Security is provided in SQL systems using the GRANT and REVOKE commands.

- Entity integrity is the property that states that no field that is part of the primary key can accept null values.

- Referential integrity is the property that states that the value in any foreign key field must either be null or match an actual value in the primary key field of another table. Referential integrity is specified in SQL using the FOREIGN KEY clause. In Access, foreign keys are specified by creating relationships.

- Legal-values integrity is the property that states that the value entered in a field must be one of the legal values; that is, it must be one of the values that satisfies some particular condition. Legal-values integrity is specified in SQL using the CHECK clause. In Access, legal-values integrity is specified using validation rules.

- The ALTER TABLE command allows you to add fields to a table, delete fields, or change the characteristics of fields. In Access, you can change the structure of a table by making the desired changes in the table design.

- The DROP TABLE command lets you delete a table from a database. In Access, you can delete a table by selecting the Delete command on the table's shortcut menu.

- The system catalog is a feature of many relational DBMSs that stores information about the structure of a database. The system updates the catalog automatically. Each DBMS includes facilities to produce documentation of the database structure using the information in the catalog.

■ KEY TERMS

cascade delete
cascade update
catalog
defining query
Documenter
entity integrity
foreign key
index
index key
legal-values integrity
multiple-column index
multiple-field index
null

referential integrity
row-and-column subset view
security
single-column index
single-field index
Syscolumns
Sysindexes
Systables
system catalog
Sysviews
validation rule
validation text
view

■ REVIEW QUESTIONS

1. What is a view? How is it defined? Do the data described in a view definition ever exist in that form? What happens when a user accesses a database through a view?

2. Using data from the Premiere Products database, define a view named TopLevelCust. It consists of the number, name, address, balance, and credit limit for all customers with credit limits that are greater than or equal to $10,000.
   a. Using SQL, write the view definition for SmallCust.
   b. Write an SQL query to retrieve the number and name of all customers in the SmallCust view with balances that exceed their credit limits.
   c. Convert the query you wrote in Question 2b to the query that the DBMS will actually execute.

3. Define a view named PartOrder. It consists of the part number, description, price, order number, order date, number ordered, and quoted price for all order lines currently on file.
   a. Using SQL, write the view definition for PartOrder.
   b. Write an SQL query to retrieve the part number, description, order number, and quoted price for all orders in the PartOrder view for parts with quoted prices that exceed $100.
   c. Convert the query you wrote in Question 3b to the query that the DBMS will actually execute.

4. What is an index? What are the advantages and disadvantages of using indexes? How do you use SQL to create an index?

5. Describe the GRANT statement and explain how it relates to security. What types of privileges may be granted? How are they revoked?

6. Write the SQL commands to grant the following privileges:
   a. User Stillwell must be able to retrieve data from the Part table.

   b. Users Webb and Bradley must be able to add new orders and order lines.

7. Write the SQL command to revoke user Stillwell's privilege.

8. What is the system catalog? Name three items about which the catalog maintains information.

9. Write the SQL commands to obtain the following information from the system catalog:
   a. List every table that you created.
   b. List every field in the Customer table and its associated data type.
   c. List every table that contains a field named PartNum.

10. Why is it a good idea for the DBMS to update the catalog automatically when a change is made in the database structure? Could users cause problems by updating the catalog themselves?

11. What are nulls? Which field cannot accept null values, and why?

12. State the three integrity rules. Indicate the reasons for enforcing each rule.

13. The Orders table contains a foreign key, CustomerNum, that must match the primary key of the Customer table. What type of update to the Orders table would violate referential integrity? If deletes do not cascade, what type of update to the Customer table would violate referential integrity? If deletes do cascade, what would happen when a customer is deleted?

14. How would you use SQL to change a table's structure? What general types of changes are possible? Which commands are used to implement these changes?

In the following exercises, you will use the data in the Premiere Products database shown in Figure 2.1 in Chapter 2. (If you use a computer to complete these exercises, use a copy of the original Premiere Products database so your data will not reflect the changes you made in Chapter 3.) If you have access to a DBMS, use the DBMS to perform the tasks and explain the steps you used in the process. If not, explain how you would use SQL to obtain the desired results. Check with your instructor if you are not certain about which approach to take.

1. Create the TopLevelCust view described in Review Question 2. Display the data in the view.

2. Create the PartOrder view described in Review Question 3. Display the data in the view.

3. Create a view named OrdTot. It consists of the order number and order total for each order currently on file. (The order total is the sum of the number ordered multiplied by the quoted price on each order line for each order.) Display the data in the view.

4. Create the following indexes. If it necessary to name the index in your DBMS, use the indicated name.

   a. Create an index named PartIndex1 on the PartNum field in the OrderLine table.
   b. Create an index named PartIndex2 on the Warehouse field in the Part table.
   c. Create an index named PartIndex3 on the Warehouse and Class fields in the Part table.
   d. Create an index named PartIndex4 on the Warehouse and OnHand fields in the Part table and list units on hand in descending order.

5. Drop the PartIndex3 index.

6. Assume that the Part table has been created, but that there are no integrity constraints. Create the necessary integrity constraint to ensure that the only allowable values for the Class field are AP, HW, and SG. Ensure that the PartNum field is the primary key and that the PartNum field in the OrderLine table is a foreign key that must match the primary key of the Part table.

7. Add a field named Allocation to the Part table. The allocation is a number representing the number of units of each part that have been allocated to each customer. Set all values of Allocation to zero. Calculate the number of units of part number KV29 currently on order. Change the value of Allocation for part number KV29 to this number. Display all the data in the Part table.

8. Increase the length of the Warehouse field in the Part table to two characters. Change the warehouse number for warehouse 1 to 1a. Display all the data in the Part table.

9. Delete the Allocation field from the Part table. Display all the data in the Part table.

10. What command would delete the Part table from the Premiere Products database? (Do not delete the Part table.)

Ray Henry would like you to complete the following tasks to help him maintain his database. In the following exercises, you will use the data in the Henry Books database shown in Figures 1.17 through 1.20 in Chapter 1. (If you use a computer to complete these exercises, use a copy of the original Henry Books database so your data will not reflect the changes you made in Chapter 3.) If you have access to a DBMS, use the DBMS to perform the tasks and explain the steps you used in the process. If not, explain how you would use SQL to obtain the desired results. Check with your instructor if you are uncertain about which approach to take.

1. Create a view named PenguinBooks. It consists of the book code, book title, book type, and book price for every book published by Penguin USA. Display the data in the view.

2. Create a view named Paperback. It consists of the book code, book title, publisher name, and book price for every book that is available in paperback. Display the data in the view.

3. Create a view named BookInventory. It consists of the book code and the total number of units of the book that are on hand at any branch. Display the data in the view.

4. Create the following indexes. If it is necessary to name the index in your DBMS, use the indicated name.

   a. Create an index named BookIndex1 on the PublisherName field in the Publisher table.
   b. Create an index named BookIndex2 on the Type field in the Book table.
   c. Create an index named BookIndex3 on the Type and Price fields in the Book table and list the prices in descending order.

5. Drop the BookIndex3 index.

6. Specify the integrity constraint that the price of any book must be less than $90.00.

7. Ensure that the following are foreign keys (that is, specify referential integrity) within the Henry Books database:

   a. PublisherCode is a foreign key in the Book table.
   b. BranchNum is a foreign key in the Inventory table.
   c. AuthorNum is a foreign key in the Wrote table.

8. Add to the Book table a new character field named Classic that is one character in length.

9. Change the Classic field in the Book table to Y for the book titled *The Grapes of Wrath*.

10. Change the length of the Title field in the Book table to 60.

11. What command would delete the Books table from the Henry Books database? (Do not delete the Book table.)

CHAPTER 5

Database Design 1: Normalization

OBJECTIVES

- Discuss functional dependence.

- Discuss primary keys.

- Define first normal form, second normal form, and third normal form.

- Describe the problems associated with tables (relations) that are not in first normal form, second normal form, or third normal form, along with the mechanism for converting to all three.

- Discuss the problems associated with incorrect conversions to third normal form.

- Define fourth normal form.

- Describe the problems associated with tables (relations) that are not in fourth normal form and describe the mechanism for converting to fourth normal form.

- Understand how normalization is used in the database design process.

Introduction

You have examined the basic relational model, its structure, and the various ways of manipulating data within a relational database. In this chapter, you will learn about the normalization process and its underlying concepts and features. The **normalization process** enables you to identify the existence of potential problems, called **update anomalies**, in the design of a relational database. This process also supplies methods for correcting these problems.

To correct update anomalies in a database, you must convert tables into various types of **normal forms**. A table in a particular normal form possesses a certain desirable collection of properties. The most common normal forms are first normal form (1NF), second normal form (2NF), third normal form (3NF), and fourth normal form (4NF). Normalization is a progression in which a table that is in first normal form is better than a table that is not in first normal form, a table that is in second normal form is better than one that is in first normal form, and so on. The goal of normalization is to allow you to take a table or collection of tables and produce a new collection of tables that represents the same information but that is free of update anomalies.

In this chapter, you will learn about two crucial concepts that are fundamental to understanding the normalization process: functional dependence and keys. You will learn about first normal form, second normal form, third normal form, and fourth normal form. After mastering this material, you will be ready to begin your study of the database design process in Chapter 6.

Many of the examples in this chapter use data from the Premiere Products database, which is shown in Figure 5.1.

FIGURE 5.1
....
Premiere Products
data

Rep

RepNum	LastName	FirstName	Street	City	State	Zip	Commission	Rate
20	Kaiser	Valerie	624 Randall	Grove	FL	33321	$20,542.50	0.05
35	Hull	Richard	532 Jackson	Sheldon	FL	33553	$39,216.00	0.07
65	Perez	Juan	1626 Taylor	Fillmore	FL	33336	$23,487.00	0.05

(continued)

140

Customer

CustomerNum	CustomerName	Street	City	State	Zip	Balance	CreditLimit	RepNum
148	Al's Appliance and Sport	2837 Greenway	Fillmore	FL	33336	$6,550.00	$7,500.00	20
282	Brookings Direct	3827 Devon	Grove	FL	33321	$431.50	$10,000.00	35
356	Ferguson's	382 Wildwood	Northfield	FL	33146	$5,785.00	$7,500.00	65
408	The Everything Shop	1828 Raven	Crystal	FL	33503	$5,285.25	$5,000.00	35
462	Bargains Galore	3829 Central	Grove	FL	33321	$3,412.00	$10,000.00	65
524	Kline's	838 Ridgeland	Fillmore	FL	33336	$12,762.00	$15,000.00	20
608	Johnson's Department Store	372 Oxford	Sheldon	FL	33553	$2,106.00	$10,000.00	65
687	Lee's Sport and Appliance	282 Evergreen	Altonville	FL	32543	$2,851.00	$5,000.00	35
725	Deerfield's Four Seasons	282 Columbia	Sheldon	FL	33553	$248.00	$7,500.00	35
842	All Season	28 Lakeview	Grove	FL	33321	$8,221.00	$7,500.00	20

Orders

OrderNum	OrderDate	CustomerNum
21608	10/20/2007	148
21610	10/20/2007	356
21613	10/21/2007	408
21614	10/21/2007	282
21617	10/23/2007	608
21619	10/23/2007	148
21623	10/23/2007	608

OrderLine

OrderNum	PartNum	NumOrdered	QuotedPrice
21608	AT94	11	$21.95
21610	DR93	1	$495.00
21610	DW11	1	$399.99
21613	KL62	4	$329.95
21614	KT03	2	$595.00
21617	BV06	2	$794.95
21617	CD52	4	$150.00
21619	DR93	1	$495.00
21623	KV29	2	$1,290.00

Part

PartNum	Description	OnHand	Class	Warehouse	Price
AT94	Iron	50	HW	3	$24.95
BV06	Home Gym	45	SG	2	$794.95
CD52	Microwave Oven	32	AP	1	$165.00
DL71	Cordless Drill	21	HW	3	$129.95
DR93	Gas Range	8	AP	2	$495.00
DW11	Washer	12	AP	3	$399.99
FD21	Stand Mixer	22	HW	3	$159.95
KL62	Dryer	12	AP	1	$349.95
KT03	Dishwasher	8	AP	3	$595.00
KV29	Treadmill	9	SG	2	$1,390.00

Functional Dependence

Understanding functional dependence is crucial to learning the material in the rest of this chapter. **Functional dependence** is a formal name for what is basically a simple idea. To illustrate functional dependence, suppose that the Rep table for Premiere Products contains an additional column named PayClass, as shown in Figure 5.2.

FIGURE 5.2
....
Rep table with
additional column,
PayClass

Rep

RepNum	LastName	FirstName	Street	City	State	Zip	Commission	PayClass	Rate
20	Kaiser	Valerie	624 Randall	Grove	FL	33321	$20,542.50	1	0.05
35	Hull	Richard	532 Jackson	Sheldon	FL	33553	$39,216.00	2	0.07
65	Perez	Juan	1626 Taylor	Fillmore	FL	33336	$23,487.00	1	0.05

Suppose further that one of the policies at Premiere Products is that all sales reps in any given pay class earn the same commission rate. How might you convey this fact to someone else? You might say that a sales rep's pay class *determines* his or her commission rate. Another way to convey this fact is to say that a sales rep's commission rate *depends on* his or her pay class. This phrasing uses the words *determines* and *depends on* in exactly the way that you will use them in connection with database design. If you wanted to be more formal, you would precede either expression with the word *functionally*. Thus, you might say, "A sales rep's pay class *functionally determines* his or her commission rate," or "A sales rep's commission rate *functionally depends on* his or her pay class."

The formal definition of functional dependence is as follows:

Definition: A column (attribute) B is **functionally dependent** on another column A (or possibly a collection of columns) if each value for A in the database is associated with exactly one value of B.

You can think of functional dependence as follows: If you are given a value for A in the database, do you know that it will be associated with exactly one value of B? If so, B is functionally dependent on A (written as A → B). If B is functionally dependent on A, you can also say that A **functionally determines** B.

In the Rep table, LastName is functionally dependent on RepNum. If you are given a value of 20 for RepNum, for example, you know that you will find a *single* LastName, in this case Kaiser, associated with it. (**Note:** You only need to be concerned with actual values of RepNum in the database. If you are given a value of 21 for RepNum, for example, you will not find any names associated with it because there is no row in the Rep table on which the rep number is 21.)

● ●

QUESTION

ANSWER

In the Customer table, is CustomerName functionally dependent on RepNum?

No. Rep number 20, for example, occurs on a row in which the customer name is Al's Appliance and Sport, on a row in which the customer name is Kline's, and on a row in which the customer name is All Season. Thus, a rep number can be associated with more than one customer name.

QUESTION

In the OrderLine table, is QuotedPrice functionally dependent on OrderNum?

ANSWER

No. Order number 21617, for example, occurs on a row in which the quoted price is $794.95 and on a row in which the quoted price is $150.00. Thus, an order number can be associated with more than one quoted price.

QUESTION

Is QuotedPrice functionally dependent on PartNum?

ANSWER

No. A given part number, such as DR93, can occur on more than one row in the OrderLine table. The quoted price can be different on each row. With current data, the two rows on which the part number is DR93 actually have the same price, $495.00. What is important, however, is not that they happen to be the same now, but whether there is a guarantee that they will always be the same. That is certainly not the case. In fact, the reason for storing quoted prices in the OrderLine table is to allow two rows with the same part number to have different prices.

QUESTION

On which columns is QuotedPrice functionally dependent?

ANSWER

For any combination of an order number and a part number, there can be only one row in the OrderLine table. Thus, any combination of an order number and part number in the OrderLine table is associated with exactly one quoted price. Consequently, QuotedPrice is functionally dependent on the combination (formally called the **concatenation**) of OrderNum and PartNum.

At this point, a question naturally arises: How do you determine functional dependencies? Can you determine them by looking at sample data, for example? The answer is no.

Consider the Rep table shown in Figure 5.3, in which all last names are unique. It is very tempting to say that LastName functionally determines Street, City, State, and Zip (or equivalently that Street, City, State, and Zip are all functionally dependent on LastName). After all, given the last name of a rep, you can find his or her address.

FIGURE 5.3

Rep table

Rep

RepNum	LastName	FirstName	Street	City	State	Zip	Commission	Rate
20	Kaiser	Valerie	624 Randall	Grove	FL	33321	$20,542.50	0.05
35	Hull	Richard	532 Jackson	Sheldon	FL	33553	$39,216.00	0.07
65	Perez	Juan	1626 Taylor	Fillmore	FL	33336	$23,487.00	0.05

What happens when you add rep 85, whose last name also is Kaiser, to the database? You then have the situation illustrated in Figure 5.4. If the last name you are given is Kaiser, you no longer can find a single address. Thus, you were misled by the original sample data. The only way to really determine the functional dependencies that exist is to examine the user's policies through discussions with users, an examination of user documentation, and so on.

FIGURE 5.4

Rep table with
second rep named
Kaiser added

Rep

RepNum	LastName	FirstName	Street	City	State	Zip	Commission	Rate
20	Kaiser	Valerie	624 Randall	Grove	FL	33321	$20,542.50	0.05
35	Hull	Richard	532 Jackson	Sheldon	FL	33553	$39,216.00	0.07
65	Perez	Juan	1626 Taylor	Fillmore	FL	33336	$23,487.00	0.05
85	Kaiser	Wil	172 Bahia	Norton	FL	39281	$0.00	0.05

QUESTION

Assume the following columns exist in a relation named Student:

- StudentNum (student number)
- StudentLast (student last name)
- StudentFirst (student first name)
- HighSchoolNum (number of the high school from which the student graduated)
- HighSchoolName (name of the high school from which the student graduated)
- AdvisorNum (number of the student's advisor)
- AdvisorLast (last name of the student's advisor)
- AdvisorFirst (first name of the student's advisor)

Student numbers, high school numbers, and advisor numbers are unique: no two students have the same number, no two high schools have the same number, and no two advisors have the same number. Use this information to determine the functional dependencies in the Student relation.

ANSWER

Because student numbers are unique, any given student number in the database is associated with a single last name, first name, high school number, high school name, advisor number, advisor last name, and advisor first name. Thus, all the other columns in the Student relation are functionally dependent on StudentNum, which is represented as:

```
StudentNum → StudentLast, StudentFirst, HighSchoolNum, HighSchoolName,
    AdvisorNum, AdvisorLast, AdvisorFirst
```

Because two students can have the same first and last names, StudentFirst and StudentLast do not determine anything else. Because high school numbers are unique, any given high school number is associated with exactly one high school name. If high school 128 is Robbins High, for example, any student whose high school number is 128 *must* have the high school name Robbins High. Thus, HighSchoolName is functionally dependent on HighSchoolNum, which is represented as:

```
HighSchoolNum → HighSchoolName
```

Because advisor numbers are unique, any given advisor number is associated with exactly one advisor first name and exactly one advisor last name. If advisor 20 is Mary Webb, for example, any student whose advisor number is 20 *must* have the advisor's first name Mary and advisor's last name Webb. Thus, AdvisorFirst and AdvisorLast are functionally dependent on AdvisorNum, which is represented as:

```
AdvisorNum → AdvisorLast, AdvisorFirst
```

As with students, an advisor's first and last names are not necessarily unique, so AdvisorFirst and AdvisorLast do not determine anything. The complete collection of functional dependencies is:

```
StudentNum → StudentLast, StudentFirst, HighSchoolNum, HighSchoolName,
     AdvisorNum, AdvisorLast, AdvisorFirst
HighSchoolNum → HighSchoolName
AdvisorNum → AdvisorLast, AdvisorFirst
```

■ Keys

A second underlying concept of the normalization process is that of the primary key. You already encountered the basic concept of the primary key in earlier chapters. In this chapter, however, you will need a more precise definition.

Definition: Column A (or a collection of columns) is the **primary key** for a relation (table) R, if

> Property 1. *All* columns in R are functionally dependent on A.
> Property 2. No subcollection of the columns in A (assuming A is a collection of columns and not just a single column) also has Property 1.

QUESTION

Is Class the primary key for the Part table?

ANSWER

No, because the other columns are not functionally dependent on the class. The item class HW, for example, appears on a row in the Part table in which the part number is AT94, a row in which the part number is DL71, and a row in which the part number is FD21. The item class HW is associated with three part numbers, so the part number is not functionally dependent on the class.

QUESTION

Is CustomerNum the primary key for the Customer table?

ANSWER

Yes, because customer numbers are unique. A given customer number cannot appear on more than one row. Thus, each customer number is associated with a single name, a single street, a single city, a single state, a single zip code, a single balance, a single credit limit, and a single rep number. In other words, all columns in the Customer table are functionally dependent on CustomerNum.

QUESTION

Is OrderNum the primary key for the OrderLine table?

ANSWER

No, because it does not uniquely determine NumOrdered or QuotedPrice. The order number 21617, for example, appears on a row in the OrderLine table in which the number ordered is 2 and the quoted price is $794.95 and on a row in which the number ordered is 4 and the quoted price is $150.00.

QUESTION

Is the combination of OrderNum and PartNum the primary key for the OrderLine table?

ANSWER

Yes, because all columns are functionally dependent on this combination. Any combination of an order number and a part number occurs on only one row in the OrderLine table and is associated with only one value for NumOrdered and only one value for QuotedPrice. Further, neither OrderNum nor PartNum alone has this property. For example, order number 21617 appears on more than one row, as does part DR93.

QUESTION
Is the combination of PartNum and Description the primary key for the Part table?

ANSWER
No. It is true that this combination functionally determines all columns in the Part table. PartNum alone, however, also has this property.

QUESTION
You already determined the functional dependencies in a Student relation containing the following columns: StudentNum, StudentLast, StudentFirst, HighSchoolNum, HighSchoolName, AdvisorNum, AdvisorLast, and AdvisorFirst. The functional dependencies you determined were:

```
StudentNum → StudentLast, StudentFirst, HighSchoolNum, HighSchoolName,
      AdvisorNum, AdvisorLast, AdvisorFirst
HighSchoolNum → HighSchoolName
AdvisorNum → AdvisorLast, AdvisorFirst
```

What is the primary key for the Student relation?

ANSWER
The only column that determines all the other columns is StudentNum, so it is the primary key for the Student relation.

Occasionally (but not often) there might be more than one possibility for the primary key. For example, if the Premiere Products database included an Employee table to store employee numbers and Social Security numbers, either the employee number or the Social Security number could serve as the primary key. In this case, both columns are referred to as candidate keys. Like a primary key, a **candidate key** is a column or collection of columns on which all columns in the table are functionally dependent; the definition for primary key also defines a candidate key. From all the candidate keys, one is chosen to be the primary key. The candidate keys that are not chosen as the primary key are often referred to as **alternate keys**.

Note: The primary key is often called the "key" in other studies on database management and the relational model. We will continue to use the term "primary key" in order to clearly distinguish among the several different concepts of a key that you will encounter throughout this text.

First Normal Form

A relation (table) that contains a **repeating group** (or multiple entries for a single record) is called an **unnormalized relation**. Removal of repeating groups is the starting point in the quest to create tables that are as free of problems as possible. Tables without repeating groups are said to be in first normal form.

Definition: A table (relation) is in **first normal form (1NF)** if it does not contain repeating groups.

As an example, consider the Orders table shown in Figure 5.5, in which there is a repeating group consisting of PartNum and NumOrdered. The notation for describing the Orders table is:

```
Orders (OrderNum, OrderDate, (PartNum, NumOrdered) )
```

This notation indicates a table named Orders, consisting of a primary key (OrderNum) and a column named OrderDate. The inner parentheses indicate that there is a repeating group. The repeating group contains two columns, PartNum and NumOrdered. This means that for a single order, there can be multiple combinations of a part number and a corresponding number of units ordered, as illustrated in Figure 5.5. The row for order 21617, for example, contains two such combinations. In the first combination, the part number is BV06 and the number ordered is 2. In the second, the part number is CD52 and the number ordered is 4.

FIGURE 5.5
....
Sample unnormal-
ized table

Orders

OrderNum	OrderDate	PartNum	NumOrdered
21608	10/20/2007	AT94	11
21610	10/20/2007	DR93	1
		DW11	1
21613	10/21/2007	KL62	4
21614	10/21/2007	KT03	2
21617	10/23/2007	BV06	2
		CD52	4
21619	10/23/2007	DR93	1
21623	10/23/2007	KV29	2

To convert the Orders table to first normal form, you remove the repeating group as follows:

```
Orders (OrderNum, OrderDate, PartNum, NumOrdered)
```

Figure 5.6 shows the new table, which is now in first normal form.

FIGURE 5.6
....
Result of normaliza-
tion (conversion to
first normal form)

Orders

OrderNum	OrderDate	PartNum	NumOrdered
21608	10/20/2007	AT94	11
21610	10/20/2007	DR93	1
21610	10/20/2007	DW11	1
21613	10/20/2007	KL62	4
21614	10/20/2007	KT03	2
21617	10/20/2007	BV06	2
21617	10/20/2007	CD52	4
21619	10/20/2007	DR93	1
21623	10/20/2007	KV29	2

Note that the fifth row of the unnormalized table (see Figure 5.5) indicates that part BV06 and part CD52 are both present for order 21617. In the normalized table (see Figure 5.6), this information is represented by *two* rows, the sixth and seventh. The primary key to the unnormalized Orders table was OrderNum alone. The primary key to the normalized table is now the combination of OrderNum and PartNum.

In general, when converting a non-first normal form table to first normal form, the primary key will usually include the original primary key concatenated with the key to the repeating group, which is the column that distinguishes one occurrence of the repeating group from another on a given row in the table. In this case, PartNum is the key to the repeating group and thus PartNum becomes part of the primary key of the first normal form table.

◼ Second Normal Form

A table that is in first normal form still might contain problems that will require you to restructure it. Consider the following table:

```
Orders (OrderNum, OrderDate, PartNum, Description, NumOrdered, QuotedPrice)
```

This table has the following functional dependencies:

```
OrderNum → OrderDate
PartNum → Description
OrderNum, PartNum → NumOrdered, QuotedPrice, OrderDate, Description
```

This notation indicates that OrderNum alone determines OrderDate, and PartNum alone determines Description, but *both* an OrderNum *and* a PartNum are required to determine either NumOrdered or QuotedPrice. (The combination of OrderNum and PartNum also determines both OrderDate and Description because OrderNum determines OrderDate and PartNum determines Description.) Consider the sample of this table shown in Figure 5.7.

FIGURE 5.7
....
Sample Orders
table

Orders

OrderNum	OrderDate	PartNum	Description	NumOrdered	QuotedPrice
21608	10/20/2007	AT94	Iron	11	$21.95
21610	10/20/2007	DR93	Gas Range	1	$495.00
21610	10/20/2007	DW11	Washer	1	$399.99
21613	10/21/2007	KL62	Dryer	4	$329.95
21614	10/21/2007	KT03	Dishwasher	2	$595.00
21617	10/23/2007	BV06	Home Gym	2	$794.95
21617	10/23/2007	CD52	Microwave Oven	4	$150.00
21619	10/23/2007	DR93	Gas Range	1	$495.00
21623	10/23/2007	KV29	Treadmill	2	$1290.00

The description of a specific part, DR93 for example, occurs twice in the table. This redundancy causes several problems. It is certainly wasteful of space, but that is not nearly as serious as some of the other problems. These other problems are called update anomalies and they fall into four categories:

1. **Update.** A change to the description of part DR93 requires not one change to the table, but two changes—you have to change each row on which part DR93 appears. Changing multiple rows certainly makes the update process much more cumbersome; it also is more complicated logically and takes more time to update.

2. **Inconsistent data.** There is nothing about the design that would prohibit part DR93 from having two different descriptions in the database. In fact, if part DR93 occurs on 20 rows, it could potentially have 20 *different* descriptions in the database!

3. **Additions.** You have a real problem when you try to add a new part and its description to the database. Because the primary key for the table consists of both OrderNum and PartNum, you need values for both columns to add a new row. If you have a part to add but there are no orders for it yet, what do you use for an order number? The only solution would be to make up a dummy order number and then replace it with a real order number after an order for the new part was actually received. Certainly, this is not an acceptable solution.

4. **Deletions.** If you delete order 21608 from the database, you would also *lose* all information about part AT94. For example, you would no longer know that part AT94 is an iron.

These problems occur because you have a column, Description, that is dependent on only a portion of the primary key (PartNum) and *not* on the complete primary key. This problem leads to the definition of second normal form. Second normal form represents an improvement over first normal form because it eliminates update anomalies in these situations. To understand second normal form, you need to understand the term "nonkey column."

Definition: A column is a **nonkey column** (also called a **nonkey attribute**) if it is not a part of the primary key.

Definition: A table (relation) is in **second normal form (2NF)** if it is in first normal form and no nonkey column is dependent on only a portion of the primary key.

Note: If the primary key of a table contains only a single column, the table is automatically in second normal form because there would be no way for a column to be dependent on only a portion of the key.

For another perspective on second normal form, consider Figure 5.8. This type of diagram, sometimes called a **dependency diagram**, uses arrows to indicate all the functional dependencies present in the Orders table. The arrows above the boxes indicate the normal dependencies that should be present; in other words, the primary key functionally determines all other columns. (In the Orders table, the concatenation of OrderNum and PartNum determines all other columns.) The arrows below the boxes prevent the table from being in second normal form. These arrows represent types of dependencies that are often called **partial dependencies**, which are dependencies on only a portion of the primary key. In fact, another definition for second normal form is that a table is in second normal form if it is in first normal form but contains no partial dependencies.

FIGURE 5.8
....
Dependences in
the Orders table

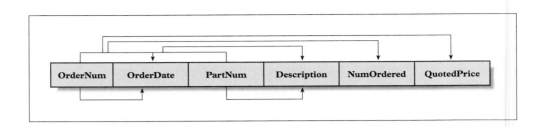

Regardless of which definition of second normal form you use, you now can identify the fundamental problem with the Orders table: It is *not* in second normal form. Although it may be pleasing to have a name for the problem, what you really need is a method to *correct* it; you want to be able to convert tables to second normal form. To do so, first take each subset of the set of columns that makes up the primary key, and then begin a new table with this subset as the primary key. For the Orders table, this would give:

```
(OrderNum,
(PartNum,
(OrderNum, PartNum,
```

Next, place each of the other columns with its appropriate primary key; that is, place each one with the minimal collection of columns on which it depends. For the Orders table, this would yield:

```
(OrderNum, OrderDate)
(PartNum, Description)
(OrderNum, PartNum, NumOrdered, QuotedPrice)
```

Now you can give each new table a name that is descriptive of the table's contents, such as Orders, Part, and OrderLine. Figure 5.9 shows the original Orders table on top; the resulting Orders, Part, and OrderLine tables created after converting the Orders table to second normal form appear below it.

FIGURE 5.9
....
Conversion to
second normal form

Orders

OrderNum	OrderDate	PartNum	Description	NumOrdered	QuotedPrice
21608	10/20/2007	AT94	Iron	11	$21.95
21610	10/20/2007	DR93	Gas Range	1	$495.00
21610	10/20/2007	DW11	Washer	1	$399.99
21613	10/21/2007	KL62	Dryer	4	$329.95
21614	10/21/2007	KT03	Dishwasher	2	$595.00
21617	10/23/2007	BV06	Home Gym	2	$794.95
21617	10/23/2007	CD52	Microwave Oven	4	$150.00
21619	10/23/2007	DR93	Gas Range	1	$495.00
21623	10/23/2007	KV29	Treadmill	2	$1290.00

Orders

OrderNum	OrderDate
21608	10/20/2007
21610	10/20/2007
21613	10/21/2007
21614	10/21/2007
21617	10/23/2007
21619	10/23/2007
21623	10/23/2007

Part

PartNum	Description
AT94	Iron
BV06	Home Gym
CD52	Microwave Oven
DL71	Cordless Drill
DR93	Gas Range
DW11	Washer
FD21	Stand Mixer
KL62	Dryer
KT03	Dishwasher
KV29	Treadmill

OrderLine

OrderNum	PartNum	NumOrdered	QuotedPrice
21608	AT94	11	$21.95
21610	DR93	1	$495.00
21610	DW11	1	$399.99
21613	KL62	4	$329.95
21614	KT03	2	$595.00
21617	BV06	2	$794.95
21617	CD52	4	$150.00
21619	DR93	1	$495.00
21623	KV29	2	$1290.00

Now you have eliminated the update anomalies. A description appears only once for each part, so you do not have the redundancy that you did in the previous design. Changing the description for part DR93 from Gas Range to Deluxe Range, for example, now is a simple process involving a single change. Because the description for a part occurs in a single place, it is not possible to have multiple descriptions for a single part in the database at the same time.

To add a new part and its description, you create a new row in the Part table and thus there is no need to have an existing order for that part. Also, deleting order 21608 does not delete part AT94 from the Part table, and thus, you still have its description (Iron) in the database. Finally, you have not lost any information in the process—you can reconstruct the data in the original design from the data in the new design.

Third Normal Form

Problems can still exist with tables that are in second normal form. Consider the following Customer table:

Customer (<u>CustomerNum</u>, CustomerName, Balance, CreditLimit, RepNum,
     LastName, FirstName)

The functional dependencies in this table are:

CustomerNum → CustomerName, Balance, CreditLimit, RepNum,
     LastName, FirstName
RepNum → LastName, FirstName

CustomerNum determines all the other columns. In addition, RepNum determines LastName and FirstName.

If the primary key of a table is a single column, the table is automatically in second normal form. (If the table were not in second normal form, some columns would be dependent on only a *portion* of the primary key, which is impossible when the primary key is just one column.) Thus, the Customer table is in second normal form.

The sample Customer data shown in Figure 5.10 illustrates that this table possesses problems similar to those encountered earlier, even though it is in second normal form. In this case, the name of a sales rep can occur many times in the table; see sales rep 65 (Juan Perez), for example.

FIGURE 5.10

Sample Customer data

Customer

CustomerNum	CustomerName	Balance	CreditLimit	RepNum	LastName	FirstName
148	Al's Appliance and Sport	$6,550.00	$7,500.00	20	Kaiser	Valerie
282	Brookings Direct	$431.50	$10,000.00	35	Hull	Richard
356	Ferguson's	$5,785.00	$7,500.00	65	Perez	Juan
408	The Everything Shop	$5,285.25	$5,000.00	35	Hull	Richard
462	Bargains Galore	$3,412.00	$10,000.00	65	Perez	Juan
524	Kline's	$12,762.00	$15,000.00	20	Kaiser	Valerie
608	Johnson's Department Store	$2,106.00	$10,000.00	65	Perez	Juan
687	Lee's Sport and Appliance	$2,851.00	$5,000.00	35	Hull	Richard
725	Deerfield's Four Seasons	$248.00	$7,500.00	35	Hull	Richard
842	All Season	$8,221.00	$7,500.00	20	Kaiser	Valerie

This redundancy creates the same set of problems that you examined in the first normal form Orders table. In addition to the problem of wasted space, you have similar update anomalies, as follows:

1. ***Updates.*** A change to the name of a sales rep requires not one change to the table, but several, making the update process cumbersome.

2. ***Inconsistent data.*** There is nothing about the design that would prohibit a sales rep from having two different names in the database. In fact, if the same sales rep represents 20 customers (and thus would be found on 20 different rows), he or she could have 20 *different* names in the database.

3. ***Additions.*** In order to add sales rep 87 (Mary Daniels) to the database, she must already represent at least one customer. If she has not yet been assigned any customers, you must add her record and create a fictitious customer for her to represent. Again, this is not a desirable solution to the problem.

4. ***Deletions.*** If you deleted all the customers of sales rep 35 from the database, you would then lose all information concerning sales rep 35.

These update anomalies are due to the fact that RepNum determines LastName and FirstName, but RepNum is not the primary key. As a result, the same RepNum and consequently the same LastName and FirstName can appear on many different rows.

You've seen that second normal form is an improvement over first normal form, but to eliminate second normal form problems, you need an even better strategy for creating tables in the database. Third normal form provides that strategy. Before looking at third normal form, however, you need to become familiar with the special name that is given to any column that determines another column (as RepNum does in the Customer table).

Definition: Any column (or collection of columns) that determines another column is called a **determinant**.

Certainly the primary key in a table will be a determinant. In fact, by definition, any candidate key will be a determinant. (Remember that a candidate key is a column or collection of columns that could function as the primary key.) In this case, RepNum is a determinant, but it is not a candidate key, and that is the problem.

Definition: A table is in **third normal form (3NF)** if it is in second normal form and the only determinants it contains are candidate keys.

Note: The previous definition is not the original definition of third normal form. This more recent definition, which is preferable to the original, is often referred to as **Boyce-Codd normal form (BCNF)** when it is important to make a distinction between this definition and the original. This text will not make such a distinction, but will take this to be the definition of third normal form.

Again, for an additional perspective, you can use a dependency diagram, as shown in Figure 5.11. The arrows above the boxes represent the normal dependencies of all columns on the primary key. The arrows below the boxes represent the problem—these arrows make RepNum a determinant. If there were arrows from RepNum to all the columns, RepNum would be a candidate key and you would not have a problem. The absence of these arrows indicates that this table contains a determinant that is not a candidate key. Thus, the table is not in third normal form.

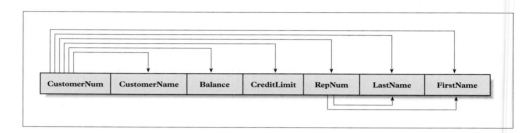

You now have identified the problem with the Customer table: It is not in third normal form. The following method corrects the deficiency in the Customer table and in all tables having similar deficiencies.

First, for each determinant that is not a candidate key, remove from the table the columns that depend on this determinant (but don't remove the determinant). Next, create a new table containing all the columns from the original table that depend on this determinant. Finally, make the determinant the primary key of this new table. In the Customer table, for example, you would remove LastName and FirstName because they depend on the determinant RepNum, which is not a candidate key. A new table is formed, consisting of RepNum (as the primary key), LastName, and FirstName.

```
Customer (CustomerNum, CustomerName, Balance, CreditLimit, RepNum)
Rep (RepNum, LastName, FirstName)
```

Figure 5.12 shows samples of the revised Customer table and the new Rep table.

FIGURE 5.12
....
Conversion to third
normal form

Customer

CustomerNum	CustomerName	Balance	CreditLimit	RepNum	LastName	FirstName
148	Al's Appliance and Sport	$6,550.00	$7,500.00	20	Kaiser	Valerie
282	Brookings Direct	$431.50	$10,000.00	35	Hull	Richard
356	Ferguson's	$5,785.00	$7,500.00	65	Perez	Juan
408	The Everything Shop	$5,285.25	$5,000.00	35	Hull	Richard
462	Bargains Galore	$3,412.00	$10,000.00	65	Perez	Juan
524	Kline's	$12,762.00	$15,000.00	20	Kaiser	Valerie
608	Johnson's Department Store	$2,106.00	$10,000.00	65	Perez	Juan
687	Lee's Sport and Appliance	$2,851.00	$5,000.00	35	Hull	Richard
725	Deerfield's Four Seasons	$248.00	$7,500.00	35	Hull	Richard
842	All Season	$8,221.00	$7,500.00	20	Kaiser	Valerie

Customer

CustomerNum	CustomerName	Balance	CreditLimit	RepNum
148	Al's Appliance and Sport	$6,550.00	$7,500.00	20
282	Brookings Direct	$431.50	$10,000.00	35
356	Ferguson's	$5,785.00	$7,500.00	65
408	The Everything Shop	$5,285.25	$5,000.00	35
462	Bargains Galore	$3,412.00	$10,000.00	65
524	Kline's	$12,762.00	$15,000.00	20
608	Johnson's Department Store	$2,106.00	$10,000.00	65
687	Lee's Sport and Appliance	$2,851.00	$5,000.00	35
725	Deerfield's Four Seasons	$248.00	$7,500.00	35
842	All Season	$8,221.00	$7,500.00	20

Rep

RepNum	LastName	FirstName
20	Kaiser	Valerie
35	Hull	Richard
65	Perez	Juan

Have you now corrected all previously identified problems? A sales rep's name appears only once, thus avoiding redundancy and simplifying the process of changing a sales rep's name. With this design, it is not possible for the same sales rep to have different names in the database. To add a new sales rep to the database, you can add a row in the Rep table without requiring the rep to have at least one assigned customer. Finally, deleting all the customers of a given sales rep will not remove the sales rep's record from the Rep table, retaining the sales rep's name. In addition, you can reconstruct all the data in the original table from the data in the new collection of tables. All previously mentioned problems have indeed been solved.

Incorrect Decompositions

It is important to note that the decomposition of a table into two or more third normal form tables *must* be accomplished by the method indicated in the previous section, even though there are other possibilities that at first glance might seem to be legitimate. For example, you can examine two other decompositions of the Customer table into third normal form tables to understand the difficulties they pose. Assume in the decomposition process that

```
Customer (CustomerNum, CustomerName, Balance, CreditLimit, RepNum,
     LastName, FirstName)
```

is replaced by

```
Customer (CustomerNum, CustomerName, Balance, CreditLimit, RepNum)
Rep (CustomerNum, LastName, FirstName)
```

Samples of these tables appear in Figure 5.13. Both new tables are in third normal form. In addition, by joining these two tables together on CustomerNum, you can reconstruct the original Customer table. The result, however, still suffers from some of the same kinds of problems as the original Customer table.

FIGURE 5.13
••••••••••••••••••••
Incorrect decom-
position of the
Customer table

Customer

CustomerNum	CustomerName	Balance	CreditLimit	RepNum	LastName	FirstName
148	Al's Appliance and Sport	$6,550.00	$7,500.00	20	Kaiser	Valerie
282	Brookings Direct	$431.50	$10,000.00	35	Hull	Richard
356	Ferguson's	$5,785.00	$7,500.00	65	Perez	Juan
408	The Everything Shop	$5,285.25	$5,000.00	35	Hull	Richard
462	Bargains Galore	$3,412.00	$10,000.00	65	Perez	Juan
524	Kline's	$12,762.00	$15,000.00	20	Kaiser	Valerie
608	Johnson's Department Store	$2,106.00	$10,000.00	65	Perez	Juan
687	Lee's Sport and Appliance	$2,851.00	$5,000.00	35	Hull	Richard
725	Deerfield's Four Seasons	$248.00	$7,500.00	35	Hull	Richard
842	All Season	$8,221.00	$7,500.00	20	Kaiser	Valerie

Customer

CustomerNum	CustomerName	Balance	CreditLimit	RepNum
148	Al's Appliance and Sport	$6,550.00	$7,500.00	20
282	Brookings Direct	$431.50	$10,000.00	35
356	Ferguson's	$5,785.00	$7,500.00	65
408	The Everything Shop	$5,285.25	$5,000.00	35
462	Bargains Galore	$3,412.00	$10,000.00	65
524	Kline's	$12,762.00	$15,000.00	20
608	Johnson's Department Store	$2,106.00	$10,000.00	65
687	Lee's Sport and Appliance	$2,851.00	$5,000.00	35
725	Deerfield's Four Seasons	$248.00	$7,500.00	35
842	All Season	$8,221.00	$7,500.00	20

Rep

CustomerNum	LastName	FirstName
148	Kaiser	Valerie
282	Hull	Richard
356	Perez	Juan
408	Hull	Richard
462	Perez	Juan
524	Kaiser	Valerie
608	Perez	Juan
687	Hull	Richard
725	Hull	Richard
842	Kaiser	Valerie

Consider, for example, the redundancy in the storage of sales reps' names, the problem encountered in changing the name of a sales rep, and the difficulty of adding a new sales rep who represents no customers. In addition, because the rep number and names are in different tables, you have actually *split a functional dependence across two different tables.* Thus, this seemingly valid decomposition is definitely not a desirable way to create third normal form tables.

There is another decomposition that you might choose, and that is to replace

```
Customer (CustomerNum, CustomerName, Balance, CreditLimit, RepNum,
     LastName, FirstName)
```

with

```
Customer (CustomerNum, CustomerName, Balance, CreditLimit,
     LastName, FirstName)
Rep (RepNum, LastName, FirstName)
```

Samples of these tables appear in Figure 5.14.

FIGURE 5.14
••••••••••••••••••
Second incorrect
decomposition of
the Customer table

Customer

CustomerNum	CustomerName	Balance	CreditLimit	RepNum	LastName	FirstName
148	Al's Appliance and Sport	$6,550.00	$7,500.00	20	Kaiser	Valerie
282	Brookings Direct	$431.50	$10,000.00	35	Hull	Richard
356	Ferguson's	$5,785.00	$7,500.00	65	Perez	Juan
408	The Everything Shop	$5,285.25	$5,000.00	35	Hull	Richard
462	Bargains Galore	$3,412.00	$10,000.00	65	Perez	Juan
524	Kline's	$12,762.00	$15,000.00	20	Kaiser	Valerie
608	Johnson's Department Store	$2,106.00	$10,000.00	65	Perez	Juan
687	Lee's Sport and Appliance	$2,851.00	$5,000.00	35	Hull	Richard
725	Deerfield's Four Seasons	$248.00	$7,500.00	35	Hull	Richard
842	All Season	$8,221.00	$7,500.00	20	Kaiser	Valerie

Customer

CustomerNum	CustomerName	Balance	CreditLimit	LastName	FirstName
148	Al's Appliance and Sport	$6,550.00	$7,500.00	Kaiser	Valerie
282	Brookings Direct	$431.50	$10,000.00	Hull	Richard
356	Ferguson's	$5,785.00	$7,500.00	Perez	Juan
408	The Everything Shop	$5,285.25	$5,000.00	Hull	Richard
462	Bargains Galore	$3,412.00	$10,000.00	Perez	Juan
524	Kline's	$12,762.00	$15,000.00	Kaiser	Valerie
608	Johnson's Department Store	$2,106.00	$10,000.00	Perez	Juan
687	Lee's Sport and Appliance	$2,851.00	$5,000.00	Hull	Richard
725	Deerfield's Four Seasons	$248.00	$7,500.00	Hull	Richard
842	All Season	$8,221.00	$7,500.00	Kaiser	Valerie

Rep

RepNum	LastName	FirstName
20	Kaiser	Valerie
35	Hull	Richard
65	Perez	Juan

This new design seems to be a possibility. Not only are both tables in third normal form, but joining them together based on LastName and FirstName seems to reconstruct the data in the original table. Or does it? Suppose that the name of sales rep 65 is also Valerie Kaiser. In that case, when you join the two new tables together, there would be no way to correctly identify which Valerie Kaiser represents which customers. You would thus get a row on which customer 148 (Al's Appliance and Sport) is associated with sales rep 20 (Valerie Kaiser) and *another* row on which customer 148 is associated with sales rep 65 (the other Valerie Kaiser). Because you obviously want decompositions that preserve the original information, this design is not appropriate.

∙ ∙

QUESTION

Using the types of entities found in a college environment (faculty, students, departments, courses, and so on), create an example of a table that is in first normal form but not in second normal form and an example of a table that is in second normal form but not in third normal form. In each case, justify your solutions and show how to convert to the higher forms.

ANSWER

There are many possible solutions. If your answer differs from the following solution, this does not mean that it is an unsatisfactory solution.

To create a first normal form table that is not in second normal form, you need a table that has no repeating groups and that has at least one column that is dependent on only a portion of the primary key. For a column to be dependent on a portion of the primary key, the primary key must contain at least two columns. The following is a picture of what you need:

(___1___, ___2___, 3 , 4)

This table contains four columns, numbered 1, 2, 3, and 4, in which the combination of columns 1 and 2 functionally determines both columns 3 and 4. In addition, neither column 1 nor column 2 can determine *all* other columns; if either one could, the primary key would contain only this one column. Finally, you want part of the primary key, say, column 2, to determine another column, say, column 4.

Now that you know the pattern you need, you would like to find columns from within the college environment to fit it. One example would be:

(StudentNum, CourseNum, Grade, CourseDescription)

In this example, the concatenation of StudentNum and CourseNum determines both Grade and CourseDescription. Both columns are required to determine Grade, and thus the primary key consists of their concatenation. CourseDescription, however, is dependent only on CourseNum, which violates second normal form. To convert this table to second normal form, you would replace it with two tables:

(StudentNum, CourseNum, Grade)
(CourseNum, CourseDescription)

You would of course now give these tables appropriate names.

To create a table that is in second normal form but not in third normal form, you need a second normal form table in which there is a determinant that is *not* a candidate key. If you choose a table that has a single column as the primary key, it is automatically in second normal form, so the real problem is the determinant. You need a table like the following:

(_____1_____,   2   ,   3   )

This table contains three columns, numbered 1, 2, and 3, in which column 1 determines each of the others and thus is the primary key. If column 2 also determines column 3, it is a determinant. If column 2 does not also determine column 1, then column 2 is not a candidate key. One example that fits this pattern would be:

```
(StudentNum, AdvisorNum, AdvisorName)
```

Here the StudentNum determines both the student's AdvisorNum and AdvisorName. AdvisorNum determines AdvisorName, but AdvisorNum does not determine StudentNum because one advisor can have many advisees. This table is in second normal form but not in third normal form. To convert it to third normal form, you would replace it with:

```
(StudentNum, AdvisorNum)
(AdvisorNum, AdvisorName)
```

QUESTION

Convert the following table to third normal form. In this table, StudentNum determines StudentName, NumCredits, AdvisorNum, and AdvisorName. AdvisorNum determines AdvisorName. CourseNum determines Description. The combination of StudentNum and CourseNum determines Grade.

```
Student (StudentNum, StudentName, NumCredits, AdvisorNum,
     AdvisorName, (CourseNum, Description, Grade) )
```

ANSWER

Step 1. Remove the repeating group to convert it to first normal form, yielding:

```
Student (StudentNum, StudentName, NumCredits, AdvisorNum,
     AdvisorName, CourseNum, Description, Grade)
```

This table is now in first normal form because it has no repeating groups. It is not, however, in second normal form, because StudentName, for example, is dependent only on StudentNum, which is only a portion of the primary key.

Step 2. Convert the first normal form table to second normal form. First, for each subset of the primary key, start a table with that subset as its key, yielding:

```
(StudentNum,
(CourseNum,
(StudentNum, CourseNum,
```

Next, place the rest of the columns with the smallest collection of columns on which they depend, giving:

```
(StudentNum, StudentName, NumCredits, AdvisorNum, AdvisorName)
(CourseNum, Description)
(StudentNum, CourseNum, Grade)
```

Finally, assign names to each of the newly created tables:

```
Student (StudentNum, StudentName, NumCredits, AdvisorNum, AdvisorName)
Course (CourseNum, Description)
StudentCourse (StudentNum, CourseNum, Grade)
```

Although these tables are all in second normal form, Course and StudentCourse are also in third normal form. The Student table is not in third normal form, however, because it contains a determinant (AdvisorNum) that is not a candidate key.

Step 3. Convert the second normal form Student table to third normal form by removing the column that depends on the determinant AdvisorNum and placing it in a separate table.

```
(StudentNum, StudentName, NumCredits, AdvisorNum)
(AdvisorNum, AdvisorName)
```

Step 4. Name these tables and put the entire collection together, giving:

```
Student (StudentNum, StudentName, NumCredits, AdvisorNum)
Advisor (AdvisorNum, AdvisorName)
Course (CourseNum, Description)
StudentCourse (StudentNum, CourseNum, Grade)
```

. . . .

■ Multivalued Dependencies and Fourth Normal Form

By converting a given collection of tables to an equivalent third normal form collection, you remove any problems arising from functional dependencies. Usually this means that you eliminate the types of previously discussed update anomalies. Converting to third normal form doesn't always avoid all problems related to dependencies, however. There is a different kind of dependency that also can lead to the same types of problems.

To illustrate the problem, suppose you are interested in faculty members at Marvel College. In addition to faculty members, you are interested in the students they advise and the committees on which the faculty members serve. A faculty member can advise many students. Because students can have more than one major, a student can have more than one faculty member as an advisor. A faculty member can serve on zero, one, or more committees. Suppose that, as an initial relational design for this situation, you chose the following unnormalized table:

```
Faculty (FacultyNum, (StudentNum), (CommitteeCode) )
```

The single Faculty table has a primary key of FacultyNum (the number that identifies the faculty member) and two separate repeating groups, StudentNum (the number that identifies the student) and CommitteeCode (the code that identifies the committee, such as ADV for Advisory committee, PER for Personnel committee, CUR for Curriculum committee, and so on). To convert this table to first normal form, you might be tempted to remove the two repeating groups and expand the primary key to include both StudentNum and CommitteeCode. This solution would give the table:

```
Faculty (FacultyNum, StudentNum, CommitteeCode)
```

Samples of the table with repeating groups and the table with the repeating groups removed appear in Figure 5.15.

Faculty

FacultyNum	StudentNum	CommitteeCode
123	12805 24139	ADV HSG PER
444	57384	HSG
456	24139 36273 37573	CUR

Faculty

FacultyNum	StudentNum	CommitteeCode
123	12805	ADV
123	12805	HSG
123	12805	PER
123	24139	ADV
123	24139	HSG
123	24139	PER
444	57384	HSG
456	24139	CUR
456	36273	CUR
456	37573	CUR

You already may have suspected that there are some problems with this approach. If so, you are correct. It is a strange way to normalize the original table. Yet, it is precisely this approach to removing repeating groups that leads to the problems concerning multivalued dependencies. In a moment, you will see how this table should have been normalized to avoid the problems altogether. For now, however, you'll examine this table to see what kinds of problems are present.

The first thing you should observe about this table is that it is in third normal form because no groups repeat, no column is dependent on only a portion of the primary key, and no determinants exist that are not candidate keys. There are several problems, however, with this third normal form table.

1. *Update.* Changing the CommitteeCode for faculty member 123 requires more than one change. If this faculty member changes from an Advisory committee member to a Curriculum committee member, you would need to change the CommitteeCode from ADV to CUR in rows 1 and 4 of the table. After all, it doesn't make sense to say that the committee is ADV when associated with student 12805 and CUR when associated with student 24139. It is the same committee served on by the same faculty member. The faculty member does not serve on one committee when advising one student and a different committee when advising another.

2. ***Additions.*** Suppose that faculty member 555 joins the faculty at Marvel College. Also suppose that this faculty member does not yet serve on any committee. When this faculty member begins advising student 44332, you have a problem because CommitteeCode is part of the primary key. You would need to enter a fictitious CommitteeCode in this situation.

3. ***Deletions.*** If faculty member 444 no longer advises student 57384 and you delete the appropriate row from the table, you lose the information that faculty member 444 serves on the Housing committee (HSG).

These problems are certainly similar to those encountered in the discussions of both second normal form and third normal form, but there are no functional dependencies among the columns in this table. A given faculty member is not associated with one student, as he or she would be if this were a functional dependence. Each faculty member, however, is associated with a specific collection of students. More importantly, this association is *independent* of any association with committees. It is this independence that causes the problem. This type of dependency is called a multivalued dependency.

Definition: In a table with columns A, B, and C, there is a **multivalued dependence** of column B on column A (also read as "B is **multidependent** on A" or "A **multidetermines** B") if each value for A is associated with a specific collection of values for B and, further, this collection is independent of any values for C. This is usually written as

A →→ B

Definition: A table is in **fourth normal form (4NF)** if it is in third normal form and there are no multivalued dependencies.

As you might expect, converting a table to fourth normal form is similar to the normalization process encountered in the treatments of second normal form and third normal form. You split the third normal form table into separate tables, each containing the column that multidetermines the others, which, in this case, is FacultyNum. This means you replace

Faculty (<u>FacultyNum</u>, <u>StudentNum</u>, <u>CommitteeCode</u>)

    with

FacStudent (<u>FacultyNum</u>, <u>StudentNum</u>)
FacCommittee (<u>FacultyNum</u>, <u>CommitteeCode</u>)

Figure 5.16 shows samples of these tables. As before, the problems have disappeared. There is no problem with changing the CommitteeCode ADV to CUR for faculty member 123 because the committee code occurs only in one place. To add the information that faculty member 555 advises student 44332, you need only add a row to the FacStudent table—it doesn't matter whether this faculty member serves on a committee. Finally, to delete the information that faculty member 444 advises student 57384, you need only remove a row from the FacStudent table. In this case, you do not lose the information that this faculty member serves on the Housing committee.

FIGURE 5.16
• • • • • • • • • • • • • • • • • •
Conversion to
fourth normal form

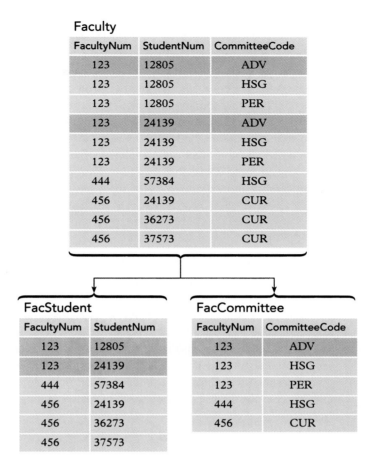

Figure 5.17 summarizes the four normal forms.

FIGURE 5.17
• • • • • • • • • • • • • • • •
Normal forms

Normal Form	Meaning/Required Conditions	Notes
First normal form	No repeating groups exist	
Second normal form	First normal form and no non-key column is dependent on only a portion of the primary key	Automatically second normal form if the primary key contains only a single column
Third normal form	Second normal form and the only determinants are candidate keys	Actually Boyce-Codd normal form (BCNF)
Fourth normal form	Third normal form and no multivalued dependencies exist	

Avoiding the Problem with Multivalued Dependencies

Any table that is not in fourth normal form suffers some serious problems, but there is a way to avoid dealing with the issue. It's better to have a design methodology for normalizing tables that will prevent this situation from occurring in the first place. You already have most of such a methodology in place from the discussion of the first normal form, second normal form, and third normal form normalization processes. All you need is a slightly more sophisticated method for converting an unnormalized table to first normal form.

The conversion of an unnormalized table to first normal form requires the removal of repeating groups. When this was first demonstrated, you merely removed the repeating group symbol and expanded the primary key. You will recall, for example, that

```
Orders (OrderNum, OrderDate, (PartNum, NumOrdered) )
```

became

```
Orders (OrderNum, OrderDate, PartNum, NumOrdered)
```

The primary key was expanded to include the primary key of the original table together with the key to the repeating group.

What if there are two or more repeating groups? The method you used earlier is inadequate for such situations. Instead, you must place each repeating group in a separate table. Each table will contain all the columns that make up the given repeating group, as well as the primary key of the original unnormalized table. The primary key to each new table will be the concatenation of the primary key of the original table and the primary key of the repeating group. For example, consider the following unnormalized table containing two repeating groups:

```
Faculty (FacultyNum, FacultyName, (StudentNum, StudentName),
    (CommitteeCode, CommitteeName) )
```

In this example, FacultyName is the name of the faculty member and StudentName is the name of the student. The columns CommitteeCode and CommitteeName refer to the committee's code and name. (For example, one row in this table would have PER in the CommitteeCode column and "Personnel Committee" in the CommitteeName column.) Applying this new method to create first normal form tables would produce:

```
Faculty (FacultyNum, FacultyName)
FacStudent (FacultyNum, StudentNum, StudentName)
FacCommittee (FacultyNum, CommitteeCode, CommitteeName)
```

As you can see, this collection of tables avoids the problems with multivalued dependencies. At this point, you have a collection of first normal form tables and you still need to convert them to third normal form. By using the above process, however, you are guaranteed that the result will also be in fourth normal form.

Application to Database Design

The normalization process used to convert a relation or collection of relations to an equivalent collection of third normal form tables is a crucial part of the database design process. By following a careful and appropriate normalization methodology, you should not need to worry about normal forms higher than third normal form. There are three aspects concerning normalization that you need to keep in mind in the process, however.

First, you should carefully convert tables to third normal form. Suppose the following columns exist in a Coach relation (the ellipsis (…) represents additional columns that exist but are not included in this example):

```
Coach (CoachNum, LastName, FirstName, Street, City,
    State, Zip, ...)
```

In addition to the functional dependencies that all the columns have on CoachNum, there are two other functional dependencies. As originally designed by the United States Postal Service, Zip determines State and Zip also determines City.

Does this mean that you should replace the Coach relation with

```
Coach (CoachNum, LastName, FirstName, Street, Zip, . . .)
ZipCode (Zip, City, State)
```

If you are determined to ensure that every relation is in third normal form, then you would replace the Coach relation with the revised Coach relation and the new Zip relation, but this approach is probably unnecessary. If you review the list of problems normally associated with relations that are not in third normal form, you will see that they really don't apply here. Are you likely to need to change the state in which the zip code 49428 is located? Do you need to add the fact that zip code 49401 corresponds to Allendale, Michigan if you have no customers who live in Allendale? In this case, the design of the original Coach relation is sufficient.

Second, there are currently situations where the same zip code corresponds to more than one city or to even more than one state. This situation illustrates both the wisdom in not making the change and the fact that requirements and, consequently, the functional dependencies, can change over time. It is critical to review assumptions and dependencies periodically to see if any changes to the design are warranted.

Third, by splitting relations to achieve third normal form tables, you create the need to express interrelation constraints. In the example given earlier for converting to third normal form, you split the Customer relation in the Premiere Products database from

```
Customer (CustomerNum, CustomerName, Balance, CreditLimit, RepNum,
    LastName, FirstName)
```

to

```
Customer (CustomerNum, CustomerName, Balance, CreditLimit, RepNum)
Rep (RepNum, LastName, FirstName)
```

There is nothing about these two relations by themselves that would force the RepNum on a row in the Customer relation to actually match a value of RepNum in the Rep relation. Requiring this to take place is an example of an **interrelation constraint**, that is, a condition that involves two or more relations. Foreign keys handle this type of interrelation constraint. You will learn more about and specify foreign keys during the database design process, which is covered in Chapter 6.

■ SUMMARY

- A column (attribute) B is functionally dependent on another column A (or possibly a collection of columns) if each value for A in the database is associated with exactly one value of B.

- The primary key is a column (or a collection of columns) A such that all other columns are functionally dependent on A and no subcollection of the columns in A also has this property.

- If there is more than one possible choice for the primary key, one of the possibilities is chosen to be *the* primary key. The others are referred to as candidate keys.

- A table (relation) is in first normal form (1NF) if it does not contain repeating groups.

- A column is a nonkey column (also called a nonkey attribute) if it is not a part of the primary key.

- A table (relation) in the second normal form (2NF) if it is in first normal form and no nonkey column is dependent on only a portion of the primary key.

- A determinant is any column that functionally determines another column.

- A table (relation) is in third normal form (3NF) if it is in second normal form and the only determinants it contains are candidate keys.

- A collection of tables (relations) that is not in third normal form has inherent problems called update anomalies. Replacing this collection by an equivalent collection of tables (relations) that is in third normal form removes these anomalies. This replacement must be done carefully, following a method like the one proposed in this text. If not, other problems, such as those discussed in this chapter, could be introduced.

- A table is in fourth normal form (4NF) if it is in third normal form and there are no multivalued dependencies.

■ KEY TERMS

alternate key
Boyce-Codd normal form (BCNF)
candidate key
concatenation
dependency diagram
determinant
first normal form (1NF)
fourth normal form (4NF)
functional dependence
functionally dependent
functionally determines
interrelation constraint
multidependent

multidetermines
multivalued dependence
nonkey attribute
nonkey column
normal form
normalization process
partial dependency
primary key
repeating group
second normal form (2NF)
third normal form (3NF)
unnormalized relation
update anomaly

■ REVIEW QUESTIONS

1. Define functional dependence.

2. Give an example of a column A and a column B such that B is functionally dependent on A. Give an example of a column C and a column D such that D is *not* functionally dependent on C.

3. Define primary key.

4. Define candidate key.

5. Define first normal form.

6. Define second normal form. What types of problems would you find in tables that are not in second normal form?

7. Define third normal form. What types of problems would you find in tables that are not in third normal form?

8. Define fourth normal form. What types of problems would you find in tables that are not in fourth normal form?

9. Define interrelation constraint and give one example of such a constraint. How are interrelation constraints addressed?

10. Consider a Student table containing StudentNum, StudentName, student's StudentMajor, student's AdvisorNum, student's AdvisorName, student's AdvisorOfficeNum, student's AdvisorPhone, student's NumCredits, and student's Class (freshman, sophomore, and so on). List the functional dependencies that exist, along with the assumptions that would support these dependencies.

11. Convert the following table to an equivalent collection of tables that is in third normal form. This table contains information about patients of a dentist. Each patient belongs to a household.

```
Patient (HouseholdNum, HouseholdName, Street, City, State, Zip,
        Balance, PatientNum, PatientName, (ServiceCode, Description,
        Fee, Date) )
```

The following dependencies exist in the Patient table:

```
PatientNum → HouseholdNum, HouseholdName, Street, City, State,
        Zip, Balance, PatientName
HouseholdNum → HouseholdName, Street, City, State, Zip, Balance
ServiceCode → Description, Fee
PatientNum, ServiceCode → Date
```

12. Using your knowledge of the college environment, determine the functional dependencies that exist in the following table. After determining the functional dependencies, convert this table to an equivalent collection of tables that are in third normal form.

```
Student (StudentNum, StudentName, NumCredits, AdvisorNum,
        AdvisorName, DeptNum, DeptName, (CourseNum, Description, Term,
        Grade) )
```

13. Again using your knowledge of the college environment, determine the functional or multivalued dependencies that exist in the following table. After determining the functional dependencies, convert this table to an equivalent collection of tables that are in fourth normal form. ActivityNum and ActivityName refer to activities in which a student can choose to participate. For example, activity number 1 might be soccer, activity 2 might be band, activity 3 might be the debate team, and so on. A student can choose to participate in multiple activities. CourseNum and Description refer to courses the student is taking.

```
Student (StudentNum, StudentName, ActivityNum, ActivityName,
        CourseNum, Description)
```

The following exercises are based on the Premiere Products database.

1. Using your knowledge of Premiere Products, determine the functional dependencies that exist in the following table. After determining the functional dependencies, convert this table to an equivalent collection of tables that are in third normal form.

   ```
   Part (PartNum, Description, OnHand, Class, Warehouse, Price,
         (OrderNum, OrderDate, CustomerNum, CustomerName, RepNum,
         LastName, FirstName, NumOrdered, QuotedPrice) )
   ```

2. List the functional dependencies in the following table that concerns invoicing (an application Premiere Products is considering adding to its database), subject to the specified conditions. For a given invoice (identified by the InvoiceNum), there will be a single customer. The customer's number, name, and complete address appear on the invoice, as does the date. Also, there may be several different parts appearing on the invoice. For each part that appears, display the part number, description, price, and number shipped. Each customer that orders a particular part pays the same price. Convert this table to an equivalent collection of tables that are in third normal form.

   ```
   Invoice (InvoiceNum, CustomerNum, LastName, FirstName, Street, City,
            State, Zip, Date, (PartNum, Description, Price, NumShipped) )
   ```

3. The requirements for Premiere Products have changed. A number *and* a name now identify each warehouse. Units of each part may be stored in multiple warehouses and it is important to know precisely how many parts are stored in each warehouse. In addition, Premiere Products now wants to manage information about the suppliers from which it purchases parts. For each part, Premiere Products needs to know the number and name of each supplier as well as the expected lead time for delivering each part. (Lead time is the amount of time a supplier is expected to take to deliver the part after Premiere Products has ordered it.) Each part can have many suppliers and each supplier can supply many parts. Using this information, convert the following unnormalized relation to fourth normal form:

   ```
   Part (PartNum, Description, Class, Price (WarehouseNum, WarehouseName,
         OnHand), (SupplierNum, SupplierName, LeadTime) )
   ```

■ Henry Books Case

Ray Henry wants you to help with database design activities, and he knows that the concepts in this chapter will be critical in the database design process. To ensure you understand these concepts, he would like you to complete the following exercises.

1. Using the types of entities found in the Henry Books database (books, authors, and publishers), create an example of a table that is in first normal form but not in second normal form and an example of a table that is in second normal form but not in third normal form. In each case, justify your answers and show how to convert to the higher forms.

2. Henry Books is considering selling textbooks to the local college. To do so, it must maintain information about courses, textbooks, and instructors. Determine the multivalued dependencies in the following table, and then convert this table to an equivalent collection of tables that are in fourth normal form. Each course is associated with a specific set of textbooks independent of the instructors who are teaching the course. In other words, although many instructors may be teaching the course, they all will use the same set of textbooks.

    Course (<u>CourseNum</u>, <u>Textbook</u>, <u>InstructorNum</u>, InstructorName)

3. The following unnormalized table is similar in content to the table in Exercise 2. Note that this table has two separate repeating groups: one listing the textbooks used for the course and the other listing the instructors who teach the course. Convert it to fourth normal form. Did you encounter the table from Exercise 2 along the way?

    Course (<u>CourseNum</u>, Description, NumCredits, (Textbook),
        (InstructorNum, InstructorName))

4. Identify the functional dependencies in the following unnormalized table. Convert the table to third normal form. Is the result also in fourth normal form?

    Book (<u>BookCode</u>, Title, PublisherCode, PublisherName, (AuthorNum,
        AuthorLast, AuthorFirst))

CHAPTER 6

Database Design 2: Design Methodology

OBJECTIVES

- Discuss the general process and goals of database design.

- Define user views and explain their function.

- Define Database Design Language (DBDL) and use it to document database designs.

- Create an entity-relationship (E-R) diagram to visually represent a database design.

- Present a methodology for database design at the information level and view examples illustrating this methodology.

- Explain the physical-level design process.

- Discuss top-down and bottom-up approaches to database design and examine the advantages and disadvantages of both methods.

- Use a survey form to obtain information from users prior to beginning the database design process.

- Review existing documents to obtain information prior to beginning the database design.

- Discuss special issues related to implementing one-to-one relationships and many-to-many relationships involving more than two entities.

- Discuss entity subtypes and their relationships to nulls.

- Learn how to avoid potential problems when merging third normal form relations.

- Examine the entity-relationship model for representing and designing databases.

Introduction

Now that you have learned how to identify and correct poor table designs, you will turn your attention to the design process itself; that is, you will learn about the process of determining the tables (relations) and columns (attributes) that will make up the database. In addition, you will determine the relationships between the various tables.

Most designers tackle database design using a two-step process. In the first step, the database designers design a database that satisfies the organization's requirements as cleanly as possible. This step is called **information-level design**, and it is completed *independently* of any particular database management system (DBMS) that the organization will ultimately use. In the second step, which is called the **physical-level design**, designers transform the information-level design into a design for the specific DBMS that the organization will use. During the physical-level design, designers must consider the characteristics of the particular DBMS that the organization will use.

After examining the information-level design process, you will explore the general database design methodology and view examples illustrating this methodology. You will construct entity-relationship diagrams to visually represent the database design. You'll then learn about the physical-level design process. You will also compare top-down and bottom-up approaches to database design.

You will explore special issues related to database design, including survey forms and their use in database design and how to obtain important information from existing documents. You will examine issues related to the implementation of some special types of relationships. You will learn about entity subtypes and their relationship to nulls. You will examine issues related to merging third normal form relations. Finally, you will learn about the entity-relationship model.

User Views

No matter which approach an organization adopts to implement its database design, a complete database design that will satisfy all the organization's requirements is rarely a one-step process. Unless the requirements are simple, an organization will usually divide the overall job of database design into many smaller tasks by identifying the individual pieces of the design problem, called user views. A **user view** is the set of requirements that is necessary to support the operations of a particular database user. For example, at Premiere Products, the database must be capable of storing the part's number, description, units on hand, item class, the number of the warehouse in which the part is located, and the price for each part.

For each user view, designers must design the database structure to support the view and then merge it into a **cumulative design**, which is a design that supports all the user views encountered thus far in the design process. Each user view is generally much simpler than the total collection of requirements, so working on individual user views is usually more manageable than attempting to turn the design of the entire database into one large task.

Information-Level Design Methodology

The information-level design methodology in this text involves representing individual user views, refining them to eliminate any problems, and then merging them into a cumulative design. After you have represented and merged all user views, you can complete the cumulative design for the entire database.

When creating user views, a "user" can be a person or a group that will use the system, a report that the system must produce, or a type of transaction that the system must support. In the last two instances, you might think of the user as the person who will use the report or enter the transaction. In fact, if the same user requires three separate reports, for example, it is more efficient to consider each report as a separate user view, even though only one "user" is involved, because smaller user views are easier to construct.

For each user view, the information-level design methodology requires you to complete the following steps:

1. Represent the user view as a collection of tables.

2. Normalize these tables.

3. Identify all keys in these tables.

4. Merge the result of Steps 1 through 3 into the cumulative design.

In the following sections, you will examine each of these steps in detail.

Represent the User View as a Collection of Tables

When provided with a user view or some sort of stated requirement, you must develop a collection of tables that will support it. In some cases, the collection of tables may be obvious to you. For example, suppose that a requested user view involves departments and employees, that each department can hire many employees, and that each employee can work in only one department (a typical restriction). A design similar to the following may have naturally occurred to you and is appropriate to represent the circumstances:

```
Department (DepartmentNum, Name, Location)
Employee (EmployeeNum, LastName, FirstName, Street, City, State,
     Zip, WageRate, SocSecNum, DepartmentNum)
```

You will undoubtedly find that the more designs you complete, the easier it will be for you to develop such a collection without resorting to any special procedure. The real question is this: What procedure should you follow when the correct design is not so obvious? In this case, you can complete the following four steps.

Step 1. Determine the entities involved and create a separate table for each type of entity. At this point, you do not need to do anything more than name the tables. For example, if a user view involves departments and employees, you can create a Department table and an Employee table. At this point, you will write something like this:

```
Department (
Employee (
```

That is, you will write the name of a table and an opening parenthesis, *and that is all.* You will assign columns to these tables in later steps.

Step 2. Determine the primary key for each table. In this step, you can add one or more columns, depending on how many columns are required for the primary key. You'll add additional columns later. Even though you have yet to determine the columns in the table, you can usually determine the primary key. For example, the primary key in an Employee table will probably be EmployeeNum, and the primary key in a Department table will probably be DepartmentNum.

The primary key is the unique identifier, so the essential question is this: What does it take to uniquely identify an employee or a department? Even if you are in the process of trying to automate a previously designed manual system, you usually can find a unique identifier in that system. If no unique identifier is available, you'll need to assign one. For example, in a manual system, customers may not have been assigned numbers because the customer base was small and the organization did not require or use customer numbers. Because the organization is computerizing its records, however, now is a good time to assign customer numbers to become the unique identifiers you are seeking.

After creating unique identifiers, you add these primary keys to what you have written already. At this point, you will have something like the following:

```
Department (DepartmentNum,
Employee (EmployeeNum,
```

Now you have the name of the table and the primary key, but that is all. In later steps, you will add the other columns.

Step 3. Determine the properties for each entity. You can look at the user requirements and then determine the other required properties of each entity. These properties, along with the primary key identified in Step 2, will become columns in the appropriate tables. For example, an Employee entity may require columns for LastName, FirstName, Street, City, State, Zip, WageRate, and SocSecNum (Social Security number). The Department entity may require columns for Name (department name) and Location (department location). Adding these columns to what is already in place would produce the following:

```
Department (DepartmentNum, Name, Location
Employee (EmployeeNum, LastName, FirstName, Street, City, State,
     Zip, WageRate, SocSecNum
```

Step 4. Determine relationships among the entities. The basic relationships are one-to-many, many-to-many, and one-to-one. You'll see how to handle each type of relationship next.

To create a one-to-many relationship, include the primary key of the "one" table as a foreign key in the "many" table. For example, assume that each employee works in a single department but that a department can have many employees. Thus *one* department is related to *many* employees. In this case, you would include the primary key of the Department table (the "one" part) as a foreign key in the Employee table (the "many" part). The tables would now look like this:

```
Department (DepartmentNum, Name, Location)
Employee (EmployeeNum, LastName, FirstName, Street, City, State,
     Zip, WageRate, SocSecNum, DepartmentNum)
```

You create a **many-to-many relationship** by creating a new table whose primary key is the combination of the primary keys of the original tables. Assume that each employee can work in multiple departments and that each department can have many employees. In this case, you would create a new table whose primary key is the combination of EmployeeNum and DepartmentNum. Because the new table represents the fact that an employee *works in* a department, you might choose to call it WorksIn, in which case, the collection of tables is as follows:

```
Department (DepartmentNum, Name, Location)
Employee (EmployeeNum, LastName, FirstName, Street, City, State,
     Zip, WageRate, SocSecNum)
WorksIn (EmployeeNum, DepartmentNum)
```

In some situations, no other columns will be required in the new table. The other columns in the WorksIn table would be those columns that depend on both the employee and the department, if such columns existed. One possibility, for example, would be the date the department hired the employee because it depends on *both* the employee *and* the department.

If each employee works in a single department and each department has only one employee, the relationship between employees and departments is **one-to-one**. (In practice, such relationships are rare.) The simplest way to implement a one-to-one relationship is to treat it as a one-to-many relationship. Which table is the "one" part of the relationship and which table is the "many" part? Sometimes looking ahead helps. For example, you might ask this question: If the relationship changes in the future, is it more likely that one employee will work in many departments or that one department might hire several employees rather than just one? If your research determines that it is more likely that a department might hire more than one employee, you would make the Employee table the "many" part of the relationship. If both situations might happen, you could treat the relationship as many-to-many. If neither situation is likely to occur, you could arbitrarily choose the "many" part of the relationship.

Normalize the Tables

After establishing the relationships among the entities, the next task is to normalize each table, with the target being third normal form. (The target is actually fourth normal form, but careful planning in the early phases of the normalization process will usually avoid the need to consider fourth normal form.)

Represent All Keys

For each table, you must identify the primary key and any alternate keys, secondary keys, and foreign keys. In the database containing information about employees and departments, you already determined the primary keys for each table in an earlier step.

Recall that an alternate key is a column or collection of columns that could have been chosen as a primary key, but was not. It is not common to have alternate keys; if they do exist and the system must enforce their uniqueness, however, you should note them. If there are any **secondary keys** (columns that are of interest strictly for the purpose of retrieval), you should represent them at this point. If a user were to indicate, for example, that rapidly retrieving an employee record on the basis of his or her last name were important, you would designate the LastName column as a secondary key.

In many ways, the foreign key is the most important key because it is through foreign keys that you can create relationships among tables and enforce certain types of integrity constraints in a database. Remember that a foreign key is a column (or collection of columns) in one table that is required either to match the value of the primary key for some row in another table or to be null. (This property is called referential integrity.) Consider, for example, the following tables:

```
Department (DepartmentNum, Name, Location)
Employee (EmployeeNum, LastName, FirstName, Street, City, State, Zip,
     WageRate, SocSecNum, DepartmentNum)
```

As before, the DepartmentNum column in the Employee table indicates the department in which the employee works. You would say that the DepartmentNum column in the Employee table is a foreign key that *identifies* Department. Thus, the number in this column on any row in the Employee table must be either a department number that is already in the database or null. (Null would indicate that, for whatever reason, the employee is not assigned to a department.)

Types of Primary Keys

There are three types of primary keys that you can use in your database design. A **natural key** (or a **logical key** or an **intelligent key**) is a primary key that consists of a column that uniquely identifies an entitiy, such as a person's Social Security number, a book's ISBN (International Standard Book Number), a product's UPC (Universal Product Code), or a vehicle's VIN (Vehicle Identification Number). These characteristics are inherent to the entity and visible to users. If a natural key exists for an entity, you usually can select it as the primary key.

If a natural key does not exist for an entity, it is common to create a primary key column that will be unique and accessible to users. The primary keys in the Premiere Products database (RepNum, CustomerNum, OrderNum, and PartNum) were created to serve as the primary keys. A column that you create for an entity to serve solely as the primary key and that is visible to users is called an **artificial key**.

The final type of primary key, which is called a **surrogate key** (or a **synthetic key**), is a system-generated primary key that is usually hidden from users. When a DBMS creates a surrogate key, it is usually an automatic numbering data type, such as the Access AutoNumber data type. Suppose, for example, you have the following relation for Customer payments:

```
Payment (CustomerNum, PaymentDate, PaymentAmount)
```

Because a customer can make multiple payments, CustomerNum cannot be the primary key. Assuming it is possible for a customer to make more than one payment on a particular day, the combination of CustomerNum and PaymentDate also cannot be the primary key. Adding an artificial key, such as PaymentNum, means you would have to assign a PaymentNum every time the customer makes a payment. Adding a surrogate key, such as PaymentId, would make more sense because the DBMS will automatically assign a unique value to each payment. Users do not need to be aware of the Payment Id value, however.

178

Database Design Language (DBDL)

In order to carry out the design process, you must have a mechanism for representing tables and keys. The standard notation you have used thus far for representing tables is fine, but it does not go far enough—there is no way to represent alternate, secondary, or foreign keys. Because the information-level design methodology is based on the relational model, it is desirable to represent tables with the standard notation. To do so, you will add additional features capable of representing additional information. The end result is **Database Design Language (DBDL)**. Figure 6.1 shows sample DBDL documentation for the Employee table.

FIGURE 6.1
·················
DBDL for the
Employee table

```
Employee (EmployeeNum, LastName, FirstName, Street, City, State, Zip,
     WageRate, SocSecNum, DepartmentNum)
   AK    SocSecNum
   SK    LastName
   FK    DepartmentNum → Department
```

In DBDL, you represent a table by listing all columns and then underlining the primary key. Below the table definition, you list any alternate keys, secondary keys, and foreign keys using the abbreviations AK, SK, and FK, respectively. For alternate and secondary keys, you can list the column or collection of columns by name. In the case of foreign keys, however, you must also represent the table whose primary key the foreign key must match. In DBDL, you write the foreign key followed by an arrow pointing to the table that the foreign key identifies.

The rules for defining tables and their keys using DBDL are as follows:

- Tables (relations), columns (attributes), and primary keys are written by first listing the table name, and then in parentheses listing the columns that comprise the table. The column(s) that comprise the primary key are underlined.

- Alternate keys are identified by the abbreviation AK, followed by the column(s) that comprise the alternate key.

- Secondary keys are identified by the abbreviation SK, followed by the column(s) that comprise the secondary key.

- Foreign keys are identified by the abbreviation FK, followed by the column(s) that comprise the foreign key. Foreign keys are followed by an arrow pointing to the table identified by the foreign key. When several tables are listed, it is common practice to place the table containing the foreign key below the table that the foreign key identifies, if possible.

Figure 6.1 shows that there is a table named Employee, containing the columns EmployeeNum, LastName, FirstName, Street, City, State, Zip, WageRate, SocSecNum, and DepartmentNum. The primary key is EmployeeNum. Another possible primary key is SocSecNum, which is listed as an alternate key. The LastName column is a secondary key, which allows you to retrieve data more efficiently based on an employee's last name. (You can add additional secondary key designations later as necessary.) The DepartmentNum column is a foreign key that identifies the department number in the Department table in which the employee works.

Entity-Relationship Diagrams

A popular type of diagram that visually represents the structure of a database is the entity-relationship (E-R) diagram. In an E-R diagram, rectangles represent the entities (tables). Foreign key restrictions determine relationships between the tables, and these relationships are represented as lines joining the corresponding rectangles.

There are several different styles of E-R diagrams currently in use. In this text, the style you will use is called **IDEF1X**.

Note: IDEF stands for "Integrated Definition" and is the name for a family of modeling languages that began with a project of the U.S. Air Force called Integrated Computer Aided Manufacturing. There are languages for such areas as activity modeling (IDEF0), conceptual data modeling (IDEF1), simulation modeling (IDEF2), process modeling (IDEF3), and object-oriented software design (IDEF4). The language in this family for logical data modeling is IDEF1X.

Consider the following database design, as written in DBDL:

```
Department (DepartmentNum, Name, Location)
Employee (EmployeeNum, LastName, FirstName, Street, City, State, Zip,
    WageRate, SocSecNum, DepartmentNum)
    AK    SocSecNum
    SK    LastName, FirstName
    FK    DepartmentNum → Department
```

The E-R diagram for the preceding database design appears in Figure 6.2.

FIGURE 6.2
· · · · · · · · · · · · · · · · · · ·
E-R diagram

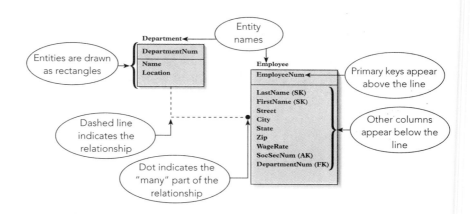

The E-R diagram shown in Figure 6.2 has the following characteristics:

- A rectangle represents each entity in the E-R diagram; that is, there is one rectangle for the Department entity and a second rectangle for the Employee entity. The name of each entity appears above the rectangle.

- The primary key for each entity appears above the line in the rectangle for each entity. DepartmentNum is the primary key of the Department entity and EmployeeNum is the primary key of the Employee entity.

- The other columns that comprise each entity appear below the line within each rectangle.

- The letters AK, SK, and FK appear in parentheses following the alternate key, secondary key, and foreign key, respectively, in the Employee entity. (The Department entity doesn't have an alternate, secondary, or foreign key.)

- For each foreign key, there is a line leading from the rectangle that corresponds to the table being identified to the rectangle that corresponds to the table containing the foreign key. The dot at the end of the line indicates the "many" part of the one-to-many relationship between the Department and Employee entities. (In Figure 6.2, *one* department is related to *many* employees, so the dot is at the end of the line connected to the Employee entity.)

When you use an E-R diagram to represent a database, it visually illustrates all the information listed in the DBDL. Thus, you would not need to also include the DBDL version of the design. There are other styles, however, that do not include such information within the diagram. In that case, you should represent the design with *both* the diagram *and* the DBDL.

■ Merge the Result into the Design

As soon as you have completed Steps 1 through 3 for a given user view, you can merge these results into the cumulative design. If the view on which you have been working is the first user view, then the cumulative design will be identical to the design for the first user. Otherwise, you merge all the tables for this user with those that are currently in the cumulative design.

Next, you combine tables that have the same primary key to form a new table. The new table has the same primary key as those tables that you have combined. The new table also contains all the columns from both tables. In the case of duplicate columns, you remove all but one copy of the column. For example, if the cumulative design already contains the following table

```
Employee (EmployeeNum, LastName, FirstName, WageRate, SocSecNum,
    DepartmentNum)
```

and the user view you just completed contains the following table

```
Employee (EmployeeNum, LastName, FirstName, Street, City, State, Zip)
```

then you would combine the two tables because they have the same primary key. All the columns from both tables would appear in the new table, but without any duplicate columns. Thus, LastName and FirstName would appear only once, even though they are in each table. The end result is:

```
Employee (EmployeeNum, LastName, FirstName, WageRate, SocSecNum,
    DepartmentNum, Street, City, State, Zip)
```

If necessary, you could reorder the columns at this point. For example, you might move the Street, City, State, and Zip columns to follow the FirstName column, which is the more traditional arrangement of this type of data. This change would give the following:

```
Employee (EmployeeNum, LastName, FirstName, Street, City, State, Zip,
     WageRate, SocSecNum, DepartmentNum)
```

At this point, you would need to check the new design to ensure that it is still in third normal form. If it is not, you should convert it to third normal form before proceeding.

Figure 6.3 summarizes the process that is repeated for each user view until all user views have been examined. At that point, the design is reviewed to resolve any problems that may remain and to ensure that it can meet the needs of all individual users. After satisfying all user view requirements, the information-level design is considered complete.

FIGURE 6.3
....
Information-level
design methodology

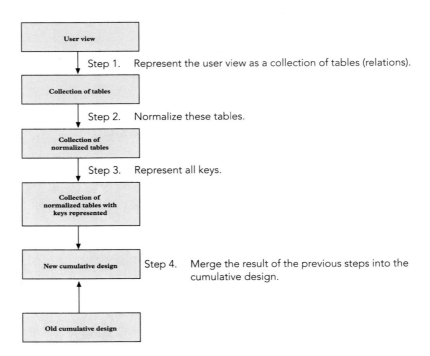

Step 1. Represent the user view as a collection of tables (relations).

Step 2. Normalize these tables.

Step 3. Represent all keys.

Step 4. Merge the result of the previous steps into the cumulative design.

◼ Database Design Examples

Now that you understand how to represent a database in DBDL and in an E-R diagram, you can examine the requirements of another database, the Premiere Products database. In the process, you will see how a set of requirements led to the database with which you have been working throughout this text.

EXAMPLE 1

Complete an information-level design for a database that satisfies the following constraints and user view requirements for a company that stores information about sales reps, customers, parts, and orders. You need to create four user views.

182

User View 1 Requirements: For a sales rep, store the sales rep's number, name, address, total commission, and commission rate.

User View 2 Requirements: For a customer, store the customer's number, name, address, balance, and credit limit. In addition, store the number and name of the sales rep who represents this customer. A sales rep can represent many customers, but a customer must have exactly one sales rep (a customer *must have* a sales rep and cannot have more than *one* sales rep).

User View 3 Requirements: For a part, store the part's number, description, units on hand, item class, the number of the warehouse in which the part is located, and the price. All units of a particular part are stored in the same warehouse.

User View 4 Requirements: For an order, store the order number; the order date; the number, name, and address of the customer that placed the order; and the number of the sales rep who represents that customer. In addition, for each line item within the order, store the part number and description, the number of the part that was ordered, and the quoted price. The user also has supplied the following constraints:

a. Each order must be placed by a customer that is already in the Customer table.

b. There is only one customer per order.

c. On a given order, there is at most one line item for a given part. For example, part DR93 cannot appear in several lines within the same order.

d. The quoted price might not match the current price in the Part table, allowing the company to sell the same parts to different customers at different prices. The user wants to be able to change the price for a part without affecting orders that are currently on file.

What are the user views in the preceding example? In particular, how should the design proceed if you are given requirements that are not specifically stated in the form of user views? Sometimes you might encounter a series of well thought-out user views in a form that you can easily merge into the design. Other times, you might be given only a set of requirements, such as those described in Example 1. In another situation, you might be given a list of reports and updates that a system must support. If, in addition to the requirements, you are able to interview users and document their needs before beginning the design process, you can make sure that you understand the specifics of their user views *prior* to starting the design process. On the other hand, you may have to take information as you get it and in whatever format it is provided.

If the user views are not clearly defined, then you should consider each stated requirement as a separate user view. Thus you can think of each report or update transaction that the system must support, as well as any other requirement as stated in the user views, as an individual user view. In fact, even if the requirements are presented as user views, you may want to split a complex user view into smaller pieces and consider each piece as a separate user view for the design process.

To transform each user view into DBDL, examine the requirements and create the necessary entities, keys, and relationships.

User View 1: For a sales rep, store the sales rep's number, name, address, total commission, and commission rate. You'll need to create only one table to support this view:

```
Rep (RepNum, LastName, FirstName, Street, City, State, Zip, Commission,
    Rate)
```

This table is in third normal form. Because there are no foreign, alternate, or secondary keys, the DBDL representation of the table is precisely the same as the relational model representation.

Notice that you have assumed that the sales rep's number (RepNum) is the Rep table's primary key—this is a reasonable assumption. Because the user didn't provide this information, however, you would need to verify its accuracy with the user. In each of the following requirements, you can assume that the obvious column (customer number, part number, and order number) is the primary key. Because you are working on the first user view, the "merge" step of the design methodology will produce a cumulative design consisting of only the Rep table, which is shown in Figure 6.4. This design is simple, so you don't need to represent it with an E-R diagram.

FIGURE 6.4

....

Cumulative design
after first user view

```
Rep (RepNum, LastName, FirstName, Street, City, State, Zip,
    Commission, Rate)
```

User View 2: Because the first user view was simple, you were able to create the necessary table without having to complete each step mentioned in the information-level design methodology section. The second user view is more complicated, however, so you'll use all the steps to determine the tables. (If you've already determined what the tables should be, you have a natural feel for the process. If so, please be patient and work through the process.)

For a customer, store the customer's number, name, address, balance, and credit limit. In addition, store the number and name of the sales rep who represents this customer. You'll take two different approaches to this requirement so that you can see how they both can lead to the same result. The only difference between the two approaches is the entities that you initially identify. In the first approach, suppose you identify two required entities for sales reps and customers. You would begin by listing the following two tables:

```
Rep (
Customer (
```

After determining the unique identifiers, you add the primary keys, which would give:

```
Rep (RepNum,
Customer (CustomerNum,
```

Adding columns for the properties of each of these entities would yield:

```
Rep (RepNum, LastName, FirstName
Customer (CustomerNum, CustomerName, Street, City, State, Zip,
    Balance, CreditLimit
```

Finally, you would deal with the relationship: *one* sales rep is related to *many* customers. To implement this one-to-many relationship, include the key of the "one" table as a foreign key in the "many" table. In this case, you would include the RepNum column in the Customer table. Thus, you would have the following:

```
Rep (RepNum, LastName, FirstName)
Customer (CustomerNum, CustomerName, Street, City, State, Zip,
    Balance, CreditLimit, RepNum)
```

184

Both tables are in third normal form, so you can move on to representing the keys. Before doing that, however, consider another approach that you could have used to determine the tables.

Suppose you didn't realize that there were really two entities, and you created only a single table for customers. You would begin by listing the table as follows:

```
Customer (
```

Adding the unique identifier as the primary key would give:

```
Customer (CustomerNum,
```

Finally, adding the other properties as additional columns would yield:

```
Customer (CustomerNum, CustomerName, Street, City, State, Zip,
     Balance, CreditLimit, RepNum, LastName, FirstName)
```

A problem appears, however, when you examine the functional dependencies that exist in the Customer entity. The CustomerNum column determines all the other columns, as it should. However, the RepNum column determines the LastName and FirstName columns; yet, RepNum is not an alternate key. This table is in second normal form because no column depends on a portion of the primary key, but it is not in third normal form. Converting the table to third normal form produces the following two tables:

```
Customer (CustomerNum, CustomerName, Street, City, State, Zip,
     Balance, CreditLimit, RepNum)
Rep (RepNum, LastName, FirstName)
```

Notice that these are precisely the same tables that you determined with the first approach—it just took a little longer to get there.

Besides the obvious primary keys, CustomerNum for Customer and RepNum for Rep, the Customer table now contains a foreign key, RepNum. There are no alternate keys, nor did the requirements state anything that would require a secondary key. If there were a requirement to retrieve the customer based on the customer's name, for example, you would probably choose to make CustomerName a secondary key.

The next step is to merge these two tables into the cumulative design. You could now represent the Rep table in DBDL in preparation for merging these two tables into the cumulative design you already have. Looking ahead, however, you see that because this table has the same primary key as the Rep table from the first user view, you can merge the two tables to form a single table that has the common column RepNum as its primary key and that contains all the other columns from both tables without duplication. For this second user view, the only columns in the Rep table besides the primary key are LastName and FirstName. These columns were already in the Rep table from the first user view that you added to the cumulative design. Thus, you won't need to add anything to the Rep table that already appears in the cumulative design. The cumulative design now contains the Rep and Customer tables shown in Figure 6.5.

FIGURE 6.5
·················
Cumulative
design after
second user view

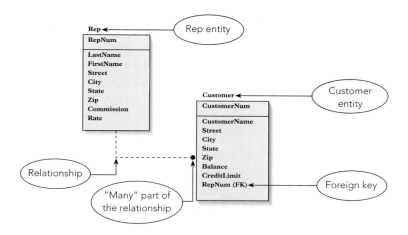

User View 3: Like the first user view, this one poses no special problems. For a part, store the part's number, description, units on hand, item class, the number of the warehouse in which the part is located, and the price. Only one table is required to support this user view:

```
Part (PartNum, Description, OnHand, Class, Warehouse, Price)
```

This table is in third normal form. The DBDL representation is identical to the relational model representation.

Because PartNum is not the primary key of any table you have already encountered, merging this table into the cumulative design produces the design shown in Figure 6.6, which contains the tables Rep, Customer, and Part.

FIGURE 6.6
·················
Cumulative design
after third user view

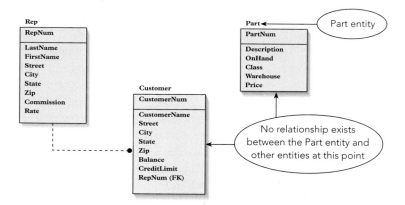

User View 4: This user view is more complicated and you can approach it in several ways. For an order, store the order number; the order date; the number, name, and address of the customer that placed the order; and the number of the sales rep who represents that customer. In addition, for each line item within the order, store the part number and description, the number of the part that was ordered, and the quoted price. Suppose that on first glance you determined that you needed to create only a single entity for orders. You might create the following table:

```
Orders (
```

Because order numbers uniquely identify orders, you would add the OrderNum column as the primary key, giving:

```
Orders (OrderNum,
```

Examining the various properties of an order, such as the date, customer number, and so on as listed in the requirements, you would add the appropriate columns, giving:

```
Orders (OrderNum, OrderDate, CustomerNum, CustomerName, Street, City,
    State, Zip, RepNum,
```

What about the fact that you are supposed to store the part number, description, number ordered, and quoted price for each order line in this order? One way of doing this would be to include all these columns within the Orders table as a repeating group (because an order can contain many order lines). This would yield:

```
Orders (OrderNum, OrderDate, CustomerNum, CustomerName, Street, City,
    State, Zip, RepNum, (PartNum, Description, NumOrdered,
    QuotedPrice) )
```

At this point, you have a table that contains all the necessary columns. Now you must convert this table to an equivalent collection of tables that are in third normal form. Because this table is not in first normal form, you would remove the repeating group and expand the primary key to produce the following:

```
Orders (OrderNum, OrderDate, CustomerNum, CustomerName, Street, City,
    State, Zip, RepNum, PartNum, Description, NumOrdered,
    QuotedPrice)
```

In the new Orders table, you have the following functional dependencies:

```
OrderNum → OrderDate, CustomerNum, CustomerName, Street, City, State,
    Zip, RepNum
CustomerNum → CustomerName, Street, City, State, Zip, RepNum
PartNum → Description
OrderNum, PartNum → NumOrdered, QuotedPrice
```

Note: Certainly the combination of OrderNum and PartNum functionally determines all attributes that OrderNum alone would determine. In addition, the combination determines all attributes that PartNum alone would determine. Adding all these other attributes after the combination of OrderNum and PartNum, while technically correct, would serve only to clutter the list of dependencies. In general, you should list an attribute after the smallest possible combination that determines it. Because you can determine Description by PartNum alone, for example, you *should* list Description after PartNum, but you *should not* list Description after the combination of OrderNum and PartNum.

From the discussion of the quoted price in the requirement, it should be noted that a quoted price depends on *both* the order number and the part number, not on the part number alone. Because some columns depend on only a portion of the primary key, the Orders table is not in second normal form. Converting to second normal form would yield the following:

```
Orders (OrderNum, OrderDate, CustomerNum, CustomerName, Street, City,
     State, Zip, RepNum)
Part (PartNum, Description)
OrderLine (OrderNum, PartNum, NumOrdered, QuotedPrice)
```

The Part and OrderLine tables are in third normal form. The Orders table is not in third normal form because CustomerNum determines CustomerName, Street, City, State, Zip, and RepNum, but CustomerNum is not an alternate key. Converting the Orders table to third normal form and leaving the other tables as written would produce the following design for this requirement:

```
Orders (OrderNum, OrderDate, CustomerNum)
Customer (CustomerNum, CustomerName, Street, City, State, Zip, RepNum)
Part (PartNum, Description)
OrderLine (OrderNum, PartNum, NumOrdered, QuotedPrice)
```

You can represent this collection of tables in DBDL and then merge them into the cumulative design. Again, however, you can look ahead and see that you can merge this Customer table with the existing Customer table and this Part table with the existing Part table. In both cases, you won't need to add anything to the Customer and Part tables already present in the cumulative design, so the Customer and Part tables for this user view will not affect the overall design. The DBDL representation for the Orders and OrderLine tables appears in Figure 6.7.

FIGURE 6.7
....
DBDL for the
Orders and
OrderLine tables

```
Orders (OrderNum, OrderDate, CustomerNum)
     FK    CustomerNum → Customer

OrderLine (OrderNum, PartNum, NumOrdered, QuotedPrice)
     FK    OrderNum → Orders
     FK    PartNum → Part
```

At this point, you have completed the process for each user view. Now it's time to review the design to make sure that it will fulfill all the stated requirements. If the design experiences problems or new information arises, you must modify the design to meet the new user views. Based on the assumption that you do not have to modify the design further, the final information-level design appears in Figure 6.8.

FIGURE 6.8
....
Final information-
level design

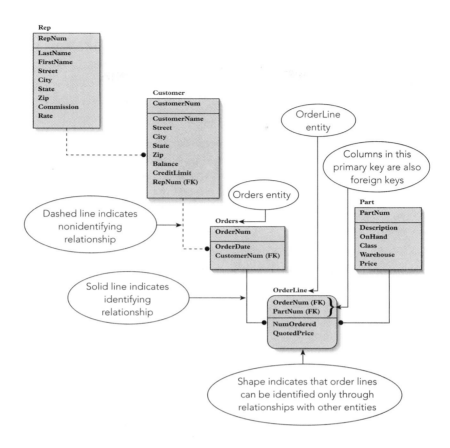

There are some differences between the E-R diagram shown in Figure 6.8 and the ones you have seen so far. The OrderLine entity appears as a rectangle with curved corners. Further, the relationships from Orders to OrderLine and from Part to OrderLine are represented with solid lines instead of with dashed lines.

Both of these differences are due to the fact that the primary key of the OrderLine entity contains foreign keys. In the OrderLine entity, both columns that comprise the primary key (OrderNum and PartNum) are foreign keys. Thus, to identify an OrderLine, you need to know the order and the part to which the order corresponds.

This situation is different from one in which the primary key does not contain one or more foreign keys. Consider the Customer table, for example, in which the primary key is CustomerNum, which is not a foreign key. The Customer table does contain a foreign key, RepNum, which identifies the Rep table. To identify a customer, however, all you need is the customer number; you do not need to know the rep number. In other words, you do not need to know the sales rep to which the customer corresponds.

An entity that does not require a relationship to another entity for identification is called an **independent entity**, and one that does require such a relationship is called a **dependent entity**. Thus, the Customer table is independent, whereas the OrderLine table is dependent. Independent entities have square corners in the diagram, and dependent entities have rounded corners.

A relationship that is necessary for identification is called an **identifying relationship**, whereas one that is not necessary is called a **nonidentifying relationship**. Thus, the relationship between the Rep and Customer tables is nonidentifying, and the relationship between the Orders and OrderLine tables is identifying. In an E-R diagram, a solid line represents an identifying relationship, and a dashed line represents a nonidentifying relationship.

EXAMPLE 2

Ray Henry, the owner of a bookstore chain named Henry Books, gathers and organizes information about branches, publishers, authors, and books. Each branch has a number that uniquely identifies the branch. In addition, Ray tracks the branch's name, location, and number of employees. Each publisher has a code that uniquely identifies the publisher. In addition, Ray tracks the publisher's name and city. The only user for the Book database is Ray, but you don't want to treat the entire project as a single user view. Ray has provided you with all the reports the system must produce, and you will treat each report as a user view. Ray has given you the following requirements:

User View 1 Requirements: For each publisher, list the publisher code, publisher name, and the city in which the publisher is located.

User View 2 Requirements: For each branch, list the number, name, location, and number of employees.

User View 3 Requirements: For each book, list its code, title, publisher code and name, price, and whether it is a paperback.

User View 4 Requirements: For each book, list its code, title, price, and type. In addition, list the book's authors and their names. If a book has more than one author, all names must appear in the order in which they are listed on the book's cover. The author order is not always alphabetical.

User View 5 Requirements: For each branch, list its number and name. In addition, list the code and title of each book currently in the branch as well as the number of copies the branch has available.

User View 6 Requirements: For each book, list its code and title. In addition, for each branch that currently has the book in stock, list the number and name of the branch along with the number of copies available.

To transform each user view into DBDL, examine the requirements and create the necessary entities, keys, and relationships.

User View 1: For each publisher, list the publisher code, publisher name, and the city in which the publisher is located. The only entity in this user view is Publisher, as follows:

```
Publisher (PublisherCode, PublisherName, City)
```

This table is in third normal form; the primary key is PublisherCode. There are no alternate or foreign keys. Assume that Ray wants to be able to access a publisher rapidly on the basis of its name. You'll need to specify the PublisherName column as a secondary key.

Because this is the first user view, there is no previous cumulative design. Thus, at this point, the new cumulative design will consist only of the design for this user view, as shown in Figure 6.9. There is no need for an E-R diagram at this point.

FIGURE 6.9
· · · · · · · · · · · · · · · · ·
DBDL for Book
database after first
user view

```
Publisher (PublisherCode, PublisherName, City)
     SK    PublisherName
```

User View 2: For each branch, list the number, name, location, and number of employees. The only entity in this user view is Branch, and it is written as follows:

```
Branch (BranchNum, BranchName, BranchLocation, NumEmployees)
```

This table is in third normal form. The primary key is BranchNum, and there are no alternate or foreign keys. Ray wants to be able to access a branch rapidly on the basis of its name, so you'll make the BranchName column a secondary key.

Because there is no table in the cumulative design with the BranchNum column as its primary key, you can add the Branch table to the cumulative design during the merge step, as shown in Figure 6.10. Again, there is no need for an E-R diagram with this simple design.

FIGURE 6.10
· · · · · · · · · · · · · · · · ·
DBDL for Book
database after
second user view

```
Publisher (PublisherCode, PublisherName, City)
     SK    PublisherName

Branch (BranchNum, BranchName, BranchLocation, NumEmployees)
     SK    BranchName
```

User View 3: For each book, list its code, title, publisher code and name, price, and whether it is paperback. To satisfy this user requirement, you'll need to create entities for publishers and books and establish a one-to-many relationship between them. This leads to the following:

```
Publisher (PublisherCode, PublisherName)
Book (BookCode, Title, PublisherCode, Price, Paperback)
```

The PublisherCode column in the Book table is a foreign key identifying the publisher. Merging these tables with the ones you already created does not add any new columns to the Publisher table, but it does add columns to the Book table. The result of merging the Book table with the cumulative design is shown in Figure 6.11. Assuming that Ray will need to access books based on their titles, you'll designate the Title column as a secondary key.

FIGURE 6.11
....
Cumulative design
after third user view

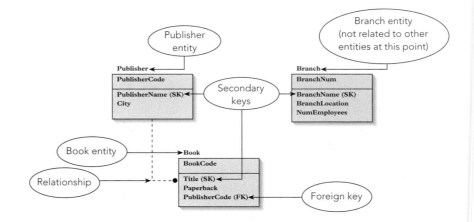

User View 4: For each book, list its code, title, price, and type. In addition, list the book's authors and their names. If a book has more than one author, all names must appear in the order in which they are listed on the book's cover. There are two entities in the user view for books and authors. The relationship between them is many-to-many (one author can write many books and one book can have many authors). Creating tables for each entity and the relationship between them gives:

```
Author (AuthorNum, AuthorLast, AuthorFirst)
Book (BookCode, Title, Price, Type)
Wrote (BookCode, AuthorNum)
```

The third table is named Wrote because it represents the fact that an author *wrote* a particular book. In this user view, you need to be able to list the authors for a book in the appropriate order. To accomplish this goal, you will add a sequence number column to the Wrote table. This completes the tables for this user view, which are:

```
Author (AuthorNum, AuthorLast, AuthorFirst)
Book (BookCode, Title, Price)
Wrote (BookCode, AuthorNum, Sequence)
```

The Author and Wrote tables are new; merging the Book table adds nothing new. Because it may be important to find an author based on the author's last name, the AuthorLast column is a secondary key. The result of the merge step is shown in Figure 6.12.

FIGURE 6.12

....

Cumulative design
after fourth user view

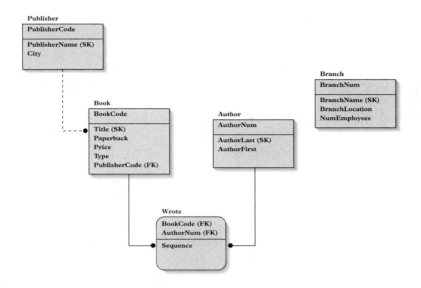

User View 5: For each branch, list its number and name. In addition, list the code and title of each book currently in the branch as well as the number of copies the branch has available. Suppose you decide that the only entity mentioned in this requirement was for information about branches. You would create the following table:

```
Branch (
```

You would then add the BranchNum column as the primary key, producing the following:

```
Branch (BranchNum,
```

The other columns include the branch name as well as the book code, book title, and number of units on hand. Because a branch will have several books, the last three columns will form a repeating group. Thus, you have the following:

```
Branch (BranchNum, BranchName, (BookCode, Title, OnHand) )
```

You convert this table to first normal form by removing the repeating group and expanding the primary key. This gives:

```
Branch (BranchNum, BranchName, BookCode, Title, OnHand)
```

In this table, you have the following functional dependencies:

```
BranchNum → BranchName
BookCode → Title
BranchNum, BookCode → OnHand
```

The table is not in second normal form because some columns depend on just a portion of the primary key. Converting to second normal form gives:

```
Branch (BranchNum, BranchName)
Book (BookCode, Title)
Inventory (BranchNum, BookCode, OnHand)
```

You can name the new table Inventory because it represents each branch's inventory. In the Inventory table, the BranchNum column is a foreign key that identifies the Branch table, and the BookCode column is a foreign key that identifies the Book table. In other words, in order for a row to exist in the Inventory table, *both* the branch number *and* the book code must already be in the database.

You can merge this Branch table with the existing Branch table without adding any new columns or relationships to the database. After adding the Inventory table to the existing cumulative design, you have the design shown in Figure 6.13.

FIGURE 6.13
·················
Cumulative design
after fifth user view

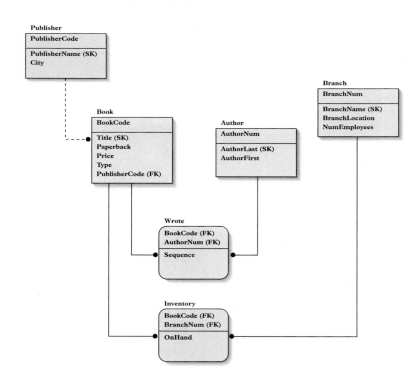

Note: When using a software tool to produce these diagrams, it is possible the software will reverse the order of the columns that make up the primary key from the order you intended. For example, the diagram in Figure 6.13 indicates that the primary key for the Inventory table is BookCode and BranchNum, even though you intended it to be BranchNum and BookCode. This change in order is not a problem. Indicating the fields that comprise the primary key is what is significant, not the order in which they appear.

QUESTION How would the design for this user view have turned out if you began with two entities, Branch and Book, instead of just the single entity Branch?

ANSWER In the first step, you would create the following tables:

```
Branch (
Book (
```

Adding the primary keys would give:

```
Branch (BranchNum,
Book (BookCode,
```

Adding the other columns would give:

```
Branch (BranchNum, BranchName)
Book (BookCode, Title)
```

Finally, you have to implement the relationship between the Branch and Book tables. Because a branch can have many books and a book can be in stock at many branches, the relationship is many-to-many. To implement a many-to-many relationship, you add a new table whose primary key is the combination of the primary keys of the other tables. Doing this, you produce the following:

```
Branch (BranchNum, BranchName)
Book (BookCode, Title)
Inventory (BranchNum, BookCode)
```

Finally, you add any column that depends on both the BranchNum and PublisherCode columns to the Inventory table, giving:

```
Branch (BranchNum, BranchName)
Book (BookCode, Title)
Inventory (BranchNum, BookCode, OnHand)
```

Thus you end up with exactly the same collection of tables, which illustrates a point made earlier: There's more than one way of arriving at the correct result.

User View 6: For each book, list its code and title. In addition, for each branch currently having the book in stock, list the number and name of each branch along with the number of copies available. This user view leads to precisely the same set of tables that were created for User View 5.

You have satisfied all the requirements, and the design shown in Figure 6.13 represents the complete information-level design.

Physical-Level Design

After the information-level design is complete, you are ready to begin the physical-level design process by implementing the design for the specific DBMS selected by the organization.

Because most DBMSs are relational and the final information-level design already exists in a relational format, producing the design for the chosen DBMS is usually an easy task—you simply use the same tables and columns. At this point, you also need to supply format details, such as specifying that the CustomerNum field will store characters and that its length is three.

Most DBMSs support primary, alternate, secondary, and foreign keys. If you are using a system that supports these keys, you simply use these features to implement the various types of keys that are listed in the final DBDL version of the information-level design. When working in DBMSs that don't support these keys, you need to devise a scheme for handling them to ensure the uniqueness of primary and alternate keys. In addition, you must ensure that values in foreign keys are legitimate; they must match the value of the primary key in some row in another table. For secondary keys, you must ensure that it is possible to retrieve data rapidly on the basis of a value of the secondary key.

For instance, suppose you are implementing the Employee table shown in Figure 6.1 and that has the following DBDL:

```
Employee (EmployeeNum, LastName, FirstName, Street, City, State, Zip,
    WageRate, SocSecNum, DepartmentNum)
    AK    SocSecNum
    SK    LastName
    FK    DepartmentNum → Department
```

The Employee table uses the EmployeeNum column as its primary key, the SocSecNum column as its alternate key, the LastName column as its secondary key, and the DepartmentNum column as a foreign key that matches the DepartmentNum column in the Department table. You must find a way for the DBMS to ensure that the following conditions hold true:

- Employee numbers are unique.

- Social Security numbers are unique.

- Access to an employee on the basis of his or her last name is rapid. (This restriction differs in that it merely states that a certain type of activity must be efficient, but it is an important restriction nonetheless.)

- Department numbers must match the number of a department currently in the database.

When the DBMS can't enforce these restrictions, who should enforce them? Two choices are possible: the users of the system could enforce them, or the programmers could. If users must enforce these restrictions, they must be careful not to enter two employees with the same EmployeeNum, an employee with an invalid DepartmentNum, and so on. Clearly, this type of enforcement would put a tremendous burden on users.

When the DBMS can't enforce these restrictions, the appropriate place for the enforcement to take place is in the programs written to access the data in the database. Thus, the responsibility for this enforcement should fall on the programmers who write these programs. Incidentally, users *must* update the data through these programs and *not* through the built-in features of the DBMS in such circumstances; otherwise, they would be able to bypass all the controls that you are attempting to program into the system.

To enforce restrictions, programmers must include logic in their programs. With respect to the DBDL for the Employee table, this means the following:

1. Before an employee is added, the program should determine and process three restrictions:

   a. Determine if an employee with the same EmployeeNum is already in the database; if so, the program should reject the update.

b. Determine if an employee with the same Social Security number is already in the database; if so, the program should reject the update.

c. Determine that the inputted department number matches a department number that is already in the database; if it doesn't, the program should reject the update.

2. When a user changes the department number of an existing employee, the program should check to make sure that the new number matches a department number that is already in the database. If it doesn't, the program should reject the update.

3. When a user deletes a department number, the program should verify that no employees work in the department. If the employees do work in the department and the program allows the deletion of the department, these employees will have invalid department numbers. In this case, the program should reject the update.

Programs must perform these verifications efficiently; in most systems, this means that the database administrator will create indexes for each column (or combination of columns) that is a primary key, an alternate key, a secondary key, or a foreign key.

◼ Top-Down Versus Bottom-Up

Another way to design a database is to use a **bottom-up design methodology** in which specific user requirements are synthesized into a design. The opposite of a bottom-up design methodology is a **top-down design methodology**, which begins with a general database design that models the overall enterprise and that repeatedly refines the model to achieve a design that supports all necessary applications. The original design and refinements are often represented with E-R diagrams.

Both strategies have their advantages. The top-down approach lends a more global feel to the project; you at least have some idea where you are headed, which is not so with a strictly bottom-up approach. On the other hand, a bottom-up approach provides a rigorous way of tackling each separate requirement and ensuring that it will be met. In particular, tables are created to satisfy each user view or requirement precisely. When these tables are correctly merged into the cumulative design, you can be sure that you have satisfied the requirements for each user view.

The ideal strategy combines the best of both approaches. Assuming that the design problem is sufficiently complicated to warrant the benefits of the top-down approach, you could begin the design process for Premiere Products by completing the following steps:

1. After gathering data on all user views, review them without attempting to create any tables. In other words, try to get a general feel for the task at hand.

2. From this information, determine the basic entities of interest to the organization (sales reps, customers, orders, and parts). Do not be overly concerned that you might miss an entity. If you do, it will show up in later steps of the design methodology.

3. For each entity, start a table. For example, if the entities are sales reps, customers, orders, and parts, you will have:

```
Rep (
Customer (
Orders (
Part (
```

4. Determine and list a primary key for each table. In this example, you might have:

```
Rep (RepNum,
Customer (CustomerNum,
Orders (OrderNum,
Part (PartNum,
```

5. For each one-to-many relationship that you can identify among these entities, optionally create and document an appropriate foreign key. For example, if there is a one-to-many relationship between the Rep and Customer tables, add the foreign key RepNum to the Customer table. If you omit this step or fail to list any foreign keys, you'll usually find them when you examine the individual users views later.

After completing the steps for a top-down approach, you can apply the methodology for examining individual user views. As you design each user view, keep in mind the tables you have created and their keys. When you need to determine the primary key for a table, look for a primary key in your cumulative design. When it is time to determine a foreign key, check the entity's primary key to see if a match exists in the cumulative design. In either case, if the primary key already exists, use the existing name as a foreign key to ensure that you can merge the tables properly. At the end of the design process, you can consider removing any tables that do not contain any columns and that have no foreign keys matching them.

Adding these steps to the process brings the benefits of the top-down approach to the approach you have been using. As you proceed through the design process for the individual user views, you will have a general idea of the overall picture.

■ Survey Form

When designing a database, you might find it helpful to design a survey form to obtain the required information from users. You can ask users to complete the form or you may want to complete the form yourself during an interview with the user. Before beginning the interview, you can identify all existing data by viewing various reports, documents, and so on. In any case, it is imperative that the completed survey form contain all the information necessary for the design process.

To be truly valuable to the design process, the survey form must contain the following information:

- *Entity information.* For each entity (reps, customers, parts, and so on), record a name and description and identify any synonyms for the entity. For example, at Premiere Products, your survey might reveal that what one user calls "parts" another user calls "products." In addition, record any general information about the entity, such as its use within the organization.

- *Attribute (column) information.* For each attribute of an entity, list its name, description, synonyms, and physical characteristics (such as being a 20-character alphanumeric or a 5-digit number), along with general information concerning its use. In addition, list any restrictions on values and the place from which the values for the item originate. (For example, the values might originate from time cards or from orders placed by customers, or be computed from values from other attributes, such as when subtracting the balance from the credit limit to obtain available credit, and so on). Finally, list any security restrictions that apply to the attribute.

- **Relationships.** For any relationship, the survey form should include the entities involved, the type of relationship (one-to-one, one-to-many, or many-to-many), the significance of the relationship (that is, what determines when two objects are related), and any restrictions on the relationship.

- **Functional dependencies.** The survey form should include information concerning the functional dependencies that exist among the columns. To obtain this information, you might ask the user a question such as this: "If you know a particular employee number, can you establish other information, such as the name?" If so, you can determine that the name is functionally dependent on the employee number. Another question you might ask is this: "Do you know the number of the department to which the employee is assigned?" If so, you can determine that the department number is functionally dependent on the employee number. If a given employee can be assigned to more than one department, you then would not know the department number, and the department number would not be dependent on the employee number. Users probably won't understand that term "functional dependency," so it is important to ask the right questions so that you can identify any functional dependencies. An accurate list of functional dependencies is absolutely essential to the design process.

- **Processing information.** The survey form should include a description of the manner in which the various types of processing (updates to the database, reports that must be produced, and so on) are to take place. To obtain this information, you would pose questions such as "How exactly is the report to be produced?", "From where do the entries on the report come?", "How are the report entries calculated?", "When a user enters a new order, from where does the data come?", and "Which entities and columns must be updated, and how?" In addition, you need to obtain estimates on processing volumes by asking questions such as "How often is the report produced?", "On average, how many pages or screens is the report?", "What is the maximum length of the report?", "What is the maximum number of orders the system receives per day?", "What is the average number of orders the system receives per day?", "What is the maximum number of invoices the system prints per day?", and "What is the average number of invoices the system prints per day?"

■ Obtaining Information from Existing Documents

Existing documents can often furnish helpful information concerning the database design. You need to take an existing document, like the invoice for the company named Holt Distributors, which is shown in Figure 6.14, and determine the tables and columns that would be required to produce the document.

FIGURE 6.14
•••••••••••••••
Invoice for Holt
Distributors

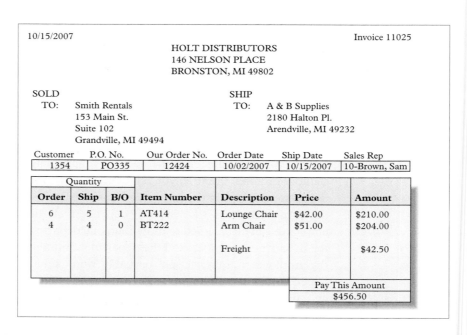

The first step in obtaining information from an existing document is to identify and list all columns and give them appropriate names. Figure 6.15 lists the columns that you can determine from the invoice shown in Figure 6.14.

FIGURE 6.15
••••••••••••••••
List of possible
attributes for the
Holt Distributors
invoice

```
Invoice Number
Invoice Date
CustomerNumber
CustomerSoldToName
CustomerSoldToAddressLine1
CustomerSoldToAddressLine2
CustomerSoldToCity
CustomerSoldToState
CustomerSoldToZip
CustomerShipToName
CustomerShipToAddress
CustomerShipToCity
CustomerShipToState
CustomerShipToZip
CustomerPONumber
OrderNumber
OrderDate
ShipDate
CustomerRepNumber
CustomerRepLastName
CustomerRepFirstName
ItemNumber
ItemDescription
ItemQuantityOrdered
ItemQuantityShipped
ItemQuantityBackordered
ItemPrice
ItemAmount
Freight
InvoiceTotal
```

The names chosen by the user for many of these columns might differ from the names you select, but this difference isn't important at this stage. After interviewing the user, you might learn that a column is required that was not apparent on the document you reviewed. For example, the shipping address for the customer shown in Figure 6.14 didn't require a second line, so you simply listed CustomerShipToAddress, rather than CustomerShipToAddressLine1 and CustomerShipToAddressLine2, in your preliminary list of columns (see Figure 6.15). Some columns that you identify may not be required. For example, if the ship date is always the same as the invoice date, then a separate ShipDate column is unnecessary. Clearly, the user's help is needed to clarify these types of issues.

Next, you need to identify functional dependencies. If you are unfamiliar with the document you are examining, you might not be able to determine its functional dependencies. In this case, you'll need to interview the user to determine the functional dependencies that exist. Sometimes you can make intelligent guesses based on your general knowledge of the type of document you are studying. You may make mistakes, of course, but you can correct them when you interview the user. After initially determining the functional dependencies shown in Figure 6.16, you may find additional information.

FIGURE 6.16
....
Tentative list of functional dependencies for the Holt Distributors invoice

```
CustomerNumber →

        CustomerSoldToName
        CustomerSoldToAddressLine1
        CustomerSoldToAddressLine2
        CustomerSoldToCity
        CustomerSoldToState
        CustomerSoldToZip
        CustomerShipToName
        CustomerShipToAddressLine1
        CustomerShipToAddressLine2
        CustomerShipToCity
        CustomerShipToState
        CustomerShipToZip
        CustomerRepNumber
        CustomerRepLastName
        CustomerRepFirstName


ItemNumber →
        ItemDescription
        ItemPrice


InvoiceNumber →
        InvoiceDate
        CustomerNumber
        OrderNumber
        OrderDate
        ShipDate
        Freight
        InvoiceTotal


InvoiceNumber, ItemNumber →
        ItemQuantityOrdered
        ItemQuantityShipped
        ItemQuantityBackordered
        ItemAmount
```

Based on your list of functional dependencies, you may learn that the shipping address for a given customer will vary from one invoice to another. In other words, the shipping address depends on the invoice number, not the customer number. A default shipping address may be defined for a given customer in case no shipping address is entered with an order. However, the address that actually appears on the invoice would depend on the invoice number. You may also determine that several columns actually depend on the order that was initially entered. The order date, customer, shipping address, and quantities ordered on each line of the invoice may have been entered as part of the initial order. At the time the invoice was printed, additional information, such as the quantities shipped, the quantities backordered, and the freight charges, may have been added. You may also find that the price is not necessarily the one stored with the item and that it can vary from one order to another. Given all these corrections, a revised list of functional dependencies might look like Figure 6.17.

FIGURE 6.17
. . . .
Revised list of functional dependencies for the Holt Distributors invoice

```
CustomerNumber  →
            CustomerSoldToName
            CustomerSoldToAddressLine1
            CustomerSoldToAddressLine2
            CustomerSoldToCity
            CustomerSoldToState
            CustomerSoldToZip
            CustomerRepNumber
            CustomerRepLastName
            CustomerRepFirstName

ItemNumber  →
            ItemDescription
            ItemPrice

InvoiceNumber  →
            InvoiceDate
            OrderNumber
            ShipDate
            Freight
            InvoiceTotal

OrderNumber  →
            OrderDate
            CustomerPONumber
            CustomerShipToName
            CustomerShipToAddressLine1
            CustomerShipToAddressLine2
            CustomerShipToCity
            CustomerShipToState
            CustomerShipToZip

OrderNumber, ItemNumber  →
            ItemQuantityOrdered (added when order is entered)
            ItemQuantityShipped (added during invoicing)
            ItemQuantityBackordered (added during invoicing)
            ItemPrice (added when order is entered)
```

After you have determined the preliminary functional dependencies, you can begin determining the tables and assigning columns. You could create tables with the determinant (the column or columns to the left of the arrow) as the primary key and with the columns to the right of the arrow as the remaining columns. This would lead to the following initial collection of tables:

```
Customer (CustomerNumber, CustomerSoldToName, CustomerSoldToAddressLine1,
     CustomerSoldToAddressLine2, CustomerSoldToCity,
     CustomerSoldToState, CustomerSoldToZip, CustomerRepNumber,
     CustomerRepLastName, CustomerRepFirstName)
Part (ItemNumber, ItemDescription, ItemPrice)
Invoice (InvoiceNumber, InvoiceDate, OrderNumber, ShipDate,
     Freight, InvoiceTotal)
Order (OrderNumber, OrderDate, CustomerPONumber,
     CustomerShipToName, CustomerShipToAddressLine1,
     CustomerShipToAddressLine2, CustomerShipToCity,
     CustomerShipToState, CustomerShipToZip)
OrderLine (OrderNumber, ItemNumber, ItemQuantityOrdered,
     ItemQuantityShipped, ItemQuantityBackordered, ItemPrice)
```

These tables would then need to be converted to third normal form and the result merged into the cumulative design.

Some people prefer not to get so specific at this point. Rather, they will examine the various columns and determine a preliminary list of entities, as shown in Figure 6.18.

FIGURE 6.18

Tentative list of entities

```
Orders
Customer
Rep
Part
```

After examining the functional dependencies, they will refine this list, producing a list similar to the one shown in Figure 6.19. At this point, they would create tables for these entities and position each column into a table in which it seems to fit best.

FIGURE 6.19

Expanded list of entities

```
Invoice
Customer
Rep
Part
Orders
OrderLine
```

Whichever approach you take, this kind of effort is certainly worthwhile; it gives you a better feel for the problem when you interact with the user. You can change your work based on your interview with the user. Even if your work proves to be accurate, you still need to ask additional questions of the user. These questions include "What names do you think are appropriate for the various entities and attributes?", "What synonyms are in use?", "What restrictions exist?", and "What are the meanings of the various entities, attributes, and relationships?"

If the organization has a computerized system, current file layouts can provide you with additional information about entities and attributes. Current file sizes can provide information on volume. Examining the logic in current programs and their operational instructions can yield processing information. Again, however, this is just a starting point. You still need further information from the user by asking questions such as "How many invoices do you expect to print?", "Exactly how are the values on the invoice calculated, and where do they come from?", "What updates must be made during the invoicing cycle of processing?", and "What fields in the Customer table will be updated?"

One-to-One Relationship Considerations

What, if anything, is wrong with implementing a one-to-one relationship by simply including the primary key of each table as a foreign key in the other table? Suppose, for example, that at Premiere Products each customer has a single sales rep and each sales rep represents a single customer. Applying the suggested technique to this one-to-one relationship produces two tables:

```
Rep (RepNum, LastName, FirstName, CustomerNum)
Customer (CustomerNum, CustomerName, RepNum)
```

In practice, these tables would contain any additional sales rep or customer columns of interest in the design problem. For the purposes of illustration, however, assume that these are the tables' only columns.

Samples of these tables are shown in Figure 6.20. This design clearly forces a sales rep to be related to a single customer. Because the customer number is a column in the Rep table, there can be only one customer for each sales rep. Likewise, this design forces a sales rep to be related to a single customer.

FIGURE 6.20
. . . .
One-to-one relationship implemented by including the primary key of each table as a foreign key in the other

Rep

RepNum	LastName	FirstName	CustomerNum
20	Kaiser	Valerie	148
35	Hull	Richard	282
65	Perez	Juan	356

Customer

CustomerNum	CustomerName	RepNum
148	Al's Appliance and Sport	20
282	Brookings Direct	35
356	Ferguson's	65

QUESTION ANSWER

What is the potential problem with this solution?

There is no guarantee that the information will match. Consider Figure 6.21, for example. The data in the first table indicates that sales rep 20 represents customer 148. The data in the second table, on the other hand, indicates that customer 148 is represented by sales rep 35! This solution may be the simplest way of implementing a one-to-one relationship from a conceptual standpoint, but it clearly suffers from this major deficiency. The programs themselves would have to ensure that the data in the two tables matched, a task that the design should be able to accomplish on its own.

FIGURE 6.21
· · · · · · · · · · · · · · · · ·
Implementation of
a one-to-one rela-
tionship in which
information does
not match

Rep

RepNum	LastName	FirstName	CustomerNum
20	Kaiser	Valerie	148
35	Hull	Richard	282
65	Perez	Juan	356

Customer

CustomerNum	CustomerName	RepNum
148	Al's Appliance and Sport	35
282	Brookings Direct	20
356	Ferguson's	65

The following are possible solutions you might consider for avoiding these problems. The first possibility is to create a single table, such as:

Customer (CustomerNum, CustomerName, RepNum, LastName, FirstName)

A sample of this table is shown in Figure 6.22. Which column should be the primary key? If it is the customer number, then there is nothing to prevent all three rows from containing the same rep number. On the other hand, if it is the rep number, the same would hold true for the customer number.

FIGURE 6.22
· · · · · · · · · · · · · · · · ·
One-to-one relation-
ship implemented in
a single table

Customer

CustomerNum	CustomerName	RepNum	LastName	FirstName
148	Al's Appliance and Sport	20	Kaiser	Valerie
282	Brookings Direct	35	Hull	Richard
356	Ferguson's	65	Perez	Juan

The solution would be to choose either the customer number or the rep number as the primary key and make the other column the alternate key. In other words, the uniqueness of both customer numbers and rep numbers should be enforced. Because each customer and each sales rep will appear in exactly one row, there is a one-to-one relationship between them.

Although this solution is workable, it has two features that are not particularly attractive. First, it combines columns of two different entities into a single table. It certainly would seem more natural to have one table with customer columns and a separate table with sales rep columns. Second, if it is possible for one entity to exist without the other (for example, if there is a customer that has no sales rep), this structure is going to cause problems.

A better solution is to create two separate tables for customers and sales reps and to include the primary key of one of them as a foreign key in the other. This foreign key would also be designated as an alternate key. Thus, you could choose either

```
Rep (RepNum, LastName, FirstName, CustomerNum)
Customer (CustomerNum, CustomerName)
```

or

```
Rep (RepNum, LastName, FirstName)
Customer (CustomerNum, CustomerName, RepNum)
```

FIGURE 6.23
••••••••••••••••••
One-to-one relationship implemented by including the primary key of one table as the foreign key (and alternate key) in the other

Samples of these two possibilities are shown in Figure 6.23. In either case, you must enforce the uniqueness of the foreign key that you added. In the first solution, for example, if customer numbers need not be unique, all three rows might contain customer 148, violating the one-to-one relationship. You can enforce the uniqueness by designating these foreign keys as alternate keys. They will also be foreign keys because they must match an actual row in the other table.

Solution 1:

Rep

RepNum	LastName	FirstName	CustomerNum
20	Kaiser	Valerie	148
35	Hull	Richard	282
65	Perez	Juan	356

Customer

CustomerNum	CustomerName
148	Al's Appliance and Sport
282	Brookings Direct
356	Ferguson's

Solution 2:

Rep

RepNum	LastName	FirstName
20	Kaiser	Valerie
35	Hull	Richard
65	Perez	Juan

Customer

CustomerNum	CustomerName	RepNum
148	Al's Appliance and Sport	20
282	Brookings Direct	35
356	Ferguson's	65

How do you make a choice between the possibilities? In some cases, it really makes no difference which arrangement you choose. Suppose, however, that you anticipate the possibility that this relationship may not always be one-to-one. Suppose there is a likelihood that in the future, a sales rep might represent more than one customer, but that each customer still will be assigned to exactly one sales rep.

The relationship would then be one-to-many, and it would be implemented with a structure similar to Solution 2. In fact, the structure would differ only in that the rep number in the Customer table would not be an alternate key. Thus, to convert from the second alternative to the appropriate structure would be a simple matter—you would remove the restriction that the rep number in the Customer table is an alternate key. This situation would lead you to favor the second alternative.

Many-to-Many Relationship Considerations

Complex issues arise when more than two entities are related in a many-to-many relationship. Suppose, for example, that Premiere Products needs to know which sales reps sold which parts to which customers. In this example, there are no restrictions on which customers a given sales rep may sell to or on the parts that a sales rep may sell. Sample data for this relationship is shown in Figure 6.24.

FIGURE 6.24

Sample Sales data

Sales

RepNum	CustomerNum	PartNum
20	148	AT94
20	282	DR93
35	148	DR93
35	148	DW11
65	282	AT94
65	282	DR93
65	356	AT94

The first row in the table indicates that sales rep 20 sold part AT94 to customer 148 (the number of units sold to the customer is not important in this example). The second row indicates that sales rep 20 sold part DR93 to customer 282.

QUESTION

ANSWER

What is the primary key of the Sales table?

Clearly none of the three columns (RepNum, CustomerNum, and PartNum) alone will uniquely identify a record. The combination of RepNum and CustomerNum does not work because there are two rows on which the rep number is 35 and the customer number is 148. The combination of RepNum and PartNum does not work because there are two rows on which the rep number is 65 and the part number is AT94. Finally, the combination of CustomerNum and PartNum does not work because there are two rows on which the customer number is 282 and the part number is DR93. Thus, the primary key for the Sales table must be the combination of all three columns:

```
Sales (RepNum, CustomerNum, PartNum)
```

Attempting to model this particular situation as two (or three) many-to-many relationships is not legitimate. Consider the following code and the data shown in Figure 6.25, for example, in which the same data is split into three tables:

```
RepCustomer (RepNum, CustomerNum)
CustomerPart (CustomerNum, PartNum)
PartRep (PartNum, RepNum)
```

FIGURE 6.25
••••••••••••••••
Result obtained
by splitting the
Sales table into
three tables

RepCustomer

RepNum	CustomerNum
20	148
20	282
35	148
65	282
65	356

CustomerPart

CustomerNum	PartNum
148	AT94
148	DR93
148	DW11
282	AT94
282	DR93
356	AT94

PartRep

PartNum	RepNum
AT94	20
AT94	65
DR93	20
DR93	35
DR93	65
DW11	35

Figure 6.26 shows the result of joining these three tables. Note that it contains inaccurate information. The second row, for example, indicates that rep 20 sold part DR93 to customer 148. If you look back to Figure 6.24, you will see this is not the case.

FIGURE 6.26
••••••••••••••••
Result obtained by
joining three
tables—the second
and third rows are
in error!

Sales

RepNum	CustomerNum	PartNum
20	148	AT94
20	148	DR93 !!!!
20	282	AT94 !!!!
20	282	DR93
35	148	DR93
35	148	DW11
65	282	AT94
65	282	DR93
65	356	AT94

The second row appears in the join because rep 20 is related to customer 148 in the RepCustomer table (rep 20 sold a part to customer 148), customer 148 is related to part DR93 in the CustomerPart table (customer 148 bought part DR93 from a rep), and part DR93 is related to rep 20 in the RepPart table (rep 20 sold part DR93 to a customer). In other words, rep 20 sold a part to customer 148, customer 148 bought part DR93, and rep 20 sold part DR93. Of course, these three facts do not imply that rep 20 sold part DR93 to customer 148 (customer 148 might have purchased this part from another rep).

The problem with the preceding relationship is that it really involves all three entities—reps, customers, and parts. Splitting the Sales table shown in Figure 6.26 any further is inappropriate. Such a relationship is called a **many-to-many-to-many relationship**.

Remember from the discussion of fourth normal form that there are examples of three-way relationships in which you must split the tables. In particular, if the relationship between sales reps and customers has nothing to do with the relationship between sales reps and parts, this table would violate fourth normal form and would need to be split.

The crucial issue in making the determination between a single many-to-many-to-many relationship and two (or three) many-to-many relationships is the independence. If all three entities are critical in the relationship, then the three-way relationship (like Sales) is appropriate. If there is independence among the individual relationships, then separate many-to-many relationships are appropriate. Incidentally, if a many-to-many-to-many relationship is created when it is not appropriate to do so, the conversion to fourth normal form will correct the problem.

Nulls and Entity Subtypes

Recall that null is a special value. A null actually represents the *absence* of a value in a field. In other words, setting a particular field to null is equivalent to not entering a value in the field. Nulls are used when a value is either unknown or inapplicable. This section focuses on the second possibility—when the value is inapplicable.

Consider, for example, a Student table in which one of the columns, DormNum, is a foreign key that identifies a Dorm (dormitory) table. The DormNum column indicates the number of the dormitory in which a student currently resides. This foreign key is allowed to be null because some students do not live in a dormitory; for these students, DormNum is inapplicable. Thus, for some rows in the Student table, the DormNum column would be null.

You can avoid this use of null values by removing the DormNum column from the Student table and creating a separate table named StudentDorm that contains the columns StudentNum (the primary key) and DormNum. Students living in a dorm would have a row in this new table. Students not living in a dorm would have a row in the Student table but not in the StudentDorm table.

This change is illustrated in Figure 6.27. Note that StudentNum, the primary key of the StudentDorm table, is also a foreign key that must match a student number in the Student table.

FIGURE 6.27
· · · · · · · · · · · · · · · · ·
Student table split
to avoid use of null
values

Student

StudentNum	LastName	FirstName	DormNum
1253	Johnson	Ann	3
1662	Anderson	Tom	1
2108	Lewis	Bill	
2546	Davis	Mary	2
2867	Albers	Cathy	2
2992	Matthew	Mark	
3011	Candela	Tim	3
3574	Talen	Sue	

Student

StudentNum	LastName	FirstName
1253	Johnson	Ann
1662	Anderson	Tom
2108	Lewis	Bill
2546	Davis	Mary
2867	Albers	Cathy
2992	Matthew	Mark
3011	Candela	Tim
3574	Talen	Sue

StudentDorm

StudentNum	DormNum
1253	3
1662	1
2546	2
2867	2
3011	3

In the process, you have created an **entity subtype**. You can say that the StudentDorm table is a subtype of the Student table. In other words, "students living in dorms" is a subtype (or subset) of "students."

Some design methodologies have specific ways of denoting entity subtypes, but it is not necessary to denote entity subtypes in DBDL. You can recognize entity subtypes by the fact that the primary key is also a foreign key, as shown in Figure 6.28.

FIGURE 6.28
· · · · · · · · · · · · · · · ·
Sample DBDL with
entity subtypes

```
Student (StudentNum, LastName, FirstName)

StudentDorm (StudentNum, DormNum)
    FK    StudentNum → Student
    FK    DormNum → Dorm
```

Most approaches to diagramming database designs have ways of representing subtypes. In IDEF1X, for example, a subtype, which is called a **category** in IDEF1X terminology, is represented in the manner shown in Figure 6.29. The circle is the symbol for category. The single line below the circle indicates that the category is an **incomplete category**; that is, there are students who do not fall into the StudentDorm category.

FIGURE 6.29
· · · · · · · · · · · · · · · · · · ·
Entity subtype

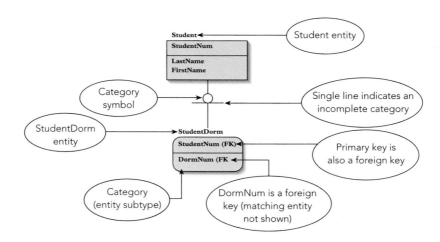

The issue is more complicated when more than one column can accept null values. Suppose the DormNum, ThesisTitle, and ThesisArea columns in the following Student table can be null:

```
Student (StudentNum, LastName, FirstName, DormNum, ThesisTitle,
    ThesisArea)
```

In this table, the dorm number is either the number of the dorm in which the student resides or null if the student does not live in a dorm. In addition, students at this college must write a senior thesis. After students attain senior standing, they must select a thesis title and the area in which they will write their thesis. Thus, seniors will have a thesis title and a thesis area, whereas other students will not. You can handle this situation by allowing the fields ThesisTitle and ThesisArea to be null.

The Student table now has three different columns—DormNum, ThesisTitle, and ThesisArea—that can be null. The DormNum column will be null for students who do not reside in a dorm. The ThesisTitle and ThesisArea columns will be null for students who have not yet attained senior standing. It wouldn't make much sense to combine all three of these columns into a single table. A better choice would be to create the following table for students living in dorms:

```
StudentDorm (StudentNum, DormNum)
```

For seniors, you could create a second table as follows:

```
SeniorStudent (StudentNum, ThesisTitle, ThesisArea)
```

FIGURE 6.30
••••••••••••••••
Student table split
to avoid the use of
null values

Samples of these tables are shown in Figure 6.30. Both tables represent entity sub-types. In both tables, the primary key (StudentNum) will also be a foreign key matching the student number in the new Student table.

Student

StudentNum	LastName	FirstName	DormNum	ThesisTitle	ThesisArea
1253	Johnson	Ann	3		
1662	Anderson	Tom	1	P.D.Q. Bach	Music
2108	Lewis	Bill		Cluster sets	Math
2546	Davis	Mary	2		
2867	Albers	Cathy	2	Rad. Treatment	Medicine
2992	Matthew	Mark			
3011	Candela	Tim	3		
3574	Talen	Sue			

Student

StudentNum	LastName	FirstName
1253	Johnson	Ann
1662	Anderson	Tom
2108	Lewis	Bill
2546	Davis	Mary
2867	Albers	Cathy
2992	Matthew	Mark
3011	Candela	Tim
3574	Talen	Sue

StudentDorm

StudentNum	DormNum
1253	3
1662	1
2546	2
2867	2
3011	3

SeniorStudent

StudentNum	ThesisTitle	ThesisArea
1662	P.D.Q. Bach	Music
2108	Cluster sets	Math
2867	Rad. treatment	Medicine

The DBDL for these tables appears in Figure 6.31. The primary key of the StudentDorm and SeniorStudent tables (StudentNum) is also a foreign key matching the student number in the new Student table.

FIGURE 6.31
····················
Sample DBDL with
entity subtypes

```
Student (StudentNum, LastName, FirstName)

StudentDorm (StudentNum, DormNum)
     FK    StudentNum → Student
     FK    DormNum →   Dorm

SeniorStudent (StudentNum, ThesisTitle, ThesisArea)
     FK StudentNum → Student
```

To represent two subtypes (categories) in IDEF1X, you use the same category symbol shown in Figure 6.29. The difference is that there will be two lines coming out of the category symbol—one to each category, as shown in Figure 6.32. Because there are students who do not live in dorms and who are not seniors, these categories are also incomplete, so there is only one line below the category symbol.

FIGURE 6.32
····················
Two entity
subtypes—
incomplete
categories

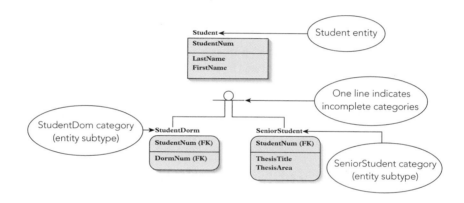

By contrast, Figure 6.33 represents a slightly different situation. There are two categories: students who live in dorms (StudentDorm) and students who do not (StudentNonDorm). For students who live in dorms, the attribute of interest is DormNum. For students who do not live in dorms, the attributes of interest are the ones that give the students' local addresses (LocalStreet, LocalCity, LocalState, and LocalZip). The difference between this example and the one shown in Figure 6.32 is that every student *must* be in one of these two categories. These are called **complete categories** and are represented by two lines below the category symbol.

FIGURE 6.33
· · · · · · · · · · · · · · · · · ·
Two entity
subtypes—
complete
categories

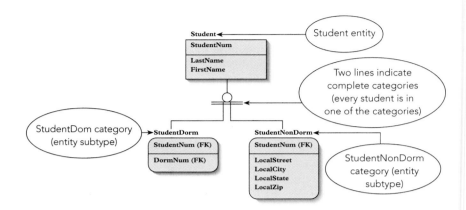

You should group columns that can be null by function. If a given subset of the entity in question can have nulls in a certain collection of columns, you should note this fact. When available, you should strongly consider splitting columns that can have nulls into a separate table (an entity subtype). If you do create an entity subtype, it's a good idea to give it a name that suggests the related entity type, as well as its relationship to it, such as SeniorStudent for students who are seniors. In addition, you should carefully document the meaning of the entity subtype, especially the conditions that will cause an occurrence of the entity itself to also be an occurrence of the entity subtype. If you do not create such an entity subtype, you must at least document precisely when the columns might take on null as a value.

■ Avoiding Problems with Third Normal Form When Merging Tables

When you combine third normal form tables, the result need not be in third normal form. For example, both of the following tables are in third normal form:

```
Customer (CustomerNum, CustomerName, RepNum)
Customer (CustomerNum, CustomerName, LastName, FirstName)
```

When you combine them, however, you get the following table:

```
Customer (CustomerNum, CustomerName, RepNum, LastName, FirstName)
```

This table is not in third normal form. You would have to convert this table to third normal form before proceeding to the next user view.

You can attempt to avoid the problem of creating a table that is not in third normal form by being cautious when representing user views. This problem occurs when a column A in one user view functionally determines a column B in a second user view. Thus, column A is a *determinant* for column B, yet column A is not a column in the second user view.

In the preceding example, the RepNum column in the first table determines the columns LastName and FirstName in the second table, yet the RepNum column is not one of the columns in the second table. If you always attempt to determine whether determinants exist and include them in the tables, you often will avoid this problem. For example, when the second user indicates that the name of a rep is part of that user's view of data, you should ask whether any special way has been provided for sales reps to be uniquely identified within the organization. Even though this user evidently does not need the rep number, he or she might very well be aware of the existence of such a number. If so, you would include this number in the table. Having done this, you would have the following table in this user view:

```
Customer (CustomerNum, CustomerName, RepNum, LastName, FirstName)
```

Now, the normalization process for this user view would produce the following two tables:

```
Customer (CustomerNum, CustomerName, RepNum)
Rep (RepNum, LastName, FirstName)
```

When you merge these two tables into the cumulative design, you will not produce any tables that are not in third normal form. Notice that the determinant RepNum has replaced the columns that it determines, LastName and FirstName, in the Customer table.

The Entity-Relationship Model

In this chapter, you examined the use of E-R diagrams to visually illustrate the relations and keys represented in DBDL. The **entity-relationship (E-R) model** is an approach to representing data in a database. This model uses E-R diagrams exclusively as the tool for representing entities, attributes, and relationships. The E-R model is widely used and also forms the basis of some computerized tools, so it is important that you understand how to use it.

In 1976, Peter Chen of the M.I.T. Sloan School of Management proposed the E-R model and since then it has been widely accepted as a graphical approach to database representation and database design. The basic constructions in the E-R model are the familiar entities, attributes, and relationships, all of which are represented in E-R diagrams. This section focuses on the standard versions of these diagrams. The versions you examined earlier in this chapter represent one of the common alternative forms of creating these diagrams that is particularly convenient for use with DBDL.

In the standard E-R diagrams, entities are drawn as rectangles and relationships are drawn as diamonds, with lines connecting the entities involved in relationships. Both entities and relationships are named in the E-R model. The lines are labeled to indicate the type of relationship. For example, in Figure 6.34, the one-to-many relationship between sales reps and customers is represented as "1" to "n."

FIGURE 6.34
One-to-many
relationship

In Figure 6.35, the many-to-many relationship between orders and parts is represented as "m" to "n."

Finally, the many-to-many-to-many relationship between sales reps, customers, and parts is represented as "m" to "n" to "p," as shown in Figure 6.36.

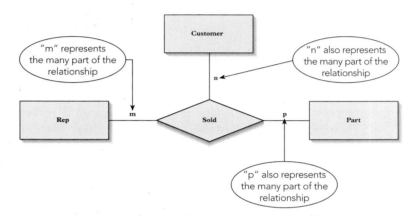

If desired, you can also indicate attributes in the E-R model by placing them in ovals and attaching them to the corresponding rectangles (entities), as shown in Figure 6.37. Just as in the relational model representation, primary keys are underlined.

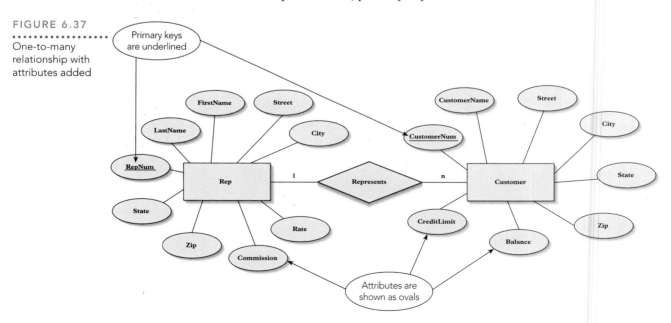

In the original version of the E-R model, attributes and relationships can have attributes. Figure 6.38 shows the many-to-many relationship between the Orders and Parts entities, in which the relationship, OrderLine, has two attributes, NumOrdered and QuotedPrice.

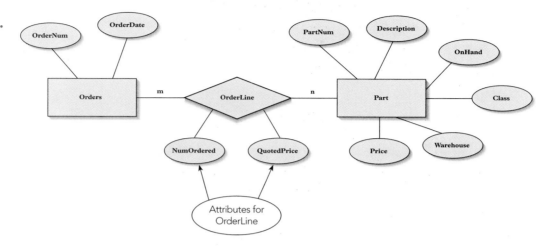

In Figure 6.38, it isn't clear whether OrderLine is an entity or a relationship. To address this possible confusion, the E-R model was changed slightly so that entities can have attributes but relationships cannot. With this change, OrderLine in Figure 6.38 would be an entity. However, making OrderLine an entity does not effectively communicate the fact that the OrderLine entity is only implementing a many-to-many relationship between orders and parts. To address this problem, an entity that exists to implement a many-to-many relationship is called a **composite entity**. A composite entity is essentially both an entity and a relationship and is represented in an E-R diagram by a diamond within a rectangle. Figure 6.39 shows this new approach.

FIGURE 6.39
. . . .
Composite entity

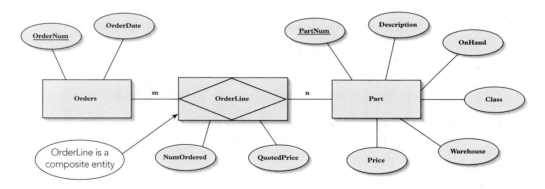

A complete E-R diagram for the Premiere Products database appears in Figure 6.40. Notice that OrderLine is represented as a composite entity.

FIGURE 6.40
•••••••••••••••••
Complete E-R
diagram for the
Premiere Products
database

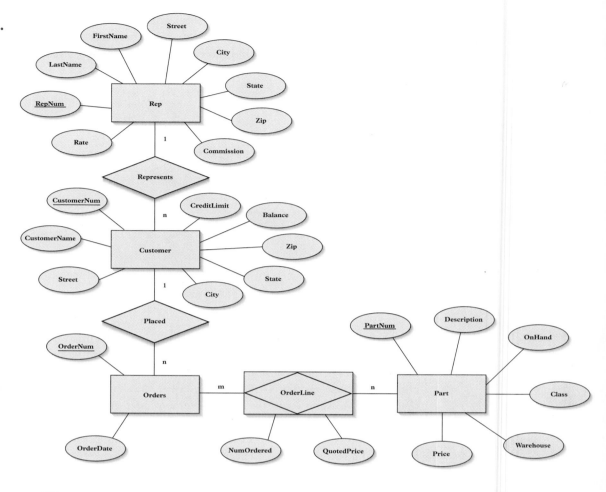

When the existence of one entity depends on the existence of another related entity, there is an **existence dependency**. For example, because an order cannot exist without a customer, the relationship between customers and orders is an existence dependency. You indicate an existence dependency by placing an "E" in the relationship diamond, as shown in Figure 6.41. An entity that depends on another entity for its own existence is called a **weak entity**. A double rectangle encloses a weak entity, as shown in Figure 6.41. A weak entity corresponds to the term dependent entity, which was previously defined in this chapter.

FIGURE 6.41
•••••••••••••••••
E-R diagram with an
existence depen-
dency and a weak
entity

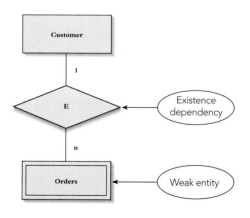

There is another popular way to indicate a one-to-many relationship. In this alternative, you do not label the "one" end of the relationship; instead, you place a crow's foot at the "many" end of the relationship. Figure 6.42 illustrates this style.

FIGURE 6.42
· · · · · · · · · · · · · · · · · · · ·
E-R diagram with a
crow's foot

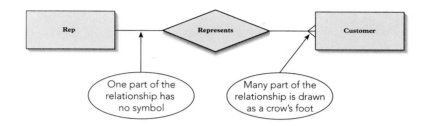

Some people represent **cardinality**, or the number of items that must be included in a relationship, in an E-R diagram. Figure 6.43 shows an E-R diagram that represents cardinality in this way. The two symbols to the right of the Rep rectangle are both the number 1. The 1 closest to the rectangle indicates that the maximum cardinality is one (a customer can have at most one sales rep). The 1 closest to the relationship is the minimum cardinality (a customer must have at least one sales rep). Together, the two symbols indicates that a customer must have exactly one sales rep. (If the minimum cardinality were zero, for example, a customer would not be required to have a sales rep.)

FIGURE 6.43
· · · · · · · · · · · · · · · · · · · ·
E-R diagram
that represents
cardinality

The crow's foot to left of the Customer rectangle indicates that the maximum cardinality is "many." The circle to the left of the crow's foot indicates that the minimum cardinality is zero; that is, a sales rep could be associated with zero customers. An entity in a relationship with minimum cardinality of zero plays an **optional role** in the relationship. An entity with a minimum cardinality of one plays a **mandatory role** in the relationship.

■ SUMMARY

- Database design is a two-part process of determining an appropriate database structure to satisfy a given set of requirements. In the information-level design, a clean database management system (DBMS) design that is not dependent on a particular DBMS is created to satisfy the requirements. In the physical-level design, the final information-level design is converted into an appropriate design for the particular DBMS that will be used.

- A user view is the set of necessary requirements to support a particular user's operations. In order to simplify the design process, the overall set of requirements is split into user views.

- The information-level design methodology involves applying the following steps to each user view: represent the user view as a collection of tables; normalize these tables (convert the collection into an equivalent collection that is in third normal form); represent all keys (primary, alternate, secondary, and foreign); and merge the results into the cumulative design.

- A database design is represented in a language called Database Design Language (DBDL).

- Designs can be represented pictorially using entity-relationship (E-R) diagrams. Such diagrams have the following characteristics: there is a rectangle for each entity; the name of the entity appears above the rectangle; the primary key appears above the line in the rectangle; the remaining columns appear below the line; alternate keys, secondary keys, and foreign keys are identified with the letters AK, SK, and FK, respectively; and for each foreign key, there is a dashed line from the rectangle that corresponds to the table being identified to the rectangle that corresponds to the table containing the foreign key. There is a dot at the end of the line indicating the "many" part of a one-to-many relationship.

- Assuming that a relational DBMS is going to be used, the physical-level design process consists of creating a table for each entity in the DBDL design. Any constraints (primary key, alternate key, or foreign key) that the DBMS cannot enforce must be enforced by the programs in the system; this fact must be documented for the programmers.

- The design methodology presented in this chapter is a bottom-up methodology. By listing potential relations before beginning the methodology, you can have the advantages of both the top-down and bottom-up approaches.

- A survey form is a useful form for documenting the information gathered for the database design process.

- To obtain information from existing documents, list all attributes present in the documents, identify potential functional dependencies, make a tentative list of tables, and then use the functional dependencies to refine the list.

- To implement a one-to-one relationship, include the primary key of one of the two tables in the other table as a foreign key, and then indicate the foreign key as an alternate key.

- If a table's primary key consists of three (or more) columns, you must determine whether there are independent relationships among pairs of these columns. If there are independent relationships, the table is not in fourth normal form and you must split it. If there are no independent relationships, you can't split the table because doing so produces incorrect information.

- If a table contains columns that can be null and the nulls represent the fact that the column is inapplicable for some rows, you can split the table, placing the null column or columns in separate tables. These new tables represent entity subtypes.

- It is possible that the result of merging third normal form tables may not be in third normal form. To avoid this problem, include determinants for columns in the individual tables before merging them.

- The entity-relationship (E-R) model is a method of representing the structure of a database using an E-R diagram. In an E-R diagram, a rectangle represents an entity, a diamond represents a relationship, and an oval represents an attribute.

■ KEY TERMS

artificial key

bottom-up design methodology

cardinality

category

complete category

composite entity

cumulative design

Database Design Language (DBDL)

dependent entity

entity-relationship (E-R) model

entity subtype

existence dependency

IDEF1X

identifying relationship

incomplete category

independent entity

information-level design

intelligent key

logical key

mandatory role

many-to-many relationship

many-to-many-to-many relationship

natural key

nonidentifying relationship

one-to-one relationship

optional role

physical-level design

secondary key

surrogate key

synthetic key

top-down design methodology

user view

weak entity

■ REVIEW QUESTIONS

1. Define the term *user view* as it applies to database design.

2. What is the purpose of breaking down the overall design problem into a consideration of individual user views?

3. Under what circumstances would you not need to break down an overall design into a consideration of individual user views?

4. The information-level design methodology presented in this chapter contains steps that must be repeated for each user view. List the steps and briefly describe the kinds of activities that must take place at each step.

5. Describe the function of each of the following types of keys: primary, alternate, secondary, and foreign.

6. Suppose that a given user view contains information about employees and projects. Suppose further that each employee has a unique EmployeeNum and that each project has a unique ProjectNum. Explain how you would implement the relationship between employees and projects in each of the following scenarios:
   a. Many employees can work on a given project, but each employee can work on only a *single* project.
   b. An employee can work on many projects but each project has a *unique* employee assigned to it.
   c. An employee can work on many projects, and a project can be worked on by many employees.

7. A database at a college is required to support the following requirements. Complete the information-level design for this set of requirements. Use your own experience to determine any constraints you need that are not stated in the problem. Represent the answer in DBDL.
   a. For a department, store its number and name.
   b. For an advisor, store his or her number and name and the number of the department to which he or she is assigned.
   c. For a course, store its code and description (for example, MTH110 or Algebra).
   d. For a student, store his or her number and name. For each course the student has taken, store the course code, course description, and grade received. In addition, store the number and name of the student's advisor. Assume that an advisor may advise any number of students but that each student has just one advisor.

8. List the changes that would need to be made in your answer to Question 7 if a student could have more than one advisor.

9. Suppose that in addition to the requirements specified in Question 7, you must store the number of the department in which the student is majoring. Indicate the changes this would cause in the design in the following two situations:
   a. The student must be assigned an advisor who is in the department in which the student is majoring.
   b. The student's advisor does not necessarily have to be in the department in which the student is majoring.

10. Illustrate the physical-level design process by means of the design shown in Question 7. List the tables, identify the keys, and list the special restrictions that programs must enforce.

11. Is the database design methodology top-down or bottom-up? How can you modify this methodology to gain the advantages to both types of design methodologies?

12. Design a survey form of your own. Fill it out as it might have been filled out during the database design for Premiere Products. For any questions that you have too little information to answer, make a reasonable guess.

13. Using a document at your own school (for example, a report card), determine the attributes present in the document. Using your knowledge of the policies at your school, determine the functional dependencies present in the document. Use these dependencies to create a set of tables and columns that you could use to produce the document.

14. Describe the different ways of implementing one-to-one relationships. Assume you are maintaining information on offices (office numbers, buildings, and phone numbers) and faculty (numbers and names). No office houses more than one faculty member; no faculty member is assigned more than one office. Illustrate the ways of implementing one-to-one relationships using offices and faculty. Which option would be best in each of the following situations?
   a. A faculty member must have an office, and each office must be occupied by a faculty member.
   b. A faculty member must have an office, but some offices are not currently occupied. You must maintain information about the unoccupied offices in an Office relation.
   c. Some faculty members do not have an office, but all offices are occupied.
   d. Some faculty members do not have an office, but some offices are not occupied.

15. For each of the following collections of relations, give the assumptions concerning the relationship between students, courses, and faculty members that are implied by the collection. In each relation, only the primary keys are shown.
   a. Student (StudentNum, CourseNum, FacultyNum)
   b. Student (StudentNum, CourseNum)
      Faculty (CourseNum, FacultyNum)
   c. Student (StudentNum, CourseNum)
      Faculty (CourseNum, FacultyNum)
      StudentFaculty (StudentNum, FacultyNum)
   d. Student (StudentNum, CourseNum, FacultyNum)
   e. Student (StudentNum, CourseNum)
      Faculty (CourseNum, FacultyNum)
      StudentFaculty (StudentNum, FacultyNum)

16. Describe the relationship between columns that can be null and entity subtypes. Under what circumstances would these columns lead to more than one entity subtype?

17. How is it possible to merge a collection of relations that is in third normal form into a cumulative design that is in third normal form but not obtain a collection of relations that is in third normal form? Give an example other than the one in the text.

18. Describe the entity-retionship model. How are entities, relationships, and attributes represented in this model? What is a composite entity? Describe the approach to diagrams that uses a "crow's foot." Describe how you would represent cardinality in an E-R diagram.

■ Premiere Products Exercises

The following exercises are based on the Premiere Products database as designed in Example 1.

1. Indicate the changes that need to be made to the design of the Premiere Products database to support the following situation: A customer is not necessarily represented by a single sales rep, but can be represented by several sales reps. When a customer places an order, the sales rep who gets the commission on the order must be one of the collection of sales reps who represents the customer.

2. Indicate the changes that need to be made to the design of the Premiere Products database to support the following situation: There is no relationship between customers and sales reps. When a customer places an order, it may be through any sales rep. On the order, identify both the customer placing the order and the sales rep responsible for the order.

3. Indicate the changes that need to be made to the design of the Premiere Products database in the event that User View 3 requirements are changed as follows. For a part, store the part's number, description, item class, and price. In addition, for each warehouse in which the part is located, store the number of the warehouse, the description of the warehouse, and the number of units of the part stored in the warehouse.

4. Indicate the changes that you need to make to the Premiere Products database design to support the following situation. The region where customers are located is divided into territories. For each territory, store the territory number (a unique identifier) and territory name. Each sales rep is assigned to a single territory. Each customer is also assigned to a single territory, but the territory must be the same as the territory to which the customer's sales rep is assigned.

5. Indicate the changes that you need to make to the the Premiere Products database design to support the following situation. The region where customers are located is divided into territories. For each territory, store the territory number (a unique identifier) and territory name. Each sales rep is assigned to a single territory. Each customer is also assigned to a single territory, which may not be the same as the territory to which the customer's sales rep is assigned.

Ray Henry is considering expanding the inventory at his book stores to include movies. He has some special ideas for how he wants to implement this change and he needs you to help with database design activities.

1. Design a database for Ray. He is interested in movies and wants to keep information on movies, actors, and directors in a database. The only user is Ray and he needs to produce the following reports:

   a. For each director, list his or her number, name, and the year he or she was born. If the director is deceased, list the year of death.
   b. For each movie, list its number, title, the year the movie was made, and its type (for example, Comedy, Drama, Science Fiction, and so on).
   c. For each movie, list its number, title, the number and name of its director, the critics' rating, the MPAA rating (G, PG, PG-13, or R), the number of awards for which the movie was nominated, and the number of awards the movie won. (The critics rate the movie with a number of "stars." Four stars is the top rating possible. Zero stars is the worst possible rating.)
   d. For each lead actor starring in each movie, list his or her number, name, birthplace, and the year he or she was born. If the actor is deceased, list the year of death.
   e. For each movie, list its number and title, along with the number and name of the actors who appeared in it.
   f. For each lead actor starring in each movie, list his or her number and name, along with the number and name of the other movies in which the actor starred.

2. Expand the database design you created in Exercise 1 so that it will also support the following situation. Ray wants to start a DVD rental program at his stores that he plans to call Henry's DVD Club. He refers to each of his customers as "members." Each member in the club is assigned a number. He also stores the members' names and addresses. In addition, he stores the number of rentals the member has made and the date the member joined the club. He periodically has promotions during which members can earn bonus units that they can later apply to the cost of renting DVDs. He needs to store the number of bonus units the member has earned.

3. Expand the database design you created in Exercise 1 and modified in Exercise 2 so that it will also support the following situation. Ray wants to store information about the DVDs the club owns. When the club purchases a DVD, Ray assigns it a number. Along with the number, he stores the number of the movie on the DVD, the date the DVD was purchased, the number of times it has been rented, and the number of the member who is currently renting it. (If the DVD is not currently being rented, the member number will be null.) Ray also needs to store the number of the branch to which the DVD is assigned. Finally, Ray would like to store the history of the rental of each particular DVD. In particular, he needs to store the DVD number, the date of the rental, the date it was returned, and the number of the member who rented the DVD. Assume that a DVD could potentially be rented more than once on the same day. (*Hint:* Review the discussion of the categories of primary keys to determine what type of primary key would be appropriate for this relation.)

DBMS Functions

OBJECTIVES

- Introduce the functions, or services, provided by a database management system (DBMS).

- Describe how a DBMS handles updating and retrieving data.

- Examine the catalog feature of a DBMS.

- Illustrate the concurrent update problem and describe how a DBMS handles this problem.

- Explain the data recovery process in a database environment.

- Describe the security services provided by a DBMS.

- Examine the data integrity features provided by a DBMS.

- Discuss the extent to which a DBMS achieves data independence.

- Define and describe data replication.

- Present the utility services provided by a DBMS.

Introduction

In Chapter 6, you learned the steps that the database designer completed to design the Premiere Products database. Management next asks the database administrator (DBA) to explain the functions that a database management system (DBMS) performs in managing a database. The DBA describes nine functions performed by a DBMS; this chapter explains these nine functions. Some of the functions have been introduced in previous chapters; however, they are re-emphasized here because they are key processing components of a DBMS. The nine functions of a DBMS are:

- **Update and retrieve data.** A DBMS must provide users with the ability to update and retrieve data in a database.

- **Provide catalog services.** A DBMS must store data about the data in a database and make this data accessible to users.

- **Support concurrent update.** A DBMS must ensure that the database is updated correctly when multiple users update the database at the same time.

- **Recover data.** A DBMS must provide methods to recover a database in the event that the database is damaged in any way.

- **Provide security services.** A DBMS must provide ways to ensure that only authorized users can access the database.

- **Provide data integrity features.** A DBMS must follow rules so that it updates data accurately and consistently.

- **Support data independence.** A DBMS must provide facilities to support the independence of programs from the structure of a database.

- **Support data replication.** A DBMS must manage multiple copies of the same data at multiple locations.

- **Provide utility services.** A DBMS must provide services that assist in the general maintenance of a database.

Update and Retrieve Data

A DBMS must provide users with the ability to update and retrieve data in a database; this is the fundamental capability of a DBMS. Unless a DBMS provides this capability, further discussion of what a DBMS does is irrelevant. In updating and retrieving data, users do not need to know how data is physically structured on disk or which processes the DBMS uses to manipulate the data. These structures and manipulations are solely the responsibility of the DBMS.

Updating data in a database includes adding new records and changing and deleting existing records. Suppose that Elena must update the Premiere Products database by adding data for part AE27, which is a new part. As shown in Figure 7.1, Elena enters the data for part AE27, and then requests that the DBMS add the data to the database. To add this data, the DBMS handles all the work to verify that part AE27 doesn't already exist in the database, stores the part AE27 data in the database, and then informs Elena that the

FIGURE 7.1

··················

Adding a new part
to the Premiere
Products database

task was completed successfully. How the DBMS performs these steps, where the DBMS stores the data in the database, how the DBMS stores the data, and all other processing details are invisible to Elena.

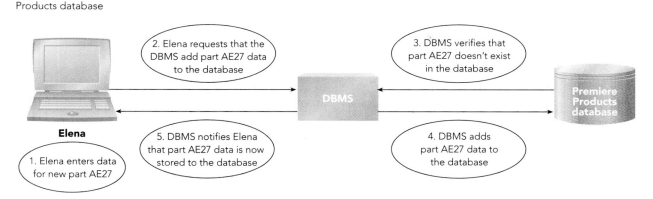

FIGURE 7.2

··················

Changing the price
of a part in the
Premiere Products
database

Suppose that Elena must update the Premiere Products database by changing the price of part KL62. As shown in Figure 7.2, Elena requests the data for the part and enters the change, but the DBMS performs the tasks of locating and reading the part data, displaying the data for Elena, and changing the price in the database. Once again, Elena does not need to be aware of the tasks that the DBMS completes, nor how the DBMS completes them.

Note: Deleting data in a database requires user and DBMS processing steps similar to those used to change data. The only differences occur in Steps 4 and 5 in Figure 7.2. In Step 4, the user requests that the DBMS delete the designated record. In Step 5, the DBMS deletes the record.

Figure 7.3 shows Elena retrieving the balance amount for All Season, a customer in the Premiere Products database. The DBMS finds the All Season record using the same strategy it used when it added the customer to the database; Elena doesn't need to know the strategy used by the DBMS to find and read the data. After finding and reading the All Season record in the database, the DBMS displays the customer's balance amount for Elena.

FIGURE 7.3
••••••••••••••••
Retrieving a balance
amount from the
Premiere Products
database

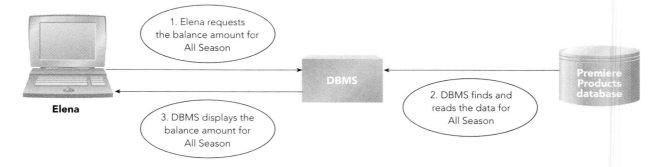

Provide Catalog Services

A DBMS must store data about the data in a database and make this data accessible to users. Data about the data in a database, or **metadata**, includes table descriptions and field definitions. As described in Chapter 4, the catalog, which is maintained automatically by the DBMS, contains table and field metadata. In addition, the catalog contains metadata about table relationships, views, indexes, users, privileges, and replicated data; the last three items are discussed later in this chapter.

The catalogs for many DBMSs consist of a set of special tables that are included in the database. The DBMS hides these special tables from the users of the database. However, the DBMS lets the DBA access and update the tables because the DBA must know the contents of the database and must create and define tables, fields, views, indexes, and other metadata. The DBA can authorize access for some catalog tables to other users as necessary.

Note: In some database systems, such as Microsoft Access, users can access and update metadata about fields, tables, relationships, and indexes. However, individuals and companies that create databases for other people usually hide this metadata so that users cannot access or update the metadata.

When the DBA uses the DBMS to access the catalog in the database, the DBA asks questions such as the following:

- What tables and fields are included in the database? What are their names?

- What are the properties of these fields? For example, is the Street field in the Customer table 15 characters long or 30? Is the CustomerNum field a numeric field or is it a character field? How many decimal places are in the Rate field in the Rep table?

- What are the possible values for the various fields? For example, are there any restrictions on the possible values for the CreditLimit field in the Customer table or for the Class field in the Part table?

- What are the meanings of the various fields? For example, what exactly is the Class field in the Part table, and what does a Class field value of HW mean?

- What relationships between tables exist in the database? Which relationships are one-to-many, many-to-many, and one-to-one? Must the relationship always exist? For example, must a customer always have a sales rep?

- Which fields and combinations of fields can you rapidly search for specific values because they are indexed? Which fields that are not indexed are candidates for indexes because they are often used in searches?

- Which users have access to the database? Which fields can Elena access for retrieval purposes but not update? Which fields can Elena update?

- Which programs or objects (queries, forms, and reports) access which data within the database? How do they access it? Do these programs merely retrieve the data, or do they update it? What kinds of updates do the programs perform? Can a certain program add a new customer, for example, or can it merely make changes regarding information about customers that are already in the database? When a program changes customer data, can it change all the fields or only some fields? Which fields?

Large, expensive DBMSs, such as Oracle and DB2, often have a catalog called a **data dictionary**, which contains answers to all these questions and more. The data dictionary serves as a super catalog containing metadata beyond what's been described previously. For example, these DBMSs let the DBA split the data in a database and store the fragmented data on multiple disks at multiple locations. In these cases, the data dictionary must track the location of the data. PC-based DBMSs do not offer a data dictionary, but they have a catalog that provides answers to most of the preceding questions.

Support Concurrent Update

A DBMS must ensure that the database is updated correctly when multiple users update the database at the same time.

Sometimes a person uses a database stored on a single computer. At other times, several people might update a database, but only one person at a time does so. For example, several people might take turns with one PC to update a database. A DBMS handles these situations easily. However, the use of networks and the use of DBMSs that are capable of running on these networks and that allow several users to update the same database raises a problem that the DBMS must address: concurrent update.

Concurrent update occurs when multiple users make updates to the same database at the same time. On the surface, you might think that a concurrent update doesn't present any problem. Why couldn't two, three, or fifty users update the database simultaneously without causing a problem?

The Concurrent Update Problem

To illustrate the problem with concurrent update, suppose that Ryan and Elena are two users who both work at Premiere Products. Ryan is currently updating the Premiere Products database to process orders and, among other actions, to increase customers' balances by the amount of their orders. For example, Ryan needs to increase the balance of customer 282 (Brookings Direct) by $100.00. Elena, on the other hand, is updating the Premiere Products database to post customer payments and, among other things, to decrease customers' balances by the amounts of their payments. Coincidentally, Elena has a $100.00 payment from Brookings Direct, so she will decrease that customer's balance by $100.00. The balance for Brookings Direct is $431.50 before the start of these updates and, because the amount of the increase exactly matches the amount of the decrease, the balance should still be $431.50 after their updates. But will it? That depends on how the database handles the updates.

How exactly does the DBMS make the required update for Ryan? First, as shown in Figure 7.4, the DBMS reads the data for Brookings Direct from the database on disk into Ryan's work area in memory (RAM). Second, Ryan enters the order data for Brookings Direct. At this point, Ryan's order entry takes place in his work area in memory, including adding the order total of $100.00 to the balance of $431.50, bringing the balance to $531.50. This change has *not* yet taken place in the database; it has taken place *only* in Ryan's work area in memory. Finally, after Ryan finishes entering the order data for Brookings Direct, the DBMS updates the database with Ryan's changes.

FIGURE 7.4
· · · · · · · · · · · · · · · · · ·
Ryan updates the database

230

Suppose that Elena begins her update at this point. As shown in Figure 7.5, the DBMS reads the data for Brookings Direct from the database, including the new balance of $531.50. Elena then enters the payment of $100.00, which decreases the customer balance to $431.50 in her work area in memory. Finally, the DBMS updates the database with Elena's change. The balance for Brookings Direct in the database is now $431.50, which is correct.

FIGURE 7.5
••••••••••••••••••
Elena updates the database

In the preceding sequence of updates, everything worked out correctly, but they don't always. Do you see how the updates to the database could occur in a way that would lead to an incorrect result?

What if the updates occur instead in the sequence shown in Figure 7.6? First, the DBMS reads the data from the database into Ryan's work area in memory, and at about the same time, the DBMS reads the data from the database into Elena's separate work area in memory. At this point, both Ryan and Elena have the correct data for Brookings Direct, including a balance of $431.50. Ryan adds $100.00 to the balance in his work area, and Elena subtracts $100.00 from the balance in her work area. At this point, in Ryan's work area in memory the balance is $531.50, while in Elena's work area in memory, it is $331.50. The DBMS now updates the database with Ryan's change. At this moment, Brookings Direct has a balance of $531.50 in the database. Finally, the DBMS updates the database with Elena's change. Her update replaces Ryan's. Now the balance for Brookings Direct in the database is $331.50! Had the DBMS updated the database in the reverse order, the final balance would have been $531.50. In either case, you now have incorrect data in the database—one of the updates has been *lost*. The DBMS must prevent these lost updates to the database.

FIGURE 7.6
Ryan's and Elena's updates to the database result in a lost update

Database before updates

Ryan   282.......431.50   Elena

Database on disk

Step 1—DBMS reads data from the database into RAM for Ryan

282.......431.50   282.......431.50   282.......531.50

Ryan   Database on disk   Elena

Step 2—DBMS reads data from the database into RAM for Elena

282.......431.50   282.......431.50   282.......431.50

Ryan   Database on disk   Elena

Step 3—Ryan changes data in RAM

282.......531.50   282.......431.50   282.......431.50

Ryan   Database on disk   Elena

(continued)

FIGURE 7.6
· · · · · · · · · · · · · · · · · · ·
Ryan's and Elena's
updates to the
database result in a
lost update
(continued)

Step 4—Elena changes data in RAM

282........531.50

Ryan

282.......431.50

Database on disk

282........331.50

Elena

Step 5—DBMS updates the database with Ryan's change

282........531.50

Ryan

282.......531.50

Database on disk

282........331.50

Elena

Step 6—DBMS updates the database with Elena's change; Ryan's update is lost!

282........531.50

Ryan

282.......331.50

Database on disk

282........331.50

Elena

Avoiding the Lost Update Problem

One way to prevent lost updates is to prohibit concurrent update. This may seem drastic, but it is not really so farfetched. You can let several users access the database at the same time, but for retrieval only; that is, they can read data from the database, but they can't update any data in the database. When these users need to update the database, such as increasing a customer's balance or changing the price of a part, the database itself is *not* updated. Instead, as shown in Figure 7.7 on the next page, a special program, which a computer programmer would create for them to use with the data in their database, adds a record to a separate file.

A record in this separate file might indicate, for example, that Premiere Products received a $100.00 payment from customer 282 on a certain date. Periodically, usually once a day, a single update program reads the *batch* of records in this file one at a time and performs the appropriate updates to the database; this processing technique is called **batch processing**. Because this program is the only way to update the database, you eliminate the problems associated with concurrent update.

FIGURE 7.7
••••••••••••••••••
Delaying updates
to the Premiere
Products database
to avoid the lost
update problem

Although this alternative approach avoids the lost update problem, it creates another problem. From the time users start updating (adding records to the special batch file) until the time the batch-processing program actually updates the database, the data in the database is out of date. If a customer's balance in the database is $4,500.00, the true balance is $5,500.00 if a user had entered an order for this customer that increased its balance by $1,000.00. If the customer has a $5,000.00 credit limit, the customer is now over its credit limit by $500.00.

The batch-processing alternative does not work in any situation that requires the data in the database to be current. These situations include credit card processing, banking, inventory control, and airline reservations. Other simple alternative solutions to the concurrent update problem, such as permitting only one user to update the database, also will not work in these situations because many users need to update the database in a timely way.

Two-Phase Locking

In most situations, you can't solve the concurrent update problem by avoiding it; you need the DBMS to have a strategy for dealing with it. One such strategy is for the DBMS to process an update completely before it begins processing the next update. For example, the DBMS can prevent Elena from beginning her update to the Brookings Direct data until the DBMS completes Ryan's update to that data, or vice versa.

To accomplish such a serial processing of updates, many DBMSs use locking. **Locking** denies access by other users to data while the DBMS processes one user's updates to the database. An example of locking using Ryan's and Elena's updates appears in Figure 7.8. After the DBMS reads the data in the database for Ryan's update, the DBMS locks the data, denying access to the data by Elena and any other user. The DBMS retains the locks until Ryan completes his change, and then the DBMS updates the database. For the duration of the locks, the DBMS rejects all attempts by Elena to access the data, and it notifies Elena that the data is locked. If she chooses to do so, she can keep attempting to access the data until the DBMS releases the locks, at which time the DBMS can process her update. In this simple case at least, the locking technique appears to solve the lost update problem.

FIGURE 7.8

· · · · · · · · · · · · · · · · ·

The DBMS uses a locking scheme to apply the updates for Ryan and Elena to the database

Database before updates

Ryan   Database on disk   Elena

Step 1—DBMS reads data from the database into RAM for Ryan and locks the record

Ryan   Database on disk   Elena

Step 2—Elena requests the same record from the database and her request fails

Ryan   Database on disk   Elena

Step 3—Ryan changes data in RAM; Elena's request for the same record again fails

Ryan   Database on disk   Elena

(continued)

FIGURE 7.8
....
The DBMS uses a
locking scheme to
apply the updates
for Ryan and Elena
to the database
(continued)

Step 4—DBMS updates the database with Ryan's change; Elena's request for the same record again fails

282.......531.50

282.......531.50

Record locked; read fails

Ryan   Database on disk   Elena

Step 5—DBMS unlocks the record; DBMS reads data from the database into RAM for Elena and locks the record

282.......531.50

282.......531.50

282.......531.50

Ryan   Database on disk   Elena

Step 6—Elena changes data in RAM

282.......531.50

282.......531.50

282.......431.50

Ryan   Database on disk   Elena

Step 7—DBMS updates the database with Elena's change

282.......531.50

282.......431.50

282.......431.50

Ryan   Database on disk   Elena

Step 8—DBMS unlocks the record

282.......531.50

282.......431.50

282.......431.50

Ryan   Database on disk   Elena

How long should the DBMS hold a lock? If the update involves changing field values in a single row in a single table, such as changing a customer's name and address, the lock no longer is necessary after this row is updated. However, sometimes an update is more involved.

Consider the task of filling an order for Premiere Products. Ryan might think that filling an order involves a single action. He simply indicates that an order currently in the database now needs to be filled. Alternatively, Ryan might need to enter data about the order. In either case, the process still feels like a single action to Ryan. Behind the scenes, though, filling an order requires that the DBMS update several records in the database. For example, suppose Ryan fills a new order for Brookings Direct that includes the sale of three washers and two dryers; Richard Hull is the rep for Brookings Direct. To fill this order, the DBMS must update the following records in the database:

- Add one record to the Orders table for the new order.

- Add one record to the OrderLine table for the sale of the three washers.

- Add one record to the OrderLine table for the sale of the two dryers.

- Change the washer record in the Part table to decrease the record's number of units on hand by three.

- Change the dryer record in the Part table to decrease the record's number of units on hand by two.

- Change the Brookings Direct record in the Customer table to increase the balance by the total amount of the order.

- Change the Richard Hull record in the Rep table to increase the commission by the commission amount owed to the rep for the order.

For this order, the DBMS updates seven records in the database; it adds three records and changes four records.

Each task that a user completes, such as filling an order, is called a transaction. A **transaction** is a set of steps completed by a DBMS to accomplish a single user task; the DBMS must successfully complete all transaction steps or none at all for the database to remain in a correct state.

For transactions such as filling an order, in which a single user task requires several updates in the database, what should the DBMS do about locks? How long does the DBMS hold each lock? For safety's sake, the DBMS should hold locks until it completes all the updates in the transaction. This approach for handling locks is called **two-phase locking**. The first phase is the **growing phase**, in which the DBMS locks more rows and releases none of the locks. After the DBMS acquires all the locks needed for the transaction and has completed all database updates, the second phase is the **shrinking phase** in which the DBMS releases all the locks and acquires no new locks. This two-phase locking approach solves the lost update problem.

Deadlock

Because each user transaction can require more than one lock, another problem can occur. Suppose that Ryan is filling the Brookings Direct order for the sale of three washers and two dryers and that Elena is filling another customer order that includes the sale of washers and dryers. Further suppose that for Ryan's transaction, the DBMS holds a lock on the washer record and is attempting to lock the dryer record, as shown in Figure 7.9. However, the DBMS has already locked the dryer record for Elena's transaction, so Ryan must wait for the DBMS to release the lock. Before the DBMS releases the lock on the dryer record for Elena's transaction, however, it needs to update, and thus lock, the washer record, which is currently locked for Ryan's transaction. Ryan is waiting for the DBMS to act for Elena (release the lock on the dryer record), while Elena is waiting for the DBMS to act for Ryan (release the lock on the washer record). Without the aid of some intervention, this dilemma could continue indefinitely. Terms used to describe such situations are **deadlock** and the **deadly embrace**. Obviously, some strategy is necessary to either prevent or handle deadlock.

One strategy is to let deadlock occur and then have the DBMS detect and break any deadlock. To detect deadlock, the DBMS must keep track of the collection of records it has locked for each transaction as well as the records it's waiting to lock. If two transactions are waiting for records held by the other, a deadlock has occurred. Actually, more than two users could be involved. Ryan could be waiting for a record held by Elena, while Elena is waiting for a record held by Pat, who in turn is waiting for a record held by Ryan.

After the DBMS detects deadlock, the DBMS must break the deadlock. To break the deadlock, the DBMS chooses one deadlocked user to be the **victim**. For the victim's transaction, the DBMS undoes all completed updates, releases all locks, and reschedules the transaction. Using this method of handling deadlocks, the user notices only a delay in the time needed to complete the transaction.

Locking on PC-Based DBMSs

Mainframe DBMSs typically offer sophisticated schemes for locking as well as for detecting and handling deadlocks. PC-based DBMSs provide facilities for the same purposes, but they usually are much more limited than the facilities provided by mainframe DBMSs. These limitations, in turn, put an additional burden on the programmers who write the programs that allow concurrent update.

Although the exact features for handling the problems associated with concurrent update vary from one PC-based DBMS to another, the following list is fairly typical of the types of facilities provided:

- Programs can lock an entire table or an individual row within a table, but only one or the other. As long as one program has a row or table locked, no other program may access that row or table.

238

- Programs can release any or all of the locks that they currently hold.

- Programs can inquire whether a given row or table is locked.

This list, although short, comprises the complete set of facilities provided by many PC-based DBMSs. Consequently, the following guidelines have been devised for writing programs for concurrent update:

- If an update transaction must lock more than one row in the same table, you must lock the entire table.

- When a program attempts to read a row that is locked, it may wait a short period of time and then try to read the row again. This process can continue until the row becomes unlocked. It usually is preferable, however, to impose a limit on the number of times a program may attempt to read the row. In this case, reading is done in a loop, which proceeds until either the read is successful or the maximum number of times that the program can repeat the operation is reached. Programs vary in terms of what action is taken should the loop be terminated without the read being successful. One possibility is to notify the user of the problem and let the user decide whether to try the same update again or to move on to something else.

- Because there is no facility to *detect and handle* deadlocks, you must try to *prevent* them. A common approach to this problem is for every program in the system to attempt to lock all the rows and/or tables it needs before beginning an update. Assuming it is successful in this attempt, each program can then perform the required updates. If any row or table that the program needs is already locked, it should immediately release all the locks that it currently holds, wait some specified period of time, and then try the entire process again. In some cases, it might be better to notify the user of the problem and see whether the user wants to try again. In effect, this means that any program that encounters a problem will immediately get out of the way of all the other programs rather than be involved in a deadlock situation.

- Because locks prevent other users from accessing a portion of the database, it is important that no user keep rows or tables locked any longer than necessary. This is especially significant for update programs. Suppose, for example, that a user is employing an update program to update information about customers. Suppose further that after the user enters the number of the customer to be updated, the customer row is locked and remains locked until the user has entered all the new data and the update has taken place. What if the user is interrupted by a phone call before he or she has finished entering the new data? What if the user goes to lunch? The row might remain locked for an extended period of time. If the update involves several rows, all of which must be locked, the problem becomes that much worse. In fact, in many DBMSs, if more than one row from the same table must be locked, the entire table must be locked, which means that entire tables might be locked for extended periods of time. Clearly, this situation must not be permitted to occur. Some variation on the timestamping technique used by some mainframe DBMSs is a programming strategy you can use to overcome this problem.

Timestamping

An alternative to two-phase locking is timestamping. With **timestamping**, the DBMS assigns to each database update the unique time when the update started; this time is called a **timestamp**. In addition, every database row includes the timestamp associated with the last update to the row. The DBMS processes updates to the database in timestamp order. If two users try to change the same row at the same time, the DBMS processes the change that has the earlier timestamp. The other transaction is restarted and assigned a new timestamp value.

Timestamping avoids the need to lock rows in the database and eliminates the processing time needed to apply and release locks and to detect and resolve deadlocks. On the other hand, additional disk and memory space are required to store the timestamp values; in addition, the DBMS uses extra processing time to update the timestamp values.

One might naturally ask at this point whether the ability to have concurrent update is worth the complexity that it adds to the DBMS. In some cases, the answer will be no. Concurrent update may be far from a necessity. In most cases, however, concurrent update is necessary to the productivity of the users of the system. In these cases, implementation of locking, timestamping, or some other strategy is essential to the proper performance of the system.

Recover Data

A DBMS must provide methods to recover a database in the event that the database is damaged in any way. A database can be damaged or destroyed in many ways. Users can enter data that is incorrect; transactions that are updating the database can end abnormally during an update; a hardware problem can occur; and so on. After any such event has occurred, the database might contain invalid or inconsistent data. It may even be totally destroyed.

Obviously, a situation in which data has been damaged or destroyed must not be allowed to go uncorrected. The database must be returned to a correct state. **Recovery** is the process of returning the database to a state that is known to be correct from a state known to be incorrect; in performing such a process, you say that you recover the database. In situations where indexes or other physical structures in the database have been damaged but the data has not, many DBMSs provide a feature that you can use to repair the database automatically to recover it.

If the data in a database has been damaged, the simplest approach to recovery involves periodically making a copy of the database (called a **backup** or a **save**). If a problem occurs, the database is recovered by copying this backup copy over it. In effect, the damage is undone by returning the database to the state it was in when the last backup was made.

Unfortunately, other activity besides that which caused the destruction also is undone. Suppose the database is backed up at 10:00 P.M. and users begin updating it at 8:00 A.M. the next day. Then suppose that at 11:30 A.M., something happens that destroys the database. If the previous night's backup is used to recover the database, the entire database is returned to the state it was in at 10:00 P.M. All updates made in the morning, not just the update or updates that were in progress at the time the problem occurred, are lost. Thus, during the final part of the recovery process, users would have to redo all the work they had done since 8:00 A.M.

Journaling

As you might expect, large, expensive DBMSs provide sophisticated features to avoid the costly and time-consuming process of having users redo their work. These features include **journaling**, which involves maintaining a **journal** or **log** of all updates to the database. The log is a file separate from the database; thus, the log is still available if a catastrophe destroys the database.

Several types of information are typically kept in the log for each transaction. This information includes the transaction ID and the date and time of each individual update. It also includes a record of what the data in the row looked like in the database before the update (called a **before image**) and a record of what the data in the row looked like in the database after the update (called an **after image**). In addition, the log contains an entry to indicate the start of a transaction and the successful completion (**commit**) of a transaction.

To illustrate the use of a log by a DBMS, consider the four sample transactions shown in Figure 7.10. Three transactions, 1, 3, and 4, require a single update to the database. The second transaction, which is Ryan's order transaction for Brookings Direct, requires seven updates to the database.

FIGURE 7.10
· · · · · · · · · · · · · · · · · ·
Sample transactions

Transaction ID	Transaction Description
1	1. Change the Price value for part number DW11 to $389.99
2	1. Add a record to the Orders table: OrderNum of 21700, OrderDate of 10/24/2007, CustomerNum of 282 2. Add a record to the OrderLine table: OrderNum of 21700, PartNum of DW11, NumOrdered of 3, QuotedPrice of $389.00 3. Add a record to the OrderLine table: OrderNum of 21700, PartNum of KL62, NumOrdered of 2, QuotedPrice of $346.50 4. Change the OnHand value for part number DW11 to 9 5. Change the OnHand value for part number KL62 to 10 6. Change the Balance value for CustomerNum 282 to $2,321.50 7. Change the Commission value for RepNum 35 to $39,346.20
3	1. Add customer 510
4	1. Delete part AT94

Suppose that these four transactions are the first transactions in a day, immediately following a backup of the database, and that they all complete successfully. In this case, the log might look like the sample log shown in Figure 7.11.

FIGURE 7.11
••••••••••••••••••
Sample log in which
all transactions
commit normally

Transaction ID	Time	Action	Record Updated	Before Image	After Image
1	8:00	Start			
2	8:01	Start			
2	8:02	Insert	Orders (21700)		(new values)
3	8:03	Start			
1	8:04	Update	Part (DW11)	(old values)	(new values)
2	8:05	Insert	OrderLine (21700, DW11)		
1	8:06	Commit			
4	8:07	Start			
3	8:08	Insert	Customer (510)		(new values)
2	8:09	Insert	OrderLine (21700, KL62)		(new values)
3	8:10	Commit			
2	8:11	Update	Part (DW11)	(old values)	(new values)
2	8:12	Update	Part (KL62)	(old values)	(new values)
4	8:13	Delete	Part (AT94)	(old values)	
2	8:14	Update	Customer (282)	(old values)	(new values)
4	8:15	Commit			
2	8:16	Update	Rep (35)	(old values)	(new values)
2	8:17	Commit			

Before studying how the log is used in the recovery process, let's examine the log itself. Each record in the log includes the ID of the transaction, as well as the time the particular action occurred. The actual time would be more precise than in the example, the DBMS would process the actions much faster, and the date would also be included in the log. For simplicity, each action occurs one minute after the preceding action. The actions are Start to indicate the start of a transaction, Commit to indicate that the transaction completed successfully, Insert to identify the addition of a record to the database, Update to identify the change of a record, and Delete to identify the deletion of a record. For an Insert action, no before image appears in the log because the data did not exist prior to the action. Similarly, for a Delete action, no after image appears in the log.

The sample log shows, for example, that transaction 2 began at 8:01. The database updates to complete the transaction occurred at 8:02 (order record 21700 inserted), 8:05 (the first order line record inserted), 8:09 (the second order line record inserted), 8:11 (the first part record updated), 8:12 (the second part record updated), 8:14 (customer record 282 updated), and 8:16 (rep record 35 updated). At 8:17, the transaction was committed. During this same time span, the other three transactions were also committed.

Forward Recovery

How is the log used in the recovery process? Suppose a catastrophe destroys the database just after 8:11. In this case, the recovery of the database begins with the most recent database backup from the previous evening at 10:00. As shown in Figure 7.12, the DBA copies the backup over the live database. Because the database is no longer current, the DBA executes a DBMS recovery program that applies the after images of committed transactions from the log to bring the database up to date. This method of recovery is called **forward recovery**.

FIGURE 7.12
. . . .
Forward recovery

In its simplest form, the recovery program copies the after image of each record in the log over the actual record in the database in chronological order. You can improve the recovery process by realizing that if a record was updated 10 times since the last backup, the recovery program copies the after image records 10 times over the database record. Thus, in reality, the first nine copies are unnecessary. The tenth after image includes all the updates accomplished in the first nine. Thus, you can improve the performance of the recovery program by having it first scan the log and then apply the last after image.

QUESTION

In the preceding scenario, which transactions in the sample log shown in Figure 7.11 does the recovery program use to update the restored database?

ANSWER

The catastrophe occurred just after 8:11. Because the recovery program applies transactions committed before the catastrophe, the program applies only transactions 1 and 3. These two transactions committed before 8:11, at which point the DBMS was still processing transactions 2 and 4.

Backward Recovery

If the database has not actually been destroyed, then the problem must involve transactions that were either incorrect or, more likely, stopped in midstream. In either case, the database is currently not in a valid state. You can use **backward recovery**, or **rollback**, to recover the database to a valid state by undoing the problem transactions. The DBMS accomplishes the backward recovery by reading the log for the problem transactions and applying the before images to undo their updates, as shown in Figure 7.13.

FIGURE 7.13
·················
Backward recovery

QUESTION

For the sample log shown in Figure 7.11, what does the DBMS do to rollback transaction 1?

ANSWER

The DBMS started transaction 1 at 8:00, changed a Part table record at 8:04 for transaction 1, and committed transaction 1 at 8:06. To rollback transaction 1, the DBMS applies the before image of the Part table record.

QUESTION

For the sample log shown in Figure 7.11, what does the DBMS do to rollback transaction 3?

ANSWER

The DBMS started transaction 3 at 8:03, added a Customer table record at 8:08 for transaction 3, and committed transaction 3 at 8:10. Because no before image exists for adding a record, to rollback transaction 3, the DBMS deletes the Customer table record.

Recovery on PC-Based DBMSs

PC-based DBMSs generally don't offer sophisticated recovery features such as journaling. Most of them simply provide users with a simple way to make backup copies and to recover the database later by copying the backup over the database.

How should you handle recovery in any application system you develop with a PC-based DBMS? You could simply use the features of the DBMS to periodically make backup copies and use the most recent backup if a recovery is necessary. The more important it is to avoid redoing work, the more often you would make backups. For example, if a backup is made every eight hours, you might have to redo up to eight hours of work. If, on the other hand, a backup is made every two hours, two hours of work at most might have to be redone.

In many situations, this approach, although not particularly desirable, is acceptable. For systems with a large number of updates made to the database between backups, however, this approach is not acceptable. In such cases, the necessary recovery features that are not supplied by the DBMS must be included in the application programs. Each of the programs that update the database could, for example, also write a record to a separate log file, indicating the update that had taken place. You could write a separate program to read the log file and re-create all the updates indicated by the records in the file. The recovery process would then consist of first copying the backup over the actual database, and then running this special program.

Although this approach does simplify the recovery process for the users of the system, it also causes some problems. First, each of the programs in the system becomes more complicated because of the extra logic involved in adding records to the special log file. Second, you must write a separate program to update the database with the information in this log file. Finally, every time a user completes an update, the system now has extra work to do, and this additional processing may slow down the system to an unacceptable level. Thus, in any application, you must determine whether the ease of recovery provided by this approach is worth the price you might have to pay for it. The answer will vary from one system to another.

■ Provide Security Services

As discussed in Chapter 4, a DBMS must provide ways to ensure that only authorized users can access the database. Security is the prevention of unauthorized access, either intentional or accidental, to a database. The most common security features used by DBMSs are encryption, authentication, authorizations, and views.

Encryption

Encryption converts the data in a database to a format that's indecipherable to a word processor or other program and stores it in an encrypted format. If unauthorized users attempt to bypass the DBMS and get to the data directly, they see only the encrypted version of the data. However, authorized users accessing the data using the DBMS have no problem viewing and working with the data.

When a user updates data in the database, the DBMS encrypts the data before updating the database. Before a legitimate user retrieves the data via the DBMS, the data is decrypted, or decoded, and presented to the user in the normal format. The entire encryption process is transparent to a legitimate user; that is, he or she is not even aware it is happening.

Access lets you encrypt a database and, after you've encrypted the database, you can use Access to decrypt it. **Decrypting** a database reverses the encryption. If your encrypted database takes longer to respond to user requests as it gets larger, you might consider decrypting it to improve its responsiveness.

Using Access to encrypt or decrypt a database is a four-step process:

1. Start Access, but don't open the database you want to encrypt or decrypt.

2. Click Tools on the menu bar, point to Security, and then click Encode/Decode Database.

3. Select the disk location for the database, select the database, and then click the OK button.

4. Select the disk location for the modified database, type its name, and then click the Save button.

Authentication

Authentication refers to techniques for identifying the person who is attempting to access the DBMS. The use of passwords is the most common authentication technique. A **password** is a string of characters assigned by the DBA to a user that the user must enter to access the database. Users also use passwords to access many operating systems, networks, and other computer and Internet resources. Increasing in use as an alternative to passwords for authentication are biometric identification techniques. **Biometrics** identify users by physical characteristics such as fingerprints, voiceprints, handwritten signatures, and facial characteristics.

Unlike individual passwords, a **database password** is a string of characters assigned by the DBA to a database that users must enter before they can access the database. As long as the database password is known only to authorized database users, unauthorized access to the database is prevented. It's best if the DBA uses a database password that's easily remembered by authorized users but that is not obvious and easily guessed by others. Access lets the DBA assign a database password, as shown in Figure 7.14. To create the database password, the DBA enters the same password twice to verify that the initial entry is the one that the DBA wants.

After the DBA creates the database password for a database, as shown in Figure 7.15, users must enter it correctly before they can open the database.

Authorizations

Using passwords is a security measure that applies to all users of a database; after users enter their passwords successfully, they can retrieve and update all the data in the database. Frequently, the security needs for a database are more individualized. For example, the DBA might need to let some users view and update all data and let other users view only certain data. In this situation, the DBA uses **authorization rules** that specify which users have what type of access to which data in the database.

Access implements authorization rules through its **user-level security**, which lets the DBA grant users specific permissions to tables, queries, and other objects in a database. A user's **permissions** specify what kind of access the user has to objects in the database. The DBA can assign permissions to individual users or to groups of users. The DBA usually creates groups of users, called **workgroups**, assigns the appropriate permissions to each group, and then assigns each user to the appropriate group based on the permissions the user requires. Access stores group, user, and group membership information in a special Access file, called a **workgroup information file**. Access stores permission information in hidden system tables in the database.

Suppose the DBA at Premiere Products uses the Access Security Wizard to implement user-level security to create two groups. As shown in Figure 7.16, the DBA has selected the Read-Only Users and Full Data Users groups. Users assigned to the Read-Only Users group can retrieve all data but can't update any data; the DBA will assign Pat to this group. Users assigned to the Full Data Users group can retrieve and update all data; the DBA will assign Ryan and Elena to this group.

FIGURE 7.16
••••••••••••••••••
Selecting the groups for the Premiere Products database

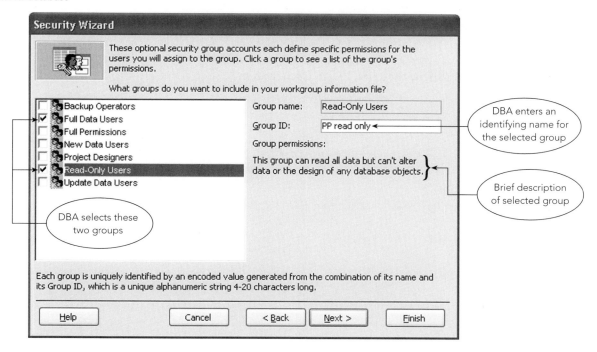

FIGURE 7.17
••••••••••••••••
Adding users to
the workgroup
information file

To create users, the DBA enters a user name, a password, and a personal ID for each user. As shown in Figure 7.17, the DBA has already created Pat and Ryan as users and is set to add Elena as a user. The DBA will use the preassigned Administrator user to access the database.

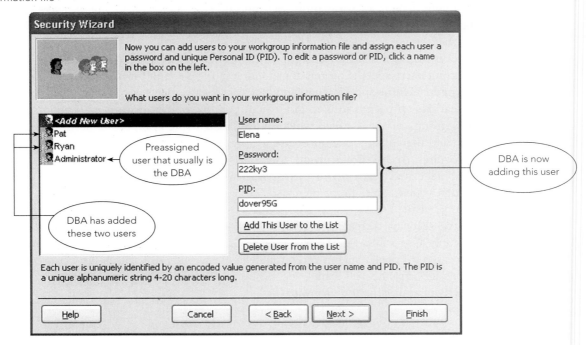

Next, the DBA assigns users to groups. As shown in Figure 7.18, the DBA has just assigned Elena to the Full Data Users group.

FIGURE 7.18
•••••••••••••••••
Assigning a user to
a group

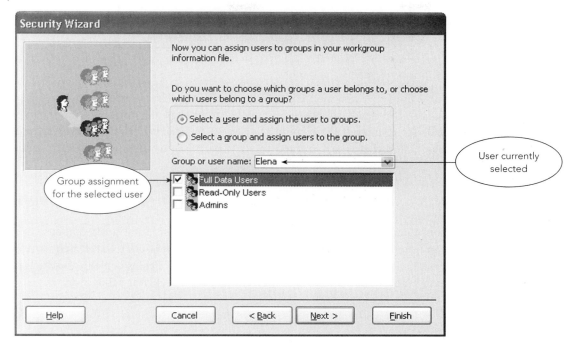

After assigning Pat and Ryan to their appropriate groups, user-level security is now in effect for the Premiere Products database. When users access the database in the future, they must first enter their user names and passwords, as assigned to them by the DBA. As shown in Figure 7.19, Elena has entered her assigned user name and password so that she can start her workday and access the database.

FIGURE 7.19
•••••••••••••••••
Elena supplying her
assigned user name
and password

The DBMS will deny access to the secured database for anyone who enters an incorrect user name or password and will inform the person of the denial, as shown in Figure 7.20.

FIGURE 7.20
· · · · · · · · · · · · · · · · ·
Message displayed
after the entry of an
invalid user name or
password

Because the DBA assigned Elena to the Full Data Users group, she can retrieve and update data in the database, but she is unable to change the design of the database. The same is true for Ryan, but Pat can only retrieve data from the database. If users attempt to perform an action that is not granted by their permissions, Access denies that action and displays an error message.

Views

Recall from Chapter 4 that a view is a snapshot of certain data in the database at a given moment in time. If a DBMS provides a facility that allows users to have their own views of a database, this facility can be used for security purposes. Tables or fields to which the user does not have access in his or her view effectively do not exist for that user.

Privacy

No discussion of security is complete without at least a brief mention of privacy. Although "security" and "privacy" are often used synonymously, they are different, although related, concepts. **Privacy** refers to the right of individuals to have certain information about them kept confidential. Privacy and security are related because it is only through appropriate *security* measures that *privacy* can be ensured.

Laws and regulations dictate some privacy rules, and companies institute additional privacy rules. Variations in what information is kept confidential occur among organizations. For example, salaries at governmental and many service organizations are public information, but salaries at many private enterprises are kept confidential.

■ Provide Data Integrity Features

A DBMS must follow rules so that it updates data accurately and consistently. These rules, called integrity constraints, are categorized as either key integrity constraints or data integrity constraints.

Key integrity constraints consist of primary key constraints and foreign key constraints. Primary key constraints, which are governed by entity integrity (as discussed in Chapter 4), enforce the uniqueness of the primary key. For example, forbidding the addition of a rep whose number matches the number of a rep already in the database is an example of a primary key constraint. Foreign key constraints, which are governed by referential integrity (as discussed in Chapter 4), enforce the fact that a value for a foreign key must match the value of the primary key for some row in a table in the database. Forbidding the addition of a customer whose rep is not already in the database is an example of a foreign key constraint.

Data integrity constraints help to ensure the accuracy and consistency of individual field values. Types of data integrity constraints include the following:

- **Data type.** The value entered for any field should be consistent with the data type for that field. For a numeric field, only numbers should be allowed to be entered. If the field is a date, only a legitimate date should be permitted. For instance, February 30, 2008, is an invalid date that should be rejected.

- **Legal values.** For some fields, not every possible value that is of the assigned data type is legitimate. For example, even though CreditLimit is a numeric field, only the values $5,000.00, $7,500.00, $10,000.00, and $15,000.00 are valid. For the OrderDate field in the Orders table, Premiere Products might insist that only the current date or a future date is an acceptable value when updating an order. In addition, you should be able to specify which fields can accept null values and which fields can't.

- **Format.** Some fields require a special entry or display format. Although the PartNum field is a character field, for example, only specially formatted strings of characters might be acceptable. Legitimate part numbers might have to consist of two letters followed by two digits; this is an example of an entry format constraint. Users might want the OrderDate field displayed with a four-digit year value instead of a two-digit year value; this is an example of a display format constraint.

Integrity constraints can be handled in one of four ways:

1. The constraint is ignored, in which case, no attempt is made to enforce the constraint.

2. The responsibility for constraint enforcement is placed on the users. This means that users must be careful that any updates they make in the database do not violate the constraint.

3. The responsibility for constraint enforcement is placed on programmers. Programmers place the logic to enforce the constraint into programs. Users must update the database only by means of these programs and not through any of the built-in entry facilities provided by the DBMS because these would allow violation of the constraint. Programmers design the programs to reject any attempt by the users to update the database in a way that violates the constraint.

4. The responsibility for constraint enforcement is placed on the DBMS. The DBA specifies the constraint to the DBMS, which then rejects any attempt to update the database in a way that violates the constraint.

QUESTION Which of these four approaches for constraint enforcement is best?

ANSWER The first approach, ignoring the constraint, is undesirable because it can lead to invalid data in the database, such as two customers with the same number, part numbers with an invalid format, invalid credit limits, and so on.

The second approach, user constraint enforcement, is a little better because at least an attempt is made to enforce the constraints. However, this approach places the burden of enforcement on users. Not only does this mean extra work for users, but any mistake on the part of a single user, no matter how innocent, can lead to invalid data in the database.

The third approach removes the burden of enforcement from users and places it on programmers. This solution is better still because it means that users can't violate the constraints. The disadvantage is that all update programs in the system become more complex. This complexity makes programmers less productive and makes programs more difficult to create and modify. This approach also makes changing an integrity constraint more difficult because this may mean changing all the programs that update the database. Furthermore, if the logic in any program used to enforce the constraints is faulty, the program could permit some constraint to be violated, and you might not realize that this had happened until a problem occurred at a later date. Finally, you would have to guard against a user bypassing the programs in the system in order to enter data directly into the database—for example, by using some built-in facility of the DBMS. If a user is able to bypass the programs and enters incorrect data, all the controls that were so diligently placed into the programs are helpless to prevent a violation of the constraints.

The best approach is the one in which the DBMS enforces. You specify the constraints to the DBMS and the DBMS ensures that they are never violated.

• •

Most DBMSs include most of the necessary capabilities to enforce the various types of integrity constraints. Consequently, you let the DBMS enforce all the constraints that it is capable of enforcing; then let application programs enforce any other constraints. You also might create a special program whose sole purpose is to examine the data in the database to determine whether any constraints have been violated. You'd run this program periodically and take corrective action to remedy any violations that the program discovers.

Access supports key constraints. Access lets you specify a primary key, and then it builds a unique index automatically for the primary key. Access also lets you specify foreign keys, and then it enforces referential integrity automatically. You can use Access to specify data integrity constraints. As shown in Figure 7.21, you can specify the data type for each field, and you can specify data format and legal-values integrity constraints.

FIGURE 7.21
• • • • • • • • • • • • • • • •
Example of integrity
constraints in Access

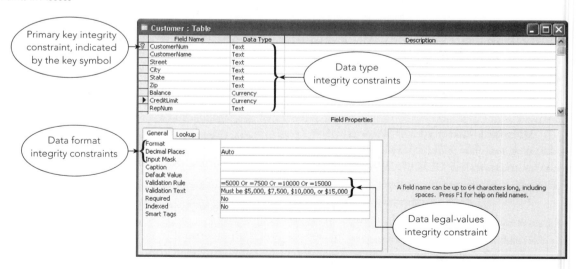

252

Support Data Independence

A DBMS must provide facilities to support the independence of programs from the structure of a database. One of the advantages of working with a DBMS is **data independence**, which is a property that lets you change the database structure without requiring you to change the programs that access the database. What types of changes could you or a DBA make to the database structure? A few of these changes are adding a field, changing a field property (such as length), creating an index, and adding or changing a relationship. The following sections describe the data independence considerations for each type of change.

Adding a Field

If you add a new field to a database, you don't need to change any program except, of course, those programs using the new field. However, if a program uses an SQL SELECT * FROM command to select all the fields from a given table, you are now presented with an extra field. To prevent this from happening, you need to change the program to restrict the output to only the desired fields. To avoid the imposition of this extra work, you should list all the required fields in an SQL SELECT command instead of using the *.

Changing the Length of a Field

In general, you don't need to change programs because you've changed the length of a field; the DBMS handles all the details concerning this change in length. However, if a program sets aside a certain portion of the screen or a report for the field, and the length of the field has increased to the point where the previously allocated space is inadequate, you'll need to change the program.

Creating an Index

To create an index, you enter a simple SQL command or select a few options. Most DBMSs use the new index automatically for all updates and queries. For some DBMSs, you might need to make minor changes in already existing programs to use the new index.

Adding or Changing a Relationship

In terms of data independence considerations, adding or changing a relationship is the trickiest of all and is best illustrated with an example. Suppose Premiere Products now has the following requirements:

- Customers are assigned to territories.

- Each territory is assigned to a single rep.

- A rep can have more than one territory.

- A customer is represented by the rep who covers the territory to which the customer is assigned.

To implement these changes, you need to restructure the database. The previous one-to-many relationship between the Rep and Customer tables is no longer valid. Instead, there's now a one-to-many relationship between the Rep table and the new Territory table, and a one-to-many relationship between the Territory table and the Customer table, as follows:

```
Rep (RepNum, LastName, FirstName, Street, City, State,
    Zip, Commission, Rate)
Territory (TerritoryNum, TerritoryDescription, RepNum)
Customer (CustomerNum, CustomerName, Street, City, State,
    Zip, Balance, CreditLimit, TerritoryNum)
```

Further suppose that a user accesses the database via the following view, which is named RepCust:

```
CREATE VIEW RepCust (RNum, RLast, RFirst, CNum, CName) AS
SELECT Rep.RepNum, LastName, FirstName, Customer.CustomerNum, CustomerName
FROM Rep, Customer
WHERE Rep.RepNum=Customer.RepNum
;
```

The defining query is now invalid because there is no RepNum field in the Customer table. A relationship still exists between reps and customers, however. The difference is that you now must go through the Territory table to relate the two tables. If users have been accessing the tables directly to form the relationship, their programs will have to change. If they are using the RepCust view, then you need to change only the definition of the view. The new definition is as follows:

```
CREATE VIEW RepCust (RNum, RLast, RFirst, CNum, CName) AS
SELECT Rep.RepNum, LastName, FirstName, Customer.CustomerNum, CustomerName
FROM Rep, Territory, Customer
WHERE Rep.RepNum=Territory.RepNum
AND Territory.TerritoryNum=Customer.TerritoryNum
;
```

The defining query is now more complicated than it was before, but this does not affect users of the view. They continue to access the database in exactly the same way that they did before, and the DBA won't need to change their programs.

Support Data Replication

A DBMS must manage multiple copies of the same data at multiple locations. Sometimes data should be duplicated—technically called **replicated**—at more than one physical location, for performance or other reasons. For example, accessing data at a local site is much more efficient than accessing data remotely. It's more efficient because using replicated data does not involve data communication and network time delays, users compete for data with fewer other users, and replicated data keeps data available to local users at times when the data might not be available at other sites.

If certain information needs to be accessed frequently from all sites, a company might choose to store the information at all its locations. At other times, users on the road—for example, reps meeting at their customers' sites—might need access to data but would not have this access unless the data was stored on their portable computers.

Replication lets users at different sites use and modify copies of a database and then share their changes with the other users. Replication is a two-step process. First, the DBMS creates copies, called **replicas**, of the database at one or more sites. For example, you could create two replicas, as shown in Figure 7.22, and give the "Replica 1 database" to one user to access at a remote location and give the "Replica 2 database" to a second user to use at a different remote location.

FIGURE 7.22
• • • • • • • • • • • • • • • • • •
The DBMS creates replicas from the master database

The master database and all replicas form a replica set. Users then update their individual replicas, just as if they were updating the master database. Periodically, the DBMS exchanges all updated data between the master database and a replica in a process called **synchronization**. For example, after the second user returns from the remote site, the DBA synchronizes the master database and the "Replica 2 database," as shown in Figure 7.23. Later, after the first user returns, the DBA synchronizes the master database and the "Replica 1 database."

FIGURE 7.23
• • • • • • • • • • • • • • • • • •
DBMS synchronizes two databases in a replica set

Ideally, the DBMS should handle all the issues associated with replication for you. The DBMS should do all the work to keep the various copies of data consistent behind the scenes; users should be unaware of the work involved. Access supports replication in the manner described. You'll learn more about replication in Chapter 9.

▓ Provide Utility Services

A DBMS must provide services that assist in the general maintenance of a database. In addition to the services already discussed, a DBMS provides a number of **utility services** that assist in the general maintenance of the database. The following is a list of services that might be provided by a PC-based DBMS:

- The DBMS lets you change the database structure—adding new tables and fields, deleting existing tables and fields, changing the name or properties of fields, and so on.

- The DBMS lets you add new indexes and delete indexes that are no longer needed.

- While using the database, the DBMS lets you use the services available from your operating system, such as Windows or Linux.

- The DBMS lets you export data to and import data from other software products. For example, you can transfer data easily between the DBMS and a spreadsheet file, a word-processing file, a graphics program file, or even another DBMS.

- The DBMS provides support for easy-to-use edit and query capabilities, screen generators, report generators, and so on.

- The DBMS provides support for both procedural and nonprocedural languages. With a **procedural language**, you must tell the computer precisely how a given task is to be accomplished; Basic, C++, and COBOL are examples of procedural languages. With a **nonprocedural language**, you merely describe the task you want the computer to accomplish. The nonprocedural language then determines how the computer will accomplish the task. SQL is an example of a nonprocedural language.

- The DBMS provides an easy-to-use, menu-driven or switchboard-driven interface that allows users to tap into the power of the DBMS without having to learn a complicated set of commands.

◼ SUMMARY

- The fundamental capability of a database management system (DBMS) is to provide users with the ability to update and retrieve data in a database without users needing to know how data is structured on disk or which processes the DBMS uses to manipulate the data.

- A DBMS must store metadata (data about the data) in a database and make this data accessible to users. The metadata is stored in a catalog or data dictionary.

- A DBMS must support concurrent update, allowing multiple users to update the same database at the same time. If concurrent update is not handled correctly, updates might be lost, causing the database to contain invalid data.

- Locking, which denies access by other users to data while the DBMS processes one user's updates, is one approach to concurrent update. Two-phase locking includes a growing phase, in which the DBMS locks more rows and releases none of the locks, followed by a shrinking phase, in which the DBMS releases all locks and acquires no new locks.

- Deadlock and deadly embrace are terms used to describe the situation in which two or more users are each waiting for the other to release a lock before they can proceed. Mainframe DBMSs have sophisticated facilities for detecting and handling deadlock. Most PC-based DBMSs do not have such facilities, which means that programs that access the database must be written in such a way that deadlocks are avoided.

- An alternative to two-phase locking is time-stamping, in which the DBMS processes updates to a database in timestamp order.

- A DBMS must provide methods to recover a database in the event that the database is damaged in any way. DBMSs provide facilities for periodically making a backup copy of the database. To recover the database when it is damaged or destroyed, your first step is to copy the backup over the damaged database.

- Large, expensive DBMSs maintain a log or journal of all database updates since the last backup. If a database is destroyed, you make the database current from the last backup by using forward recovery to apply the after images of committed transactions. If you need to remove the updates of incorrect or terminated transactions, you use backward recovery or rollback to apply the before images to undo the updates.

- A DBMS must provide security features to prevent unauthorized access, either intentional or accidental, to a database. These security features include encryption (the storing of data in an encoded form), authentication (passwords or biometrics to identify users, and database passwords assigned to the database), authorizations (assigning authorized users to groups that have permissions for accessing the database), and views (which limit a user's access to only the tables and fields included in the view).

- A DBMS must follow rules or integrity constraints so that it updates data accurately and consistently. Key integrity constraints consist of primary key and foreign constraints. Data integrity constraints help to ensure the accuracy and consistency of individual fields and include data type, legal-values, and format integrity constraints.

- A DBMS must provide facilities to support the independence of programs from the structure of a database; data independence is the term for this capability.

- A DBMS must provide a facility to handle replication by managing multiple copies of a database at multiple locations.

- A DBMS must provide a set of utility services that assist in the general maintenance of a database.

■ KEY TERMS

after image
authentication
authorization rule
backup
backward recovery
batch processing
before image
biometrics
commit
concurrent update
data dictionary
data independence
database password
deadlock
deadly embrace
decrypting
encryption
forward recovery
growing phase
journal
journaling
locking
log

metadata
nonprocedural language
password
permission
privacy
procedural language
recovery
replica
replicate
rollback
save
shrinking phase
synchronization
timestamp
timestamping
transaction
two-phase locking
user-level security
utility services
victim
workgroup
workgroup information file

■ REVIEW QUESTIONS

1. When users update and retrieve data, what tasks does a DBMS perform that are hidden from the users?

2. What is metadata, and which component of a DBMS maintains metadata?

3. How does a catalog differ from a data dictionary?

4. What is meant by concurrent update?

5. Describe a situation that could cause a lost update.

6. What is locking, and what does it accomplish?

7. What is a transaction?

8. Describe two-phase locking.

9. What is deadlock? How does it occur?

10. How do some DBMSs use timestamping to handle concurrent update?

11. What is recovery?

12. What is journaling? What two types of images does a DBMS output to its journal?

13. When does a DBA use forward recovery? What are the forward recovery steps?

14. When does a DBA use backward recovery? What does the DBMS do to perform backward recovery?

15. What is security?

16. What is encryption? How does encryption relate to security?

17. What is authentication? Describe three types of authentication.

18. What are authorization rules?

19. What are permissions? Explain the relationship between permissions and workgroups.

20 How do views relate to security?

21. What is privacy? How is privacy related to security?

22. What are integrity constraints? Describe four different ways to handle integrity constraints. Which approach is the most desirable?

23. What is data independence?

24. What is replication? What is synchronization?

25. Describe three utility services that a DBMS should provide.

26. What is a procedural language? What is a non-procedural language?

■ Premiere Products Exercises

For the following exercises, you will answer problems and questions from management at Premiere Products. You do not use the Premiere Products database for any exercise.

1. While users were updating the Premiere Products database, one of the transactions was interrupted. You need to explain to management what steps the DBMS will take to correct the database. Using the sample log in Figure 7.11, list and describe the updates that the DBMS will rollback if transaction 2 is interrupted at 8:10.

2. Occasionally, users at Premiere Products obtain incorrect results when they run queries that include built-in (aggregate, summary, or statistical) functions. The DBA told management that unrepeatable reads caused the problems. Use books, articles, or the Internet to research the unrepeatable read problem. Write a short report that explains the unrepeatable-read problem to management, and use an example with your explanation. (*Note*: Unrepeatable reads are also called inconsistent retrievals, dirty reads, and inconsistent reads.)

3. You've explained replication to management, and some managers ask you for examples of when replication could be useful to them. Describe two situations, other than the ones given in the text, when replication would be useful to an organization.

4. The staff of the marketing department at Premiere Products is scheduled to receive some statistical databases, and they need you to explain these databases to them. (A statistical database is a database that is intended to supply only statistical information to users; a census database is an example of a statistical database.) Using a statistical database, users should not be able to infer information about any individual record in the database. Use books, articles, or the Internet to research statistical databases, and then write a report that explains them, discusses the problem with using them, and gives the solution to the problem.

5. The DBA at Premiere Products wants you to investigate biometric identification techniques for potential use at the company for computer authentication purposes. Use books, articles, or the Internet to research these techniques, and then write a report that describes the advantages and disadvantages of each of these techniques. In addition, recommend one technique and provide a justification for your recommendation.

Ray Henry plans to upgrade his database from Microsoft Access and wants you to help him select a different DBMS. To help him, he would like you to complete the following exercises. You do not use the Henry Books database for any exercise.

1. Many computer magazines and Web sites present comparisons of several DBMSs. Find one such DBMS comparison and compare the functions in this chapter to the listed features and functions in the comparison. Which functions from this chapter are included in the comparison, which functions are missing from the comparison, and what additional functions are included in the comparison?

2. How well does your school's DBMS fulfill the functions of a DBMS, as described in this chapter? Which functions are fully supported, which are partially supported, and which are not supported at all?

3. Use computer magazines or the Internet to investigate one of these DBMSs: DB2, SQL Server, MySQL, Oracle, or Sybase. Then prepare a report that explains how that DBMS handles the following DBMS functions: concurrent update, data recovery, and security. (*Note*: For concurrent update, you might need to review the concurrency control features of the DBMS.)

CHAPTER 8

Database Administration

OBJECTIVES

- Discuss the need for database administration (DBA).

- Explain DBA's responsibilities in formulating and enforcing database policies for access privileges, security, disaster planning, and archiving.

- Discuss DBA's administrative responsibilities for database management system (DBMS) evaluation and selection, DBMS maintenance, data dictionary management, and training.

- Discuss DBA's technical responsibilities for database design, testing, and performance tuning.

Introduction

As you've learned in previous chapters, the database approach has many benefits. At the same time, the use of a database management system (DBMS) involves potential hazards, especially when a database serves more than one user. For example, concurrent update and security present potential problems. Whom do you allow to access various parts of the database, and in what way? How do you prevent unauthorized accesses?

Note that just managing a database involves fundamental difficulties. So that they can use the database effectively, users must be made aware of the database structure or, at least, the portion of the database that they are allowed to access. Any changes made in the database structure must be communicated to all users, along with information about how the changes will affect them. Backup and recovery must be carefully coordinated, much more so than in a single-user environment, and this coordination presents another complication.

To manage these problems, companies create an organizational group, commonly referred to as database administration (DBA), that is responsible for supervising both the database and the use of the DBMS. DBA is usually a group rather than an individual, although sometimes the term DBA is used to refer to the database administrator, who is the individual in charge of this group. Usually the context makes clear which meaning is intended.

In this chapter, you will learn about the responsibilities of DBA, which are summarized in Figure 8.1. You'll be focusing on the role of DBA in a PC environment that is similar to the environment of Premiere Products. You will learn about DBA's role in formulating and enforcing important policies with respect to the database and its use. Then you will examine DBA's other administrative responsibilities for DBMS evaluation and selection, DBMS maintenance, data dictionary management, and training. Finally, you will learn about DBA's technical responsibilities for database design, testing, and performance tuning.

FIGURE 8.1
....
DBA responsibilities

```
Database Policy Formulation and Enforcement
    Access privileges
    Security
    Disaster planning
    Archiving
Other Database Administrative Functions
    DBMS evaluation and selection
    DBMS maintenance
    Data dictionary management
    Training
Database Technical Functions
    Database design
    Testing
    Performance tuning
```

Database Policy Formulation and Enforcement

DBA formulates database policies, communicates these policies to users, and enforces them. Among the policies are those covering access privileges, security, disaster planning, and archiving.

Access Privileges

Access to every table and field in a database is not a necessity for every user. Sam, for example, is an employee at Premiere Products; his main responsibility is inventory control. Although he needs access to the entire Part table, does he also need access to the Rep table? It is unlikely. Figure 8.2 illustrates the permitted and denied access privileges for Sam.

FIGURE 8.2
••••••••••••••••••
Permitted and denied access privileges for Sam

Rep

RepNum	LastName	FirstName	Street	City	State	Zip	Commission	Rate
20	Kaiser	Valerie	624 Randall	Grove	FL	33321	$20,542.50	0.05
35	Hull	Richard	532 Jackson	Sheldon	FL	33553	$39,216.00	0.07
65	Perez	Juan	1626 Taylor	Fillmore	FL	33336	$23,487.00	0.05

Part

PartNum	Description	OnHand	Class	Warehouse	Price
AT94	Iron	50	HW	3	$24.95
BV06	Home Gym	45	SG	2	$794.95
CD52	Microwave Oven	32	AP	1	$165.00
DL71	Cordless Drill	21	HW	3	$129.95
DR93	Gas Range	8	AP	2	$495.00
DW11	Washer	12	AP	3	$399.99
FD21	Stand Mixer	22	HW	3	$159.95
KL62	Dryer	12	AP	1	$349.95
KT03	Dishwasher	8	AP	3	$595.00
KV29	Treadmill	9	SG	2	$1,390.00

Paige, whose responsibility is customer mailings at Premiere Products, clearly requires access to customers' names and addresses, but what about their balances or credit limits? Should she be able to change an address? Should she be able to retrieve customers' balances or credit limits? Figure 8.3 illustrates the permitted and denied access privileges for Paige.

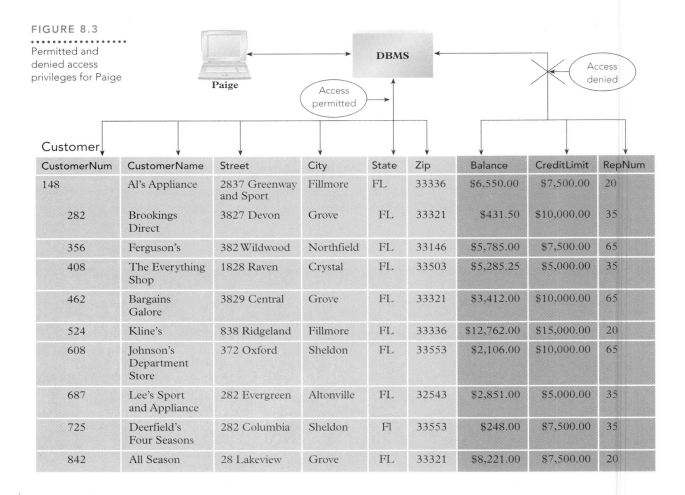

FIGURE 8.3
••••••••••••••••••
Permitted and
denied access
privileges for Paige

Customer

CustomerNum	CustomerName	Street	City	State	Zip	Balance	CreditLimit	RepNum
148	Al's Appliance and Sport	2837 Greenway	Fillmore	FL	33336	$6,550.00	$7,500.00	20
282	Brookings Direct	3827 Devon	Grove	FL	33321	$431.50	$10,000.00	35
356	Ferguson's	382 Wildwood	Northfield	FL	33146	$5,785.00	$7,500.00	65
408	The Everything Shop	1828 Raven	Crystal	FL	33503	$5,285.25	$5,000.00	35
462	Bargains Galore	3829 Central	Grove	FL	33321	$3,412.00	$10,000.00	65
524	Kline's	838 Ridgeland	Fillmore	FL	33336	$12,762.00	$15,000.00	20
608	Johnson's Department Store	372 Oxford	Sheldon	FL	33553	$2,106.00	$10,000.00	65
687	Lee's Sport and Appliance	282 Evergreen	Altonville	FL	32543	$2,851.00	$5,000.00	35
725	Deerfield's Four Seasons	282 Columbia	Sheldon	Fl	33553	$248.00	$7,500.00	35
842	All Season	28 Lakeview	Grove	FL	33321	$8,221.00	$7,500.00	20

Although rep 20 (Valerie Kaiser) should be able to obtain some of the information about her own customers, should she be able to obtain the same information about other customers? Figure 8.4 illustrates the permitted and denied access privileges for Valerie.

FIGURE 8.4
••••••••••••••••
Permitted and
denied access
privileges for Valerie

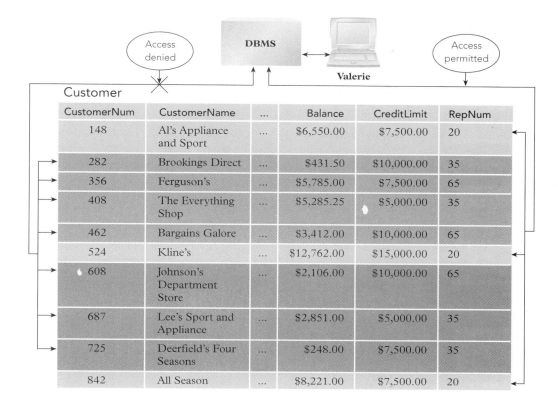

DBA determines the access privileges for all users and enters the appropriate authorization rules in the DBMS catalog to ensure that users access the database only in ways in which they are entitled to do so. DBA also documents access privilege policy, top-level management approves the policy, and DBA communicates the policy to management and to all users.

Security

As discussed in previous chapters, security is the prevention of unauthorized access, either intentional or accidental, to a database. Unauthorized access includes access by someone who has no right to access the database at all. For example, as shown in Figure 8.5, the DBMS prevents Brady, who is a programmer at Premiere Products, from accessing the database because DBA has not authorized Brady as a user.

FIGURE 8.5
••••••••••••••••
Attempted security
violation by Brady,
who's not an
authorized user

Unauthorized access also includes users who have legitimate access to some, but not all, data in a database, and who attempt to access data for which they are not authorized. For example, the DBMS prevents Paige from accessing customer balances, as shown in Figure 8.6, because DBA did not grant her access privileges to this data.

FIGURE 8.6
••••••••••••••••
Attempted security
violation by Paige,
who's authorized
to access some
customer data, but
is not authorized to
access customer
balances

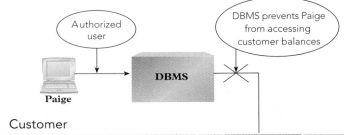

Customer

CustomerNum	CustomerName	...	Balance	CreditLimit	RepNum
148	Al's Appliance and Sport	...	$6,550.00	$7,500.00	20
282	Brookings Direct	...	$431.50	$10,000.00	35
356	Ferguson's	...	$5,785.00	$7,500.00	65
408	The Everything Shop	...	$5,285.25	$5,000.00	35
462	Bargains Galore	...	$3,412.00	$10,000.00	65
524	Kline's	...	$12,762.00	$15,000.00	20
608	Johnson's Department Store	...	$2,106.00	$10,000.00	65
687	Lee's Sport and Appliance	...	$2,851.00	$5,000.00	35
725	Deerfield's Four Seasons	...	$248.00	$7,500.00	35
842	All Season	...	$8,221.00	$7,500.00	20

DBA takes the steps necessary to ensure that the database is secure. After DBA determines the access privileges for each user, DBA creates security policies and procedures, obtains management approval of the policies and procedures, and then distributes them to authorized users.

To implement and enforce security, as shown in Figure 8.7, DBA uses the DBMS's security features, such as encryption, authentication, authorizations, and views.

If a DBMS lacks essential security features, DBA might create or purchase special security programs that provide the missing features, as shown in Figure 8.8.

In addition to relying on the security features provided by the DBMS and, if necessary, the special security programs, DBA monitors database usage to detect potential security violations. If a security violation occurs, DBA determines who breached security, how the violation occurred, and how to prevent a similar violation in the future.

Disaster Planning

The type of security discussed in the previous section concerns damage to the data in a database caused by authorized and unauthorized users. Damage to a database can also occur through a physical incident such as an abnormally terminated program, a software virus or worm, a disk problem, a power outage, a computer malfunction, a hurricane, a flood, a tornado, or other natural disaster.

To protect an organization's data against physical damage, DBA creates and implements backup and recovery procedures as part of a disaster recovery plan. A **disaster recovery plan** specifies the ongoing and emergency actions and procedures required to ensure data availability, even if a disaster occurs.

For some functions, such as credit card processing and airline reservations, data availability must be continuous. In these situations, companies can switch quickly to duplicate backup systems in the event of a malfunction in or a complete destruction of the main system. Other companies contract with firms using hardware and software similar to their own so that in the event of a catastrophe, they can temporarily use these other facilities. As part of a disaster recovery plan, DBA uses the DBMS's backup and recovery features wherever possible and supplements them where they are lacking.

Archiving

Often, users need to retain certain data in a database for only a limited time. An order that has been filled, reported on a customer's statement, and paid by the customer is in one sense no longer important. Should you keep the order in the database? If you always keep data in the database as a matter of policy, the database will continually grow. The disk space that is occupied by the database expands, and programs that access the database might take more time to perform their functions. The increased disk space and longer processing times might be good reasons to remove completed orders and all their associated order lines from the database.

On the other hand, you might need to retain orders and their associated order lines for future reference by users. Reasons for retaining the data could include customer inquiries, government regulations, and auditing requirements. If you need to retain the data, you now have a conflict on your hands: You should remove the old data from the database, yet you need to have access to it.

One solution to the conflict is to use what is known as a **data archive**, or **archive**. In ordinary usage, an archive (technically *archives*) is a place where public records and documents are kept. A data archive is similar. It is a place where a record of certain corporate data is kept. In the case of the previously mentioned completed orders and associated order lines, Figure 8.9 shows how you would remove them from the database and place them in the archive, thus storing them for future reference.

FIGURE 8.9

Movement of order
21617 from the
database to the
archive

DATABASE
Orders

OrderNum	OrderDate	CustomerNum
21608	10/20/2007	148
21610	10/20/2007	356
21613	10/21/2007	408
21614	10/21/2007	282
21617	10/23/2007	608
21619	10/23/2007	148
21623	10/23/2007	608

ARCHIVE

Orders

OrderNum	OrderDate	CustomerNum
21617	10/23/2007	608

OrderLine

OrderNum	PartNum	NumOrdered	QuotedPrice
21608	AT94	11	$21.95
21610	DR93	1	$495.00
21610	DW11	1	$399.99
21613	KL62	4	$329.95
21614	KT03	2	$595.00
21617	BV06	2	$794.95
21617	CD52	4	$150.00
21619	DR93	1	$495.00
21623	KV29	2	$1,290.00

OrderLine

OrderNum	PartNum	NumOrdered	QuotedPrice
21617	BV06	2	$794.95
21617	CD52	4	$150.00

Typically, DBA stores the archive on some mass-storage device—for example, a disk, a tape, a CD, or a DVD. Whichever medium DBA uses, DBA must store copies of both archives and database backups off site, so that recovery can take place even if a company's buildings and contents are destroyed. Once again, it is up to DBA to establish and implement procedures for the use, maintenance, and storage of the archive.

■ Other Database Administrative Functions

DBA is also responsible for DBMS evaluation and selection, DBMS maintenance, data dictionary management, and training.

DBMS Evaluation and Selection

When a company decides to purchase a new DBMS, DBA leads the DBMS evaluation and selection effort. To evaluate the DBMS candidates objectively, DBA usually prepares a checklist similar to the one shown in Figure 8.10. (This checklist applies specifically to a relational system because most DBMSs are, at least in part, relational. If DBA had not already selected a data model, such as the relational model, DBA would add a "Choice of Data Model" category to the list.) DBA evaluates each prospective purchase of a DBMS against the categories shown in the figure. An explanation of each category follows Figure 8.10.

FIGURE 8.10

DBMS evaluation
checklist

```
1.  Data Definition
    a.  Data types
        (1)  Numeric
        (2)  Character
        (3)  Date
        (4)  Logical (T/F)
        (5)  Memo
        (6)  Currency
        (7)  Binary object (pictures, drawings, sounds, and so on)
        (8)  Link to an Internet, Web, or other address
        (9)  User-defined data types
       (10)  Other
    b.  Support for nulls
    c.  Support for primary keys
    d.  Support for foreign keys
    e.  Unique indexes
    f.  Views
2.  Data Restructuring
    a.  Possible restructuring
        (1)  Add new tables
        (2)  Delete existing tables
        (3)  Add new columns
        (4)  Change the layout of existing columns
        (5)  Delete columns
        (6)  Add new indexes
        (7)  Delete existing indexes
    b.  Ease of restructuring
3.  Nonprocedural Languages
    a.  Nonprocedural languages supported
        (1)  SQL
        (2)  QBE
        (3)  Natural language
        (4)  Language unique to the DBMS. Award points on the
             basis of ease of use as well as the types of operations
             (joining, sorting, grouping, calculating various statistics,
             and so on) that are available in the language. You can
             use SQL as a standard against which you can judge the
             language.
    b.  Optimization done by one of the following:
        (1)  User (in formulating the query)
        (2)  DBMS (through built-in optimizer)
        (3)  No optimization possible; system does only sequential
             searches.
4.  Procedural Languages
    a.  Procedural languages supported
        (1)  Language unique to the DBMS. Award points on the basis
             of the quality of this language both in terms of the types
             of statements and control structures available and the database
             manipulation statements included in the language.
        (2)  Java
        (3)  C or C++
        (4)  GUI language such as Visual Basic
        (5)  COBOL
        (6)  Other
    b.  Can a nonprocedural language be used in conjunction with
        the procedural language (for example, could SQL be
        embedded in a COBOL program)?
5.  Data Dictionary
    a.  Type of entries
        (1)  Tables
        (2)  Columns
        (3)  Indexes
        (4)  Relationships
        (5)  Users
        (6)  Programs
        (7)  Other
    b.  Integration of data dictionary with other components of
        the system
```

(continued)

FIGURE 8.10
••••••••••••••••••
DBMS evaluation
checklist
(continued)

 6. Concurrent Update
    a. Level of locking
       (1) Field value
       (2) Row
       (3) Page
       (4) Table
       (5) Database
    b. Type of locking
       (1) Shared
       (2) Exclusive
       (3) Both
    c. Responsibility for handling deadlock
       (1) Programs
       (2) DBMS (automatic rollback of transaction causing
           deadlock)
 7. Backup and Recovery
    a. Backup services
    b. Journaling services
    c. Recovery services
       (1) Recover from backup copy only
       (2) Recover using backup copy and journal
    d. Rollback of individual transactions
    e. Incremental backup
 8. Security
    a. Encryption
    b. Passwords
    c. Authorization rules
       (1) Access to database only
       (2) Access/update access to any column or combination of
           columns
    d. Views
    e. Difficulty in bypassing security controls
 9. Integrity
    a. Support for entity integrity
    b. Support for referential integrity
    c. Support for data integrity
    d. Support for other types of integrity constraints
10. Replication and Distributed Databases
    a. Partial replicas
    b. Handling of duplicate updates in replicas
    c. Data distribution
    d. Procedure support
       (1) Language used
       (2) Procedures stored in database
       (3) Support for remote stored procedures
       (4) Trigger support
11. Limitations
    a. Number of tables
    b. Number of columns
    c. Length of individual columns
    d. Total length of all columns in a table
    e. Number of rows per table
    f. Number of files that can be open at the same time
    g. Sizes of database, tables, and other objects
    h. Types of hardware supported
    i. Types of LANs supported
    j. Other
12. Documentation and Training
    a. Clearly written manuals
    b. Tutorial
       (1) Online
       (2) Printed
    c. Online help available
       (1) General help
       (2) Context-sensitive help
    d. Training
       (1) Vendor or other company
       (2) Location
       (3) Types (DBA, programmers, users, others)
       (4) Cost
13. Vendor Support
    a. Type of support available
    b. Quality of support available
    c. Cost of support
    d. Reputation of support
14. Performance
    a. External benchmarking done by various organizations
    b. Internal benchmarking
    c. Includes a performance monitor

(continued)

FIGURE 8.10
....
DBMS evaluation
checklist
(continued)

```
15. Portability
    a.  Operating systems
        (1) Unix
        (2) Microsoft Windows
        (3) Linux
        (4) Other
    b.  Import/export/linking file support
        (1) Other databases
        (2) Other applications (for example, spreadsheets and
            graphics)
    c.  Internet and intranet support
16. Cost
    a.  Cost of DBMS
    b.  Cost of any additional components
    c.  Cost of any additional hardware that is required
    d.  Cost of network version (if required)
    e.  Cost and types of support
17. Future Plans
    a.  What does the vendor plan for the future of the system?
    b.  What is the history of the vendor in terms of keeping the
        system up to date?
    c.  When changes are made in the system, what is involved in
        converting to the new version?
        (1) How easy is the conversion?
        (2) What will it cost?
18. Other Considerations (Fill in your own special requirements.)
    a.  ?
    b.  ?
    c.  ?
    d.  ?
```

1. **Data definition.** What types of data does the DBMS support? Does it support nulls? What about primary and foreign keys? The DBMS undoubtedly provides support for indexes, but can you specify that an index is unique and then have the system enforce the uniqueness? Does the DBMS support views?

2. **Data restructuring.** What type of database restructuring does the DBMS allow? How easily can DBA perform the restructuring? Will the system do most of the work or will DBA have to create special programs for this purpose?

3. **Nonprocedural languages.** What types of nonprocedural language does the DBMS support? The possibilities are structured query language (SQL), Query-By-Example (QBE), natural language, and a DBMS built-in language. If the DBMS supports one of the standard languages, what's the quality of its version? If the DBMS provides its own language, how good is it? How does its functionality compare to that of SQL? How does the DBMS achieve optimization of queries? The DBMS itself optimizes each query, or the user must do so by the manner in which he or she states the query. If neither happens, no optimization occurs. Most desirable, of course, is the first alternative.

4. **Procedural languages.** What types of procedural languages does the DBMS support? Are they common languages, such as Java, C or C++, and COBOL? Is it a graphical user interface (GUI) language? Does the DBMS provide its own language? In the latter case, how complete is the language? Does it contain all the required types of statements and control structures? What facilities does the language provide for accessing the database? Does the DBMS let you use a nonprocedural language while using the procedural language?

5. **Data dictionary.** What kind of data dictionary does the DBMS provide? Is it a simple catalog, or can it contain more content, such as information about programs

and the various data items these programs access? How well is the data dictionary integrated with other components of the system—for example, the nonprocedural language?

6. **Concurrent update.** Does the DBMS support concurrent update? What is the unit that may be locked (field value, row, page, table, or database)? Are exclusive locks the only ones permitted or are shared locks also allowed? (A **shared lock** permits other users to read the data; with an **exclusive lock**, no other user may access the data in any way.) Does the DBMS resolve deadlock, or must programs resolve it?

7. **Backup and recovery.** What type of backup and recovery services does the DBMS provide? Does the DBMS maintain a journal of changes in the database and use the journal during the recovery process? If a transaction terminates abnormally, does the DBMS rollback its updates? Can the DBMS perform an incremental backup of just the data that has changed?

8. **Security.** What types of security features does the DBMS provide? Does the DBMS support encryption, password support, and authorizations rules? Does the DBMS provide a view mechanism that can be used for security? How difficult is it to bypass the security controls?

9. **Integrity.** What type of integrity constraints does the DBMS support? Does the DBMS support entity integrity (the fact that the primary key cannot be null) and referential integrity (the property that values in foreign keys must match values already in the database)? What types of data integrity does the DBMS support? Does the DBMS support any other types of integrity constraints?

10. **Replication and distributed databases.** Does the DBMS support replication? If so, does the DBMS allow partial replicas (copies of selected rows and fields from tables in a database), and how does the DBMS handle updates to the same data from two or more replicas? Can the DBMS distribute a database, that is, divide the database into segments and store the segments on different computers? If so, what types of distribution does the DBMS allow, and what types of procedure support for distribution does the DBMS provide?

11. **Limitations.** What limitations exist with respect to the number of tables and the number of fields and rows per table? How many files can you open at the same time? (For some databases, each table and each index is in a separate file. Thus, a single table with three indexes, all in use at the same time, would account for *four* files. Problems might arise if the number of files that you can open is relatively small and many indexes are in use.) On what types of operating system and hardware is the DBMS supported? What types of local area networks (LANs) can you use with the DBMS? (A **local area network (LAN)** is a configuration of several computers connected together that allow users to share a variety of hardware and software resources. One of these resources is the database. In a LAN, support for concurrent update is very important because many users might be updating the database at the same time. The relevant question here, however, is not how well the DBMS supports concurrent update, but which of the LANs you can use with the DBMS.)

12. **Documentation and training.** Does the vendor of the DBMS supply printed or online training manuals? How good are the manuals? Are they easy to use? Is there a good index? Is a tutorial, in either printed or online form, available to assist users in getting started with the system? Is online help available? If so, does the DBMS provide general help and context-sensitive help? (**Context-sensitive help** means that if a user is having trouble and asks for help, the DBMS will provide assistance for the particular feature being used at the time the user asks for the help.) Does the vendor provide training classes? Do other companies offer training? Are the classes on site or off site? Are there classes for DBA and separate classes for programmers and others? What is the cost for each type of training?

13. **Vendor support.** What type of support does the vendor provide for the DBMS, and how good is it? What is the cost? What is the vendor's reputation for support among current users?

14. **Performance.** How well does the DBMS perform, where performance is a measure of how rapidly the DBMS completes its tasks? This is a tough question to answer because each organization has a different number of users and a different mix of transactions, and both factors affect how a DBMS performs. One way to determine relative performance among DBMSs is to look into benchmark tests that have been performed on several DBMSs by various organizations. Benchmarking typically is done in areas such as sorting, indexing, and reading all rows and then changing data values in all rows. For example, the Transaction Processing Performance Council (*www.tpc.org*) provides the results of database benchmark tests to its membership. Beyond using benchmarks, if an organization has some specialized needs, it may have to set up its own benchmark tests. Does the DBMS provide a performance monitor that measures different types of performance while the DBMS is operational?

15. **Portability.** Which operating systems can you use with the DBMS? What types of files can you import or export? Can the DBMS link to other data sources, such as files and other types of DBMSs? Does the DBMS provide Internet and intranet support? (An **intranet** is an internal company network that uses software tools typically used on the Internet and the World Wide Web.)

16. **Cost.** What is the cost of the DBMS and of any additional components the organization is planning to purchase? Is additional hardware required and, if so, what is the associated cost? If the organization requires a special version of the DBMS for a network, what is the additional cost? What is the cost of vendor support, and what types of support plans are available?

17. **Future plans.** What plans has the vendor made for the future of the system? This information is often difficult to obtain, but you can get an idea by looking at the performance of the vendor with respect to how it has kept the existing system up to date. How easy has it been for users to convert to new versions of the system?

18. **Other considerations.** This is a final, catch-all category that contains any special requirements not covered in the other categories.

After DBA examines each DBMS with respect to all the preceding categories, DBA and management can compare the results. Unfortunately, this process can be difficult because of the number of categories and their generally subjective nature. To make the process more objective, DBA can assign a numerical ranking to each DBMS in each category (for example, a number between 0 and 10, where 0 is poor and 10 is excellent). Furthermore, DBA can assign weights to the categories. Weighting allows an organization to signify which categories are more critical to it than others. Then you can multiply each number used in the numerical ranking by the appropriate weight. You then add the results, producing a weighted total. You then compare the weighted totals for each DBMS, producing the final evaluation.

How does DBA arrive at the numbers to assign each DBMS in the various categories? Several methods are used. DBA can request feedback from other organizations that are currently using the DBMS being considered. DBA can read journal reviews of the various DBMSs. Sometimes DBA can obtain a trial version of the DBMS, and members of the staff can give it a hands-on test. In practice, DBA usually combines all three methods. Whichever method is used, however, DBA must carefully create the checklist and determine weights before starting the evaluation; otherwise, the findings may be inadvertently slanted in a particular direction.

DBMS Maintenance

After the organization selects and purchases the DBMS, DBA has primary responsibility for it. DBA installs the DBMS in a way that is suitable for the organization. If the DBMS configuration needs to be changed, it is DBA that makes the changes.

When the vendor releases a new version of the DBMS, DBA reviews it and determines whether the organization should upgrade to it. If the decision is made to convert to the new version or perhaps to a new DBMS, DBA coordinates the conversion. DBA also handles any fixes to problems in the DBMS that the vendor releases.

When a problem occurs that affects the database, DBA coordinates the people required to resolve the problem. Some people, such as programmers and users, are from inside the organization, and others, such as hardware and software vendors, are from outside the organization.

When users have special one-time processing needs or extensive query requirements against the database, DBA coordinates the users so that their needs are satisfied without unduly affecting other users.

Data Dictionary Management

DBA also manages the data dictionary. The data dictionary is essentially the catalog mentioned in Chapter 7, but it often contains a wider range of information, including information about tables, fields, indexes, programs, and users.

DBA establishes naming conventions for tables, fields, indexes, and so on. DBA creates the data definitions for all tables, as well as for any data integrity rules and user views. DBA also updates the contents of the data dictionary. Finally, DBA creates and distributes appropriate reports from the data dictionary to users, programmers, and other people in the organization.

Training

DBA provides training in the use of the DBMS and in how to access the database. DBA also coordinates the training of users. In those cases where the vendor of the DBMS provides training, DBA handles the scheduling to make sure that users receive the training they require.

■ Technical Functions

DBA is also responsible for database design, testing, and performance tuning.

Database Design

DBA establishes a sound methodology for database design, such as the one discussed in Chapter 6, and ensures that all database designers follow the methodology. DBA also verifies that the designers obtain all pertinent information from the appropriate users. After the database designers complete the information-level design, DBA does the physical-level design.

DBA establishes documentation standards for all the steps in the database design process. DBA also makes sure that these standards are followed, that the documentation is kept up to date, and that the appropriate personnel have access to the documentation they need.

Requirements don't remain stable over time; they change constantly. DBA reviews all changes to requirements and determines whether the changes will require modifications to the database. If so, DBA makes the changes in the design and in the data in the database. DBA also verifies that programmers modify all programs and documentation affected by the change.

Testing

The hardware, software, and database for the users is called the **production system**, or **live system**. DBA strictly controls the production system. With just two exceptions, DBA grants access and update privileges to the production system only to authorized users. The first exception is when problems occur, such as with software. DBA and others must troubleshoot the problem by accessing the production system. The second exception is when programmers complete new programs or modify existing programs for the production system. For both exceptions, DBA performs any necessary database modifications or closely controls the activities of others.

Other than for these two exceptions, DBA does not grant programmers access to the production system. Instead, DBA and the programmers create a separate system, called the **test system**, that programmers use to develop new programs and modify existing programs. After programmers complete the testing of their programs in the test system, DBA and the users review and approve the test results, and DBA reviews and approves the programs and documentation. If DBA and the users approve, DBA notifies all affected users when new or corrected features will be available. DBA then transfers the programs to the production system and makes any required database changes, as shown in Figure 8.11.

FIGURE 8.11
••••••••••••••••
DBA controls the
interaction between
the test and
production systems

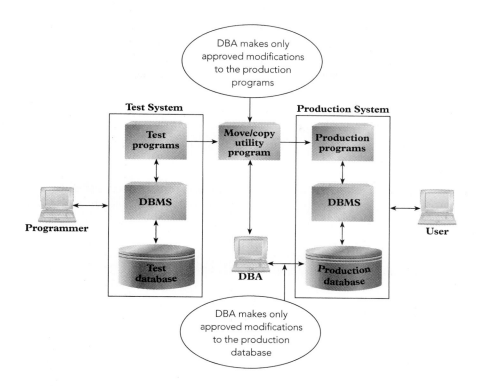

A production system with a DBMS is a complex system. A separate test system reduces the complexity of the production system and provides an extra measure of control.

Performance Tuning

Database performance deals with the ability of the production system to serve users in a timely and responsive manner. Because funding is usually a constraint, DBA's challenge is to get the best possible performance from the available funds.

Faster computers with faster disks, faster network connections, faster software, and other production system expenditures help improve performance. What can DBA do if the organization has no additional money for its production system but needs further performance improvements? DBA can change the database design to improve performance; this process is called **tuning** the design. Some of the performance tuning changes DBA can make to a database design include creating and deleting indexes, splitting tables, and changing the table design.

By default, Access and some other DBMSs automatically create indexes for primary key and foreign key fields. These indexes make accessing the fields faster than they would be without the indexes. Further, indexing common fields improves the speed of joining related tables. If a DBMS doesn't automatically index primary key and foreign key fields, DBA should create indexes for them. In addition, queries that search indexed fields run faster than do comparable queries without indexes for those fields. For example, if users frequently query the Part table to find records based on values for the Class or Warehouse fields, DBA can improve performance by adding indexes on these fields. On the other hand, a table with many indexes takes longer to update. If users experience delays when they update a table, DBA can delete some of the table's indexes to improve update performance.

If users access only certain fields in a table, you can improve performance by splitting the table into two or more tables that each have the same primary key as the original and that collectively contain all the fields from the original table. Each resulting table is smaller than the original; the lower amount of data moves faster between disk and memory and across a network. For example, suppose dozens of users at Premiere Products access the Customer table shown in Figure 8.12.

Customer

CustomerNum	CustomerName	Street	City	State	Zip	Balance	CreditLimit	RepNum
148	Al's Appliance and Sport	2837 Greenway	Fillmore	FL	33336	$6,550.00	$7,500.00	20
282	Brookings Direct	3827 Devon	Grove	FL	33321	$431.50	$10,000.00	35
356	Ferguson's	382 Wildwood	Northfield	FL	33146	$5,785.00	$7,500.00	65
408	The Everything Shop	1828 Raven	Crystal	FL	33503	$5,285.25	$5,000.00	35
462	Bargains Galore	3829 Central	Grove	FL	33321	$3,412.00	$10,000.00	65
524	Kline's	838 Ridgeland	Fillmore	FL	33336	$12,762.00	$15,000.00	20
608	Johnson's Department Store	372 Oxford	Sheldon	FL	33553	$2,106.00	$10,000.00	65
687	Lee's Sport and Appliance	282 Evergreen	Altonville	FL	32543	$2,851.00	$5,000.00	35
725	Deerfield's Four Seasons	282 Columbia	Sheldon	FL	33553	$248.00	$7,500.00	35
842	All Season	28 Lakeview	Grove	FL	33321	$8,221.00	$7,500.00	20

If some users access address data from the Customer table and other users access balances and credit limits, DBA can split the Customer table into two tables, as shown in Figure 8.13, to improve performance. Users needing data from both tables can obtain that data by joining the two split tables on the CustomerNum field.

FIGURE 8.13
··················
Result of splitting
the Customer table
into two tables

CustomerAddress

CustomerNum	CustomerName	Street	City	State	Zip
148	Al's Appliance and Sport	2837 Greenway	Fillmore	FL	33336
282	Brookings Direct	3827 Devon	Grove	FL	33321
356	Ferguson's	382 Wildwood	Northfield	FL	33146
408	The Everything Shop	1828 Raven	Crystal	FL	33503
462	Bargains Galore	3829 Central	Grove	FL	33321
524	Kline's	838 Ridgeland	Fillmore	FL	33336
608	Johnson's Department Store	372 Oxford	Sheldon	FL	33553
687	Lee's Sport and Appliance	282 Evergreen	Altonville	FL	32543
725	Deerfield's Four Seasons	282 Columbia	Sheldon	FL	33553
842	All Season	28 Lakeview	Grove	FL	33321

CustomerFinancial

CustomerNum	CustomerName	Balance	CreditLimit	RepNum
148	Al's Appliance and Sport	$6,550.00	$7,500.00	20
282	Brookings Direct	$431.50	$10,000.00	35
356	Ferguson's	$5,785.00	$7,500.00	65
408	The Everything Shop	$5,285.25	$5,000.00	35
462	Bargains Galore	$3,412.00	$10,000.00	65
524	Kline's	$12,762.00	$15,000.00	20
608	Johnson's Department Store	$2,106.00	$10,000.00	65
687	Lee's Sport and Appliance	$2,851.00	$5,000.00	35
725	Deerfield's Four Seasons	$248.00	$7,500.00	35
842	All Season	$8,221.00	$7,500.00	20

DBA can also split tables for security purposes. In Figure 8.13, the CustomerFinancial table contains customer financial data, and the CustomerAddress table contains customer address data. Those users granted access only to the CustomerAddress table have no access to customer financial data, thus providing an added measure of security.

Although you design database tables in third normal form to prevent the anomaly problems discussed in Chapter 5, DBA occasionally denormalizes tables to improve performance. **Denormalizing** converts a table that is in third normal form to a table that is no longer in third normal form. Usually, the conversion produces tables that are in first normal form or second normal form. Denormalizing introduces anomaly problems but can decrease the number of disk accesses required by certain types of transactions, thus improving performance. For example, suppose users who are processing order lines need part descriptions. DBA might include part descriptions in the OrderLine table, as shown in Figure 8.14.

FIGURE 8.14
••••••••••••••••••
Including part
descriptions in the
OrderLine table,
which creates a first
normal form table

OrderLine

OrderNum	PartNum	Description	NumOrdered	QuotedPrice
21608	AT94	Iron	11	$21.95
21610	DR93	Gas Range	1	$495.00
21610	DW11	Washer	1	$399.99
21613	KL62	Dryer	4	$329.95
21614	KT03	Dishwasher	2	$595.00
21617	BV06	Home Gym	2	$794.95
21617	CD52	Microwave Oven	4	$150.00
21619	DR93	Gas Range	1	$495.00
21623	KV29	Treadmill	2	$1,290.00

The OrderLine table in Figure 8.14 is in first normal form because there are no repeating groups. Because a part description depends only on the part number, which itself is just a portion of the primary key for the table, it is not in second normal form and, consequently, it is not in third normal form either. As a result, the table has redundancy and anomaly problems that are inherent in tables that are not in third normal form. However, users processing order lines no longer need to join the OrderLine and Part tables to obtain part descriptions, thus improving performance.

Large databases with thousands of users often suffer periodic performance problems as users change their transaction mix. In these cases, DBA must tune the databases to provide improved performance to all users.

■ SUMMARY

- Database administration (DBA) is the person or group that is responsible for supervising the database and the use of the database management system (DBMS).

- DBA formulates and enforces policies about those users who can access the database, the portions of the database they may access, and in what manner they can access the database.

- DBA formulates and enforces policies about security (which is the prevention of unauthorized access, either intentional or accidental, to a database). DBA uses the DBMS's security features and special security programs, if necessary, and monitors database usage to detect potential security violations.

- DBA creates and implements backup and recovery procedures as part of a disaster recovery plan to protect an organization's data against physical damage.

- DBA formulates and enforces policies that govern the management of an archive for data that is no longer needed in the database but that must be retained for reference purposes.

- DBA leads the effort to evaluate and select a new DBMS. DBA develops a checklist of desirable features for a DBMS and evaluates each prospective purchase of a DBMS against this checklist.

- DBA installs and maintains the DBMS after it has been selected and procured.

- DBA maintains the data dictionary, establishes naming conventions for its contents, and provides information from it to others in the organization.

- DBA provides database and DBMS training and coordinates and schedules training by outside vendors.

- DBA verifies all information-level database designs, completes all physical-level database designs, and creates documentation standards. DBA also evaluates changes in requirements to determine whether it needs to change the database design and the data in the database.

- DBA controls the production system, which is accessible only to authorized users. Other than when authorized by DBA to access the production system in exceptional situations, programmers access a separate test system. DBA migrates tested programs to the production system and makes any required database changes.

- DBA tunes the database design to improve performance. Included among the performance tuning changes DBA makes are creating and deleting indexes, splitting tables, and denormalizing tables.

■ KEY TERMS

archive
context-sensitive help
data archive
denormalizing
disaster recovery plan
exclusive lock
intranet

live system
local area network (LAN)
production system
shared lock
test system
tuning

■ REVIEW QUESTIONS

1. What is DBA? Why is it necessary?

2. What are DBA's responsibilities regarding access privileges?

3. What are DBA's responsibilities regarding security?

4. What is a disaster recovery plan?

5. What are data archives? What purpose do they serve? What is the relationship between a database and its data archives?

6. Name five categories that you usually find on a DBMS evaluation and selection checklist.

7. What is a shared lock? What is an exclusive lock?

8. What is a LAN?

9. What is context-sensitive help?

10. What is an intranet?

11. After a DBMS has been selected, what is DBA's role in DBMS maintenance?

12. What are DBA's responsibilities with regard to the data dictionary?

13. Who trains computer users in an organization? What is DBA's role in this training?

14. What are DBA's database design responsibilities?

15. What is the difference between production and test systems?

16. What is meant by "tuning a design?"

17. How can splitting a table improve performance?

18. What is denormalization?

■ Premiere Products Exercises

For the following exercises, you will answer problems and questions from management at Premiere Products. You do not use the Premiere Products database for any exercise.

1. DBA asks for your help in planning its data archive for the following Premiere Products database:

    ```
    Rep (RepNum, LastName, FirstName, Street, City, State,
        Zip, Commission, Rate)
    Customer (CustomerNum, CustomerName, Street, City, State,
        Zip, Balance, CreditLimit, RepNum)
    Orders (OrderNum, OrderDate, CustomerNum)
    OrderLine (OrderNum, PartNum, NumOrdered, QuotedPrice)
    Part (PartNum, Description, OnHand, Class, Warehouse, Price)
    ```

    Determine which data from the database to archive; that is, for each table, specify whether data needs to be archived. If it does, specify which data, when it should be archived, and whether it should be archived with data from another table.

2. DBA denormalized some of the data in the Premiere Products database to improve performance, and one of the resulting tables is the following:

    ```
    Customer (CustomerNum, CustomerName, Street, City, State,
        Zip, Balance, CreditLimit, RepNum, RepName)
    ```

    Which field or fields cause the table to no longer be in third normal form? In which normal form is the denormalized table?

3. Does your school have a formal disaster recovery plan? If it does, describe the general steps in the plan. If it does not, describe the informal steps that would be taken if a disaster occurs.

■ Henry Books Case

Ray Henry plans to upgrade his database from Microsoft Access and wants you to help him select a different DBMS. To help him, he would like you to complete the following exercises. You do not use the Henry Books database for any exercise.

1. DBA asks for your help in planning its data archive for the following Henry Books database:

   ```
   Branch (BranchNum, BranchName, BranchLocation, NumEmployees)
   Publisher (PublisherCode, PublisherName, City)
   Author (AuthorNum, AuthorLast, AuthorFirst)
   Book (BookCode, Title, PublisherCode, Type, Price, Paperback)
   Wrote (BookCode, AuthorNum, Sequence)
   Inventory (BookCode, BranchNum, OnHand)
   ```

   Determine which data from the database to archive; that is, for each table, specify whether data needs to be archived. If it does, specify which data, when it should be archived, and whether it should be archived with data from another table.

2. DBA denormalized some of the data in the Henry Books database to improve performance, and one of the resulting tables is the following:

   ```
   Wrote (BookCode, AuthorNum, Sequence, PublisherCode,
       PublisherName)
   ```

   Which field or fields cause the table to no longer be in third normal form? In which normal form is the denormalized table?

3. Interview DBA at your school or at a local business to determine the safeguards used to segregate the production system from the test system.

CHAPTER 9

Database Management Approaches

OBJECTIVES

- Describe distributed database management systems (DBMSs).

- Discuss client/server systems.

- Define data warehouses and explain their structure and access.

- Discuss the general concepts of object-oriented DBMSs.

- Summarize the impact of Web access to databases.

- Provide a brief history of database management.

- Describe the network and hierarchical data models.

Introduction

In previous chapters, you learned about **relational DBMSs (RDBMSs)**, which dominate the database market today. In this chapter, you will examine several database management topics, most of which are applicable to relational systems.

The centralized approach to processing data, in which users access a central computer through personal computers and workstations, dominated organizations from the late 1960s through the mid-1980s because there was no alternative approach to compete with it. The introduction of reasonably priced personal computers during the 1980s, however, facilitated the placement of computers at various locations within an organization; users could access a database directly at those various locations. Networks connected these computers, so users could access data not only from their local computers, but also from anywhere along the entire network. Thus, distributed processing began. In the next section, you will study the issues involved in distributed databases—the database component of distributed processing.

Organizations often off-load, or shift, data communications functions from central computers to smaller computers to improve processing speed. Similarly, organizations often use client/server systems to off-load database access functions from central computers to other computers; you'll study these client/server systems. Next, you'll learn about special database systems, called data warehouses, that allow you to retrieve data rapidly. Then you will study object-oriented systems, which treat data as objects and include the actions that operate on the objects. You also will study the impact of the Web on database access. The last two sections present a brief history of DBMSs and the network and hierarchical data models.

Distributed Databases

Premiere Products has several locations nationwide. Each location has its own sales reps and customer base, and each location maintains its own inventory. Instead of using a single, centralized mainframe computer accessed by all the separate locations, Premiere Products is considering installing a computer at each site. If it does so, each site would maintain its own data about its sales reps, customers, parts, and orders. Occasionally, an order at one site might involve parts from another site. In addition, a customer serviced at one site might place orders for its subsidiaries that are located closer to other sites. Consequently, the computer at a particular site would need to communicate with the computers at all the other sites. The computers would have to be connected in a **communications network**, or **network**, as illustrated in Figure 9.1.

FIGURE 9.1
••••••••••••••••••
Communications
network

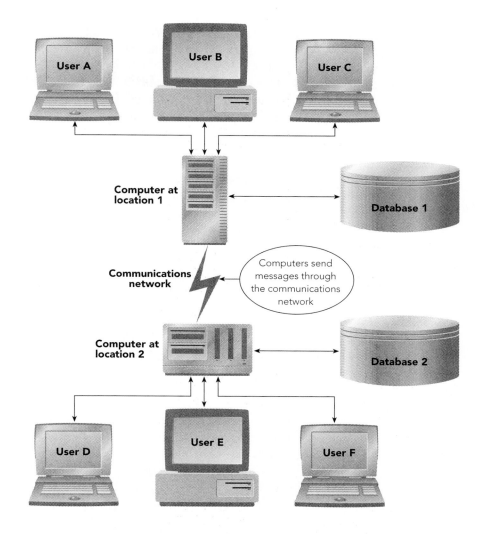

Premiere Products would also divide its existing database and distribute to each site the data needed at that site. In doing so, Premiere Products would be creating a distributed database. A **distributed database** is a single logical database that is physically divided among computers at several sites on a network. To make such a distributed database work properly, Premiere Products needs to purchase a **distributed database management system (DDBMS)**, which is a DBMS capable of supporting and manipulating distributed databases.

Computers in a network communicate through **messages**; that is, one computer sends a message to another. The word "message" is used in a fairly broad way here. A computer might send a message to request data from another computer, or a computer might send a message to indicate a problem. For example, one computer might send a message to another computer to indicate that the requested data is not available. Finally, a computer might send the requested data as a message to another computer.

Accessing data using messages over a network is substantially slower than accessing data on a disk. To access data rapidly in a centralized database, you make design decisions to minimize the number of disk accesses. In general, to access data rapidly in a distributed database, you must attempt to minimize the number of messages. The length of time required to send one message depends on the length of the message and the characteristics of the network. There is a fixed amount of time, sometimes called the **access delay**, required for every message. In addition, the time to send a message includes the time it takes to transmit all the characters in the message. The formula for message transmission time is as follows:

```
Communication time = access delay + (data volume / transmission rate)
```

To illustrate the importance of minimizing the number of messages, suppose you have a network with an access delay of 2 seconds and a transmission rate of 30,000 bits per second. Also suppose that you send a message that consists of 1,000 records, each of which is 800 bits long. (The 1,000 records is equivalent to approximately 25 pages of single-spaced text.) In this example, you calculate the communication time as follows:

```
Communication time = 2 + ((1,000 * 800) / 30,000)
                   = 2 + (800,000 / 30,000)
                   = 2 + 26.67
                   = 28.67 seconds
```

If you send a message that is 100 bits long, your communication time calculation is:

```
Communication time = 2 + (100 / 30,000)
                   = 2 + .003
                   = 2.003 seconds or, for practical purposes,
                   = 2 seconds
```

As you can see, in short messages, the access delay becomes the dominant factor. Thus, in general, it is preferable to send a small number of lengthy messages rather than a large number of short messages.

Characteristics of Distributed DBMSs

Because a DDBMS effectively contains a local DBMS at each site, an important property of DDBMSs is that they are either homogeneous or heterogeneous. A **homogeneous DDBMS** is one that has the same local DBMS at each site. A **heterogeneous DDBMS** is one that does not; there are at least two sites at which the local DBMSs are different. Heterogeneous DDBMSs are more complex than homogeneous DDBMSs and, consequently, have more problems and are more difficult to manage.

All DDBMSs share several important characteristics. Among these characteristics are location transparency, replication transparency, and fragmentation transparency.

Location Transparency

The definition of a distributed database says nothing about the *ease* with which users access data that is stored at other sites. Systems that support distributed databases should let a user access data at a **remote site**—a site other than the one at which the user is located—just as easily as the user accesses data from the **local site**—the site where the user is located. Response times for accessing data stored at a remote site might be much slower but, except for this difference, it should feel to a user as though the entire database is stored at the local site. **Location transparency** is the characteristic of a DDBMS that states that users do not need to be aware of the location of data in a distributed database; location transparency is one of the major characteristics of a DDBMS.

Replication Transparency

As described in Chapter 7, replication lets users at different sites use and update copies of a database and then share their updates with other users. However, data replication creates update problems that can lead to data inconsistencies. If you update the record of a single part at Premiere Products, the DDBMS must make the update at every location at which data concerning this part is stored. Not only do multiple updates make the process more time consuming and complicated, but, should one of the copies of data for this part be overlooked, the database would contain inconsistent data. Ideally, the DDBMS should correctly handle the updating of replicated data. The steps taken by the DDBMS to update the various copies of data should be done behind the scenes; users should be unaware of the steps. This DDBMS characteristic is called **replication transparency**.

Fragmentation Transparency

A DDBMS supports **data fragmentation** if the DDBMS can divide and manage a logical object, such as the records in a table, among the various locations under its control. The main purpose of data fragmentation is to place data at the location where the data is most often accessed.

Suppose that Premiere Products has a local DBMS at each of its three warehouses and wants to fragment its Part table data, which is shown in Figure 9.2, by placing the data for the parts stored in a warehouse in that warehouse's local database.

FIGURE 9.2

Premiere Products
Part table data

Part

PartNum	Description	OnHand	Class	Warehouse	Price
AT94	Iron	50	HW	3	$24.95
BV06	Home Gym	45	SG	2	$794.95
CD52	Microwave Oven	32	AP	1	$165.00
DL71	Cordless Drill	21	HW	3	$129.95
DR93	Gas Range	8	AP	2	$495.00
DW11	Washer	12	AP	3	$399.99
FD21	Stand Mixer	22	HW	3	$159.95
KL62	Dryer	12	AP	1	$349.95
KT03	Dishwasher	8	AP	3	$595.00
KV29	Treadmill	9	SG	2	$1,390.00

Using SQL-type statements, you can define the following fragments:

```
DEFINE FRAGMENT Part1 AS
SELECT PartNum, Description, OnHand, Class, Warehouse,
    Price
FROM Part
WHERE Warehouse='1'

DEFINE FRAGMENT Part2 AS
SELECT PartNum, Description, OnHand, Class, Warehouse,
    Price
FROM Part
WHERE Warehouse='2'

DEFINE FRAGMENT Part3 AS
SELECT PartNum, Description, OnHand, Class, Warehouse,
    Price
FROM Part
WHERE Warehouse='3'
```

Each fragment definition indicates which Part table data to select for the fragment. Note that the Part table does not actually exist in any one place. Rather, the Part table exists in three pieces. You assign these pieces, or fragments, to the databases located at the warehouses, as shown in Figure 9.3.

Fragment Part1

PartNum	Description	OnHand	Class	Warehouse	Price
CD52	Microwave Oven	32	AP	1	$165.00
KL62	Dryer	12	AP	1	$349.95

Fragment Part2

PartNum	Description	OnHand	Class	Warehouse	Price
BV06	Home Gym	45	SG	2	$794.95
DR93	Gas Range	8	AP	2	$495.00
KV29	Treadmill	9	SG	2	$1,390.00

Fragment Part3

PartNum	Description	OnHand	Class	Warehouse	Price
AT94	Iron	50	HW	3	$24.95
DL71	Cordless Drill	21	HW	3	$129.95
DW11	Washer	12	AP	3	$399.99
FD21	Stand Mixer	22	HW	3	$159.95
KT03	Dishwasher	8	AP	3	$595.00

You assign Fragment Part1 to the database at warehouse 1, Fragment Part2 to the database at warehouse 2, and Fragment Part3 to the database at warehouse 3. The effect of these assignments is that data about each part is stored in the database at the warehouse where the part is stored. You can access the complete Part table by taking the union of the three fragments.

Users should not be aware of the fragmentation—they should feel as if they are using a single central database. If users are unaware of fragmentation, the DDBMS has **fragmentation transparency**.

Advantages of Distributed Databases

When compared with a single centralized database, distributed databases offer the following advantages:

- *Local control of data.* Because each location retains its own data, a location can exercise greater control over that data. With a single centralized database, on the other hand, the central site that maintains the database will usually not be aware of all the local issues at the various sites served by the database.

- *Increasing database capacity.* In a properly designed and installed distributed database, the process of increasing system capacity is often simpler than in a centralized database. If the size of the disk at a single site becomes inadequate for its database, you need to increase the capacity of the disk only at that site. Furthermore, you can increase the capacity of the entire database simply by adding a new site.

- *System availability.* When a centralized database becomes unavailable for any reason, *no* users can continue processing. In contrast, if one of the local databases in a distributed database becomes unavailable, only users who need data in that particular database are affected; other users can continue processing in a normal fashion. In addition, if the data has been replicated (another copy of it exists in other local databases), potentially all users can continue processing. However, processing for users at the site of the unavailable database will be much less efficient because data that was formerly obtained locally must now be obtained through communication with a remote site.

- *Added efficiency.* When data is available locally, you eliminate network communication delays and can retrieve data faster than with a remote centralized database.

Disadvantages of Distributed Databases

Distributed databases have the following disadvantages:

- *Update of replicated data.* Replicating data can improve processing speed and ensure that the overall system remains available even when the database at one site is unavailable. However, replication can cause update problems, most obviously in terms of the extra time needed to update all the copies. Instead of updating a single copy of the data, the DBMS must update several copies. Because most of these copies are at sites other than the site initiating the update, each update requires extra time to update each copy and extra time to communicate all the update messages over the network.

  Replicated data causes another, slightly more serious problem. Assume that an update transaction must update data that is replicated at five sites and that the fifth site is currently unavailable. If all updates must be made or none at all, the update transaction fails. Because the data at a single site is unavailable for update, that data is unavailable for update at *all* sites. This situation certainly contradicts the earlier advantage of increased system availability. On the other hand, if you do not require that all updates be made, the data will be inconsistent.

Often a DDBMS uses a compromise strategy. The DDBMS designates one copy of the data to be the **primary copy**. As long as the primary copy is updated, the DDBMS considers the update to be complete. The primary site and the DDBMS must ensure that all the other copies are in sync. The primary site sends update transactions to the other sites and notes whether any sites are currently unavailable. If a site is unavailable, the primary site must try to send the update again at some later time and continue trying until it succeeds. This strategy overcomes the basic problem, but it obviously uses more time. Further, if the primary site is unavailable, the problem remains unresolved.

■ ***More complex query processing.*** Processing queries is more complex in a distributed database. The complexity occurs due to the difference between the time it takes to send messages between sites and the time it takes to access a disk. As discussed earlier, you learned that minimizing message traffic is extremely important in a distributed database environment. To illustrate the complexity involved with query processing, consider the following query for Premiere Products: List all parts in item class SG with a price that is more than $500.00. For this query, assume that (1) the Part table contains 1,000 rows and is stored at a remote site; (2) each record in the Part table is 500 bits long; (3) there is no special structure, such as an index, that would be helpful in processing this query faster; and (4) only 10 of the 1,000 rows in the Part table satisfy the conditions. How would you process this query?

One query strategy involves retrieving each row from the remote site and examining the item class and price to determine whether the row should be included in the result. For each row, this solution requires two messages. The first is a message from the local site to the remote site requesting a row. It is followed by the second message, which is from the remote site to the local site containing either the data or, ultimately, an indication that there is no more data because you have retrieved every row in the table. Thus, in addition to the database accesses themselves, this strategy requires 2,000 messages. Once again, suppose that you have a network with an access delay of 2 seconds and a transmission rate of 30,000 bits per second. Based on the calculations for communication time earlier in this chapter, each message requires approximately 2 seconds. You calculate the communication time for this query strategy as follows:

```
Communication time = 2 * 2,000
                   = 4,000 seconds, or 66.7 minutes
```

A second query strategy involves sending a single message from the local site to the remote site requesting the complete answer to the query. The remote site examines each row in the table and finds the 10 rows that satisfy the query. The remote site then sends a single message back to the local site containing all 10 rows in the answer. You calculate the communication time for this query strategy as follows:

```
Communication time = 2 + (2 + ((10 * 500) / 30,000))
                   = 2 + (2 + (5000 / 30,000)
                   = 2 + (2 + 0.16)
                   = 4.16 seconds
```

Even if the second message is very lengthy, especially where many rows satisfied the conditions, this second query strategy is a vast improvement over the first strategy. A small number of lengthy messages is preferable to a large number of short messages.

Systems that are record-at-a-time oriented can create severe performance problems in distributed systems. If the only choice is to transmit every record from one site to another site as a message and then examine it at the other site, the communication time required can become unacceptably high. DDBMSs that permit a request for a set of records, as opposed to an individual record, outperform record-at-a-time systems.

■ ***More complex treatment of concurrent update.*** Concurrent update in a distributed database is treated in basically the same way as it is treated in nondistributed databases. A user transaction acquires locks, and the locking is two-phase. (Locks are acquired in a growing phase, during which time no locks are released and the DDBMS applies the updates. All locks are released during the shrinking phase.) The DDBMS detects and breaks deadlocks, and then the DDBMS rolls back interrupted transactions. The primary distinction lies not in the kinds of activities that take place but in the additional level of complexity created by the very nature of a distributed database.

If all the records to be updated by a particular transaction occur at one site, the problem is essentially the same as in a nondistributed database. However, the records in a distributed database might be stored at many different sites. Furthermore, if the data is replicated, each occurrence might be stored at several sites, each requiring the same update to be performed. Assuming each record occurrence has replicas at three different sites, an update that would affect five record occurrences in a nondistributed system might affect 20 different record occurrences in a distributed system (each record occurrence together with its three replica occurrences). These 20 different record occurrences could conceivably be stored at 20 different sites.

Having more record occurrences to update is only part of the problem. Assuming that each site keeps its own locks, the DDBMS must send many messages for each record to be updated: a request for a lock; a message indicating that either the record is already locked by another user or that the lock has been granted; a message directing that the update be performed; an acknowledgment of the update; and, finally, a message indicating that the record is to be unlocked. Because all these messages must be sent for each record and its occurrences, the total time for an update can be substantially longer in a distributed database.

A partial solution to minimize the number of messages involves the use of the primary copy mentioned earlier. Recall that one of the replicas of a given record occurrence is designated as the primary copy. Locking the primary copy, rather than all copies, is sufficient and will reduce the number of messages required to lock and unlock records. The number of messages might still be large, however, and the unavailability of the primary copy can cause an entire transaction to fail. Thus, even this partial solution presents problems.

Just as in a nondistributed database, deadlock is a possibility in a distributed database. In a distributed database, however, deadlock is more complicated because two types of deadlock, local deadlock and global deadlock, are possible. **Local deadlock** is deadlock that occurs at a single site in a distributed database. If two transactions are each waiting for a record held by the other at the same site, the local DBMS can detect and resolve the deadlock with a minimum number of messages needed to communicate the situation to the other DBMSs in the distributed system.

On the other hand, **global deadlock** involves one transaction that requires a record held by a second transaction at one site, while the second transaction requires a record held by the first transaction at a different site. In this case, neither site has information individually to allow this deadlock to be detected; this is a global deadlock, and it can be detected and resolved only by sending a large number of messages between the DBMSs at the two sites.

The various factors involved in supporting concurrent update greatly add to the complexity and the communications time in a distributed database.

- ***More complex recovery measures.*** Although the basic recovery process for a distributed database is the same as the one described in Chapter 7, there is an additional potential problem. To make sure that the database remains consistent, each database update should be made permanent or aborted and undone, in which case, *none* of its changes will be made. In a distributed database, with an individual transaction updating several local databases, it is possible—due to problems affecting individual sites—for local DBMSs to commit the updates at some sites and undo the updates at other sites, thereby creating an inconsistent state in the distributed database. The DDBMS *must not* allow this inconsistency to occur.

  A DDBMS usually prevents this potential inconsistency through the use of **two-phase commit**. The basic idea of two-phase commit is that one site, often the site initiating the update, acts as **coordinator**. In the first phase, the coordinator sends messages to all other sites requesting that they prepare to update the database; in other words, each site acquires all necessary locks. The sites do not update at this point, however, but they do send messages to the coordinator that they are ready to update. If for any reason any site cannot secure the necessary locks, or if any site must abort its updates, the site sends a message to the coordinator that all sites must abort the transaction. The coordinator waits for replies from all sites involved before determining whether to commit the update. If all replies are positive, the coordinator sends a message to each site to commit the update. At this point, each site *must* proceed with the commit process. If any reply is negative, the coordinator sends a message to each site to abort the update, and each site *must* follow this instruction. In this way, the DDBMS guarantees consistency.

  While a process similar to two-phase commit is essential to the consistency of the database, there are two problems associated with it. For one thing, many messages are sent during the process. For another, during the second phase, each site must follow the instructions from the coordinator; otherwise, the process will not accomplish its intended result. This process means that the sites are not as independent as you would like them to be.

- ***More difficult management of the data dictionary.*** A distributed database introduces further complexity to the management of the data dictionary or catalog. Where should the data dictionary entries be stored? The three possibilities are as follows: choose one site and store the complete data dictionary at this site and this site alone; store a complete copy of the data dictionary at each site; and distribute, possibly with replication, the data dictionary entries among the various sites.

Although storing the complete data dictionary at a single site is a relatively simple approach to administer, retrieving information in the data dictionary from any other site is more time-consuming because of the communication involved. Storing a complete copy of the data dictionary at every site solves the retrieval problem because a local DBMS can handle any retrieval locally. Because this second approach involves total replication (every data dictionary occurrence is replicated at every site), updates to the data dictionary are more time-consuming. If the data dictionary is updated with any frequency, the extra time needed to update all copies of the data dictionary might be unacceptable. Thus, you usually implement an intermediate strategy.

One intermediate strategy is to partition the data by storing data dictionary entries at the site at which the data they describe are located. Interestingly, this approach also suffers from a problem. If a user queries the data dictionary to access an entry not stored at the user's site, the system has no way of knowing the entry's location. Satisfying this user's query might involve sending a message to every other site, which involves a considerable amount of network and DDBMS overhead.

- ■ *More complex database design.* A distributed database adds another level of complexity to database design. Distributing data does not affect the information-level design. During the physical-level design in a nondistributed database, disk activity—both the number of disk accesses and the volumes of data to be transported—is one of the principal concerns. Although disk activity is also a factor in a distributed database, communication activity becomes another concern during the physical-level design. Because transmitting data from one site to another is much slower than transferring data to and from disk, in many situations, communication activity is the most important physical-level design factor. In addition, you must consider possible fragmentation and replication during the physical-level design.

■ Rules for Distributed Databases

C.J. Date (Date, C.J. "Twelve Rules for a Distributed Database." in *ComputerWorld* 21.23, June 8, 1987) formulated 12 rules that distributed databases should follow. The basic goal is that a distributed database should feel like a nondistributed database to users; that is, users should not be aware that the database is distributed. The 12 rules serve as a benchmark against which you can measure DDBMSs. The 12 rules are as follows:

1. *Local autonomy.* No site should depend on another site to perform its database functions.

2. *No reliance on a central site.* The DDBMS should not rely on a single central site to control specific types of operations. These operations include data dictionary management, query processing, update management, database recovery, and concurrent update.

3. *Continuous operation.* Performing functions, such as adding sites, changing versions of DBMSs, creating backups, and modifying hardware, should not require planned shutdowns of the entire distributed database.

4. ***Location transparency.*** Users should not need to be concerned with the location of any specific data in the database. Users should feel as if the entire database is stored at their location.

5. ***Fragmentation transparency.*** Users should not be aware of any data fragmentation that has occurred in the database. Users should feel as if they are using a single central database.

6. ***Replication transparency.*** Users should not be aware of any data replication. The DDBMS should perform all the work required to keep the replicas consistent; users should be unaware of the data synchronization work carried out by the DDBMS.

7. ***Distributed query processing.*** You already learned about the complexities of query processing in a distributed database. The DDBMS must process queries as rapidly as possible.

8. ***Distributed transaction management.*** You already learned about the complexities of update management in a distributed database and the need for the two-phase commit strategy. The DDBMS must effectively manage transaction updates at multiple sites.

9. ***Hardware independence.*** Organizations usually have many different types of hardware, and a DDBMS must be able to run on this hardware. Without this capability, users would be restricted to accessing data stored only on similar computers, disks, and so on.

10. ***Operating system independence.*** Even if an organization uses similar hardware, different operating systems might exist in the organization. For the same reason that it is desirable for a DDBMS to support different types of hardware, a DDBMS must be able to run on different operating systems.

11. ***Network independence.*** Because different sites within an organization might use different communications networks, a DDBMS must run on different types of networks and not be restricted to a single type of network.

12. ***DBMS independence.*** Another way of stating this requirement is that a DDBMS should be heterogeneous; that is, a DDBMS must support different local DBMSs. Supporting heterogeneous DBMSs is a difficult task. In practice, each local DBMS must "speak" a common language; this common language most likely is structured query language (SQL).

■ Client/Server Systems

Networks often include a file server, as shown in Figure 9.4. The **file server** stores the files required by the users on the network. When users need data from a file or group of files, they send requests to the file server. The file server then sends the requested file or files to the user's computer; that is, the file server sends entire files, not just the data needed by users. Although this approach works to supply data to users, sending entire files generates a high level of communication activity on the network. Adding users to the network and larger files to the file server adds higher levels of communication activity and eventually causes longer delays in supplying data to users.

FIGURE 9.4
··················
File server architecture

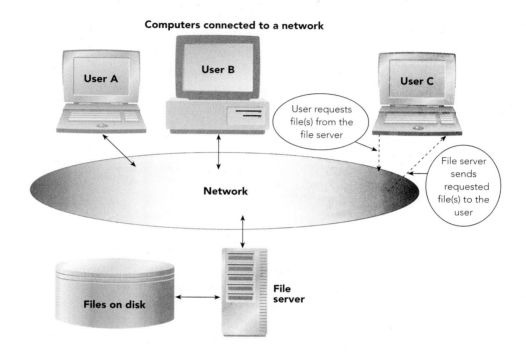

Computers connected to a network

An alternative architecture, which is called **client/server**, is illustrated in Figure 9.5. In client/server terminology, the **server** is a computer providing data to the **clients**, which are the computers that are connected to a network and that people use to access data stored on the server. A server is also called a **back-end processor** or a **back-end machine**, and a client is also called a **front-end processor** or a **front-end machine**.

With this alternative architecture, a DBMS runs on the server. A client sends a request to the server, not for entire *files*, but for specific *data*. The DBMS on the server processes the request, extracts the requested data, and then sends the data back to the client. Compared to a file server architecture, a client/server architecture reduces communication activity on a network, which reduces delays in supplying data to users. Because the clients and the server perform different functions and can run different operating systems, this arrangement of client/server architecture is called a **two-tier architecture**.

FIGURE 9.5
••••••••••••••••••
Two-tier
client/server
architecture

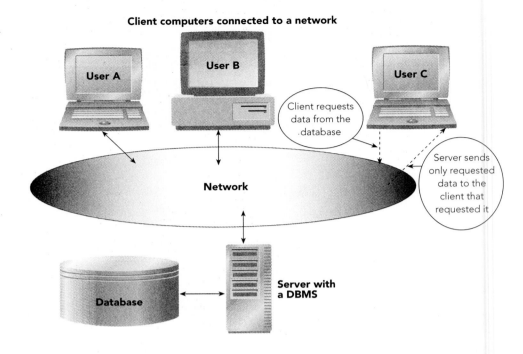

Client computers connected to a network

In a two-tier architecture, the server performs database functions and the clients perform the presentation functions (or user interface functions), such as which form to display on the screen and how to format the form's data. Which of the two tiers, server or clients, performs the business functions, such as the calculations Premiere Products uses to determine commissions, taxes, and order totals? If the clients perform the business functions—each client is called a **fat client** in this arrangement—then you have a client maintenance problem. Whenever programmers make changes to the business functions, they must make sure that they place the updated business functions on every client. For organizations with thousands of clients, updating the business functions for all clients is an almost impossible task.

To eliminate the fat client maintenance problem, you can place the business functions on the server. Because clients perform only the presentation functions in this arrangement, each client is called a **thin client**. Although you've now eliminated the fat client maintenance problem by moving the business functions to the server, you've now created a scalability problem. **Scalability** is the ability of a computer system to continue to function well as utilization of the system increases. Because the server performs both database and business functions, increasing the number of clients eventually causes a bottleneck on the server and degrades the system's responsiveness to clients. To improve a system's scalability, some organizations use a three-tier client/server architecture, as shown in Figure 9.6. In a **three-tier architecture**, the clients perform the presentation functions, a **database server** performs the database functions, and separate computers, called **application servers**, perform the business functions and serve as an interface between clients and the database server. A three-tier architecture distributes the processing functions so that you eliminate the fat client maintenance problem and maximize the scalability of the system. As the number of users increases, you can upgrade the application and database servers by adding faster processors, disks, and other hardware without changing any client computer.

FIGURE 9.6
••••••••••••••••••
Three-tier
client/server
architecture

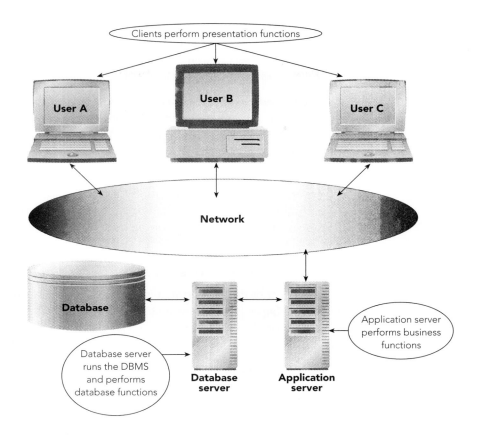

Note: A client/server system stores the database on a single server and the DBMS resides and processes on that server. Only with a DDBMS is the database itself distributed to multiple computers. However, you can combine a DDBMS with a client/server system to distribute both data and processing functions across multiple computers.

Advantages of Client/Server Systems

Compared to file server systems, a client/server system has the following advantages:

- *Lower network traffic.* A client/server system transmits only the necessary data, rather than entire files, across the network.

- *Improved processing distribution.* A client/server system lets you distribute processing functions among multiple computers.

- *Thinner clients.* Because the application and database servers handle most of the processing in a client/server system, clients do not need to be as powerful or as expensive as they would in a file-server environment.

- *Greater processing transparency.* As far as a user is concerned, all processing occurs on the client just as it does on a stand-alone system. Users do not need to learn any special commands or techniques to work in a client/server environment.

- *Increased network, hardware, and software transparency.* Because client/server systems use SQL as a common language, it is easier for users to access data from a variety of sources. A single operation could access data from different networks, different computers, and different operating systems.

- *Improved security.* Client/server systems can provide a greater level of security than do file server systems. In addition to the DBMS security features located on the database server, you can place additional security features on the application servers and on the network.

- *Decreased costs.* Client/server systems have proven to be powerful enough that organizations have replaced, at a considerable cost savings, mainframe applications and mainframe databases with PC applications and databases managed by client/server systems.

- *Increased scalability.* A three-tier client/server system is more scalable than file-server and two-tier architectures. If an application server or database server becomes a bottleneck, you can upgrade the appropriate server or add additional processors to share the processing load.

Triggers and Stored Procedures

One special aspect of the support that client/server systems provide for integrity is the use of triggers. A **trigger** is an action that occurs automatically in response to an associated database operation. Programmers create triggers with special SQL statements similar to the following:

```
CREATE AddOrder TRIGGER ON Orders
FOR INSERT, UPDATE AS
IF NOT EXISTS
(SELECT CustomerNum
FROM Inserted
WHERE CustomerNum IN
(SELECT CustomerNum FROM Customer))
BEGIN
Print("Invalid Customer Number")
END
```

This trigger, which is named AddOrder, ensures that the customer number entered for an order matches a customer number in the Customer table. The DBMS will apply this trigger on both INSERT and UPDATE actions on a row in the Orders table. The data to be updated is in a special temporary table named Inserted. If the customer number in the Inserted table is not in the collection of customer numbers in the Customer table, the DBMS will not perform the update. Instead, the DBMS will display the "Invalid Customer Number" message.

If users execute a trigger or other collection of SQL statements repeatedly, you can improve the performance of a client/server system by placing the statements in a special file, called a **stored procedure**. The DBMS compiles and optimizes the statements in a stored procedure. From that point on, users execute the compiled, optimized code in the stored procedure.

▓ Data Warehouses

Among the objectives that organizations have when they use RDBMSs are data integrity, high performance, and ample availability. The leading RDBMSs are able to satisfy these requirements. Typically, when users interact with an RDBMS, they use transactions, such as adding a new order and changing a customer's sales rep. Thus, these types of systems are called **online transaction processing (OLTP)** systems.

For each transaction, OLTP typically deals with a small number of rows from the tables in a database in a highly structured, repetitive, and predetermined way. If you need to know the status of specific customers, parts, and orders, or if you need to update data for specific customers, parts, and orders, an RDBMS and OLTP are the ideal tools to use.

When you need to analyze data from a database, however, an RDBMS and OLTP often suffer from severe performance problems. For example, finding total sales by site and by month requires the joining of all the rows in many tables; such processing takes a considerable number of database accesses and considerable time to accomplish. Consequently, many organizations continue to use RDBMSs and OLTP for their normal, day-to-day processing or for *operational purposes*, but the organizations have turned to data warehouses for the *analysis* of their data. The following definition for a data warehouse is credited to W.H. Inmon (Inmon, W.H. *Building the Data Warehouse*. QED, 1990), who originally coined the phrase.

Definition: A **data warehouse** is a subject-oriented, integrated, time-variant, nonvolatile collection of data in support of management's decision-making process.

Subject-oriented means that data is organized by entity rather than by the application that uses the data. For example, Figure 9.7 shows the databases for typical operational applications such as inventory, order entry, production, and accounts payable. When the data from these operational databases is loaded into a data warehouse, it is transformed into subjects such as product, customer, vendor, and financial. Data about products appears once in the warehouse, even though it might appear in many files and databases in the operational environment.

FIGURE 9.7

Data warehouse architecture

Note: For the operational applications shown in Figure 9.7, large organizations use a variety of DBMSs and file-processing systems that have been developed over a period of many years.

Integrated means that data is stored in one place in the data warehouse, even though the data originates from everywhere in the organization and from a variety of external sources. The data can come from recently developed applications or from legacy systems developed many years ago.

Time-variant means that data in a data warehouse represents snapshots of data at various points in time in the past, such as at the end of each month. This is unlike an operational application, which has data that is accurate as of the moment. Data warehouses also retain historical data for long periods of time, summarized to specific time periods such as daily, weekly, monthly, and annually.

Nonvolatile means that data is read-only. Data is loaded into a data warehouse periodically, but users cannot update a data warehouse directly.

In summary, a data warehouse contains read-only snapshots of highly consolidated and summarized data from multiple internal and external sources that are refreshed periodically, usually on a daily or weekly basis. Companies use data warehouses in support of their decision-making processing, which typically consists of unstructured and nonrepetitive requests for exactly the type of information contained in a data warehouse.

Data Warehouse Structure and Access

A typical data warehouse structure is shown in Figure 9.8. The central Sales table is called a fact table. A **fact table** consists of rows that contain consolidated and summarized data. The fact table contains a multipart primary key, each part of which is a foreign key to the surrounding dimension tables. Each **dimension table** contains a single-part primary key that serves as an index for the fact table and that also contains other fields associated with the primary key value. The overall structure shown in Figure 9.8 is called a **star schema** because of its conceptual shape.

FIGURE 9.8
••••••••••••••••••
A star schema with
four dimension
tables and a central
fact table

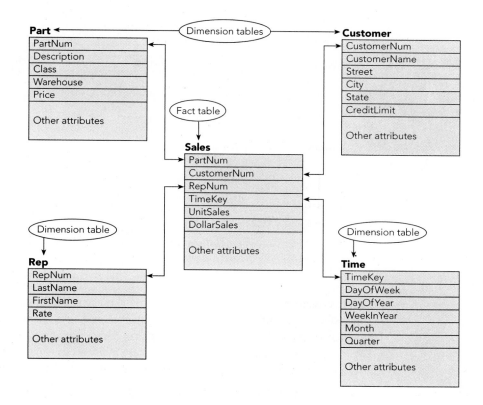

Access to a data warehouse is accomplished through the use of **online analytical processing (OLAP)** software. OLAP software is optimized to work efficiently with data warehouses.

Users access a data warehouse using OLAP software to answer questions such as the following: How has the average customer balance changed each year over the past five years? What are the total sales by month for this year, and how do they compare to last year's sales?

In posing these types of questions, users perceive the data in a data warehouse as a **multidimensional database**. For example, if users' questions pertain to the Part, Customer, and Time dimensions, which appear in Figure 9.8, they might visualize the data warehouse as a multidimensional database in the shape of a **data cube**, as shown in Figure 9.9. Each axis in the data cube (Part, Customer, and Time) represents data from a dimension table in Figure 9.8, and the cells in the data cube represent unit sales and dollar sales data from the Sales fact table in Figure 9.8.

FIGURE 9.9
....
A data cube
representation of
the Part, Customer,
and Time
dimensions

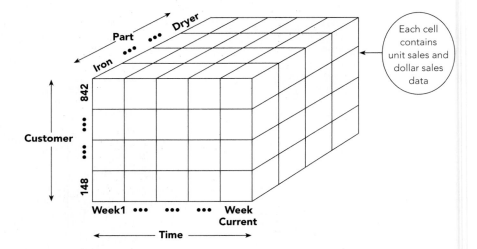

When users access a data warehouse, their queries usually involve aggregate data, such as total sales by month and average sales by customer. As users view the aggregate results from their queries, they often need to perform further analyses of the data they're viewing. OLAP software should let users perform these analyses as easily and as quickly as possible. Users' analyses typically involve actions that include the following:

- **_Slice and dice._** Instead of viewing all data in a data cube, users typically view only portions of the data. You **slice and dice** data to select portions of the available data or to reduce the data cube. For example, suppose that the Time dimension in the conceptual data cube that appears in Figure 9.9 has detailed sales data on a weekly basis for Premiere Products. Further suppose that the sales manager queries the data warehouse to view this week's total sales, both in dollars and in units sold, as shown in Figure 9.10.

FIGURE 9.10
....
Total sales query
results

TotalDollars	TotalUnits
$8,911.14	28

Conceptually, the sales manager's query slices the data cube to reduce it to the shaded "Week Current" portion, which is shown in Figure 9.11.

FIGURE 9.11
·················
Slicing the data
cube based on the
Time dimension

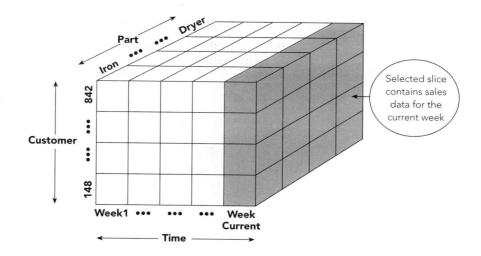

If the sales manager's next query displays this week's total sales for irons, the query dices the sliced data cube, reducing it to the shaded portion shown in Figure 9.12.

FIGURE 9.12
·················
Dicing the sliced
data cube based on
the Part dimension

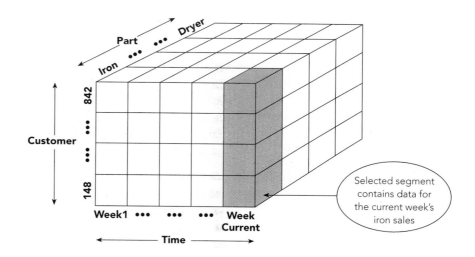

The results for the sales manager's queries for this diced portion of the data cube appear in Figure 9.13.

FIGURE 9.13
·················
Query results for
total sales of irons

TotalDollars	TotalUnits
$241.45	11

The sales manager's first query sliced the data cube to focus on the current week's sales, and the second query reduced the slice by dicing only the cells in the data cube that are for irons.

- **_Drill down._** When you view specific aggregate data, you **drill down** the data to view and analyze lower levels of aggregation; that is, you go to a more detailed view of the data. For example, suppose again that the sales manager queries the data warehouse to view this week's total sales, as shown in Figure 9.10. To analyze details of these sales, the sales manager next might drill down to view total sales by class, as shown in Figure 9.14.

Class	TotalDollars	TotalUnits
AP	$4,499.79	13
HW	$241.45	11
SG	$4,169.90	4

Finally, the sales manager might drill down to view total sales by part within class, as shown in Figure 9.15.

Class	Description	TotalDollars	TotalUnits
AP	Dishwasher	$1,190.00	2
AP	Dryer	$1,319.80	4
AP	Gas Range	$990.00	2
AP	Microwave Oven	$600.00	4
AP	Washer	$399.99	1
HW	Iron	$241.45	11
SG	Home Gym	$1,589.90	2
SG	Treadmill	$2,580.00	2

- **_Roll up._** When you view specific aggregate data, you **roll up** the data to view and analyze higher levels of aggregation. Rolling up the data is the exact opposite of drilling down the data. For example, the sales manager might start with the query results for total sales by class and part (see Figure 9.15), click the appropriate button to roll up the data for the query results for the total sales by class (see Figure 9.14), and then click another button to roll up the data for the query results for total sales (see Figure 9.10).

Data mining consists of uncovering new knowledge, patterns, trends, and rules from the data stored in a data warehouse. You use data mining software to answer questions such as the following: Which products best attract new customers? What factors best predict which customers default in making payments? What are the optimal seasonal inventory levels based on predicted economic factors? What is the optimal number of customers to assign to each sales rep? Because data warehouses often contain enormous amounts of data, users can't sift through the data in them to find answers to these questions. Instead, with minimal user interaction, data mining software attempts to answer these questions by using sophisticated analytical, mathematical, and statistical techniques.

Rules for OLAP Systems

E.F. Codd (Codd, E.F., S.B. Codd, and C.T. Salley. "Providing OLAP (On-line Analytical Processing) to User-Analysts: An IT Mandate." Arbor Software, August, 1993) formulated 12 rules that OLAP systems should follow. The 12 rules serve as a benchmark against which you can measure OLAP systems. The 12 rules are as follows:

1. ***Multidimensional conceptual view.*** Users must be able to view data in a multidimensional way, matching the way data appears naturally in an organization. For example, users can view data about the relationships among data using the dimensions of parts, customer locations, sales reps, and time.

2. ***Transparency.*** Users should not have to know they are using a multidimensional database nor need to use special software tools to access data. For example, if users usually access data using a spreadsheet, they should still be able to use a spreadsheet to access a multidimensional database.

3. ***Accessibility.*** Users should perceive data as a single user view, even though the data may be physically located in several heterogeneous locations and in different forms such as relational databases and standard files.

4. ***Consistent reporting performance.*** Retrieval performance should not degrade as the number of dimensions and the size of the warehouse grows.

5. ***Client/server architecture.*** The server component of OLAP software must be intelligent enough so that a variety of clients can be connected with minimal effort.

6. ***Generic dimensionality.*** Every dimension table must be equivalent in both its structural and operational capabilities. For example, you should be able to obtain information about parts as easily as you obtain information about sales reps.

7. ***Dynamic sparse matrix handling.*** Missing data should be handled correctly and efficiently and not affect the accuracy or speed of data retrieval.

8. ***Multiuser support.*** OLAP software must provide secure, concurrent retrieval of data. Because you don't update a data warehouse when you're using it, concurrent update is not an issue, so problems of security and access are less difficult than in an OLTP environment.

9. ***Unrestricted, cross-dimensional operations.*** Users must be able to perform the same operations across any number of dimensions. For example, you should be able to ask for statistics based on the dimensions of time, location, and part just as easily as you would ask for statistics based on the single dimension of location.

10. **Intuitive data manipulation.** Users should be able to act directly on individual data values without needing to use menus or other interfaces. Of course, these other interfaces can be used, but they should not be the required method of processing.

11. **Flexible reporting.** Users should be able to retrieve data results and view them in any way they want for analysis.

12. **Unlimited dimensions and aggregation levels.** OLAP software should allow at least 15 data dimensions and an unlimited number of aggregation (summary) levels.

Object-Oriented DBMSs

Organizations use relational databases to store and access data consisting of text and numbers. Additionally, some organizations store and access graphics, drawings, photographs, video, sound, voice mail, spreadsheets, and other complex objects in their databases. Relational DBMSs store these complex objects using special data types, generically called **binary large objects (BLOBs)**. Some applications, such as computer-aided design and manufacturing (CAD/CAM) and geographic information systems (GIS), have as their primary focus the storage and management of complex objects. For these systems, many companies use object-oriented DBMSs.

What is an Object-Oriented DBMS?

The relational model, which has a strong theoretical foundation, is the foundation for relational DBMSs. Although object-oriented DBMSs do not have a corresponding theoretical foundation, they all exhibit several common characteristics. Central to all object-oriented systems is the concept of an object. An **object** is a set of related attributes along with the actions that are associated with the set of attributes. A customer object, for example, consists of the attributes associated with customers (number, name, balance, and so on) together with the actions that are associated with customer data (add customer, change credit limit, delete customer, and so on).

In relational systems, you create the actions as part of data manipulation (in the programs that update the database), rather than as part of the data definition. In contrast, in object-oriented systems, you define the actions as part of the data definition and then use the actions whenever they are required. In an object-oriented system, the data and actions are **encapsulated**, which means that you define an object to contain both the data and its associated actions. Thus, an **object-oriented database management system (OODBMS)** is one in which data and the actions that operate on the data are encapsulated into objects.

To become familiar with OODBMSs, you should have a general understanding of the following object-oriented concepts: objects, classes, methods, messages, and inheritance.

Objects and Classes

To understand the distinction between objects and classes, you will examine an object-oriented representation of the following relational model representation of the Premiere Products database.

```
Rep (RepNum, LastName, FirstName, Street, City, State,
     Zip, Commission, Rate)
Customer (CustomerNum, CustomerName, Street, City, State,
     Zip, Balance, CreditLimit, RepNum)
Orders (OrderNum, OrderDate, CustomerNum)
OrderLine (OrderNum, PartNum, NumOrdered, QuotedPrice)
Part (PartNum, Description, OnHand, Class, Warehouse, Price,
     Allocated)
```

This version of the Premiere Products database contains an extra field, Allocated, in the Part table. The Allocated field stores the number of units of a part that are currently on order (allocated). Figure 9.16 shows a representation of this database as a collection of objects.

FIGURE 9.16

Object-oriented representation of the Premiere Products database

Rep OBJECT	
RepNum:	Sales Rep Numbers
LastName:	Last Names
FirstName:	First Names
Street:	Addresses
City:	Cities
State:	States
Zip:	Zip Codes
Commission:	Commissions
Rate:	Commission Rates
Customer:	Customer OBJECT; MV
Customer OBJECT	
CustomerNum:	Customer Numbers
CustomerName:	Customer Names
Street:	Addresses
City:	Cities
State:	States
Zip:	Zip Codes
Balance:	Balances
CreditLimit:	Credit Limits
Rep:	Rep OBJECT; SUBSET[RepNum, LastName, FirstName]
Part OBJECT	
PartNum:	Part Numbers
Description:	Part Descriptions
OnHand:	Units
Class:	Item Classes
Warehouse:	Warehouse Numbers
Price:	Prices
Allocated:	Units
OrderLine:	OrderLine OBJECT; MV
Orders OBJECT	
OrderNum:	Order Numbers
OrderDate:	Dates
Customer:	Customer OBJECT; SUBSET[CustomerNum, CustomerName, RepNum]
OrderLine:	OrderLine OBJECT; MV
OrderLine OBJECT	
OrderNum:	Order Numbers
PartNum:	Part Numbers
NumOrdered:	Units
QuotedPrice:	Prices

Note: Figure 9.16 shows just one of many approaches to representing objects. However, all techniques have the same general features.

You'll notice the following differences between the collection of objects in Figure 9.16 and the relational model representation:

- You represent each entity (Rep, Customer, and so on) as an *object* rather than as a relation.

- You list the attributes vertically below the object names. In addition, you follow each attribute by the name of the domain associated with the attribute. A **domain** is the set of values that are permitted for an attribute.

- Objects can contain other objects. For example, the Rep object contains the Customer object as one of its attributes. In the Rep object, the letters *MV* following the Customer object indicate that the Customer object is multivalued. In other words, a single occurrence of the Rep object can contain multiple occurrences of the Customer object. Roughly speaking, this is analogous to a relation containing a repeating group.

- An object can contain a portion of another object. The Customer object, for example, contains the Rep object. The word *SUBSET* indicates, however, that the Customer object contains only a subset of the Rep object. In this case, the Customer object contains three of the Rep object attributes: RepNum, LastName, and FirstName.

Notice that two objects can each appear to contain the other. The Rep object contains the Customer object, and the Customer object contains the Rep object (or at least a subset of it). The important thing to keep in mind is that users deal with *objects*. If the users of the Customer object require access to the rep's number and name, the rep number and name are part of the Customer object. If the users of the Rep object require data about all the customers of a sales rep, the Customer object is part of the Rep object. This arrangement is not to imply, of course, that the data is physically stored this way, but this is the way its users perceive the data.

Objects can contain more than one other object. Notice that the Orders object contains the Customer object and the OrderLine object, with the OrderLine object being multivalued. Nevertheless, users of the Orders object perceive it as a single unit.

Technically, the objects in Figure 9.16 are really classes. The term **class** refers to the general structure. The term "object" refers to a specific occurrence of a class. Thus, Rep is a class, whereas the data for rep 20 is an object.

Methods and Messages

Methods are the actions defined for a class. Figure 9.17 shows two methods associated with the Orders object. The first method, Add Order, adds an order to the database. In this example, users enter data, and then the program places the data temporarily in computer memory in a work area named WOrders. (In this example, the "W" prefix indicates a temporary work area record or field.) The WOrders record consists of a user-entered value for the order number stored in WOrderNum, a user-entered value for the order date stored in WOrderDate, and so on.

FIGURE 9.17
• • • • • • • • • • • • • • • • • •
Two methods for
the Premiere
Products
object-oriented
database

```
Add Order (WOrders)
        Add row to Orders table
                OrderNum      = WOrderNum
                OrderDate     = WOrderDate
                CustomerNum = WCustomerNum
        For each order line record in WOrders DO
                Add row to OrderLine table
                        OrderNum      = WOrderNum
                        PartNum       = WPartNum
                        NumOrdered    = WNumOrdered
                        QuotedPrice   = WQuotedPrice
                Update Part table (WHERE PartNum = WPartNum)
                        Allocated       = Allocated + WNumOrdered

Delete Order (WOrderNum)
        Delete row from Orders table (WHERE OrderNum = WOrderNum)
        For each OrderLine record (WHERE OrderNum = WOrderNum) DO
                Delete row from OrderLine table
                Update Part table (WHERE Part.PartNum = OrderLine.PartNum)
                        Allocated       = Allocated – NumOrdered
```

QUESTION

ANSWER

Describe the steps in the Add Order method.

The steps accomplish the following:
- Add a row to the Orders table for the new order.
- For each order line record associated with the order, add a row to the OrderLine table.
- For each matched order line record, update the Allocated value in the Part table for the corresponding part.

In Figure 9.17, the second method, Delete Order, deletes an order. The only data a user inputs to this method is the order number to be deleted, which is placed temporarily in WOrderNum.

QUESTION

ANSWER

Describe the steps in the Delete Order method.

The steps accomplish the following:
- Delete the order with the user-entered order number (WOrderNum) from the Orders table.
- For each order line record in which the order number matches the value of WOrderNum, delete the record.
- For each matched order line record, subtract the NumOrdered value from the Allocated value for the corresponding part in the Part table. (Because the method deletes the order line record, the parts are no longer allocated.)

Note: The two methods in Figure 9.17 are fairly complicated, consisting of several steps with each step involving separate updates to the database. Many methods are much simpler, although some methods are even more complicated.

You define methods during the data definition process. To execute the steps in a method, a user sends a message to the object. A message is a request to execute a method. As part of sending the message to an object, the user sends the required data (for example, full order data for the Add Order method, but only the order number for the Delete Order method). The process is similar to the process of calling a subroutine or invoking a procedure in a standard programming language.

Inheritance

A key feature of object-oriented systems is **inheritance**. For any class, you can define a **subclass**. Every occurrence of the subclass is also considered to be an occurrence of the class. The subclass *inherits* the structure of the class as well as its methods. In addition, you can define additional attributes and methods for the subclass.

As an example, suppose Premiere Products has a special type of order that has all the characteristics of other orders. In addition, it contains a freight amount and a discount that are calculated in a special way. Rather than create a new class for this type of order, you can define it as a subclass of the Orders class. In that way, the special order type automatically has all the attributes of the Orders class. The new subclass also has all the same methods of the Orders class, including the updating of the Allocated field in the Part table whenever orders are added or deleted. The only thing you would have to add would be those attributes and methods that are specific to this new type of order, thus greatly simplifying the entire process.

Unified Modeling Language

The **Unified Modeling Language (UML)** is an approach you can use to model all the various aspects of software development for object-oriented systems. UML includes a way to represent database designs.

UML includes several types of diagrams, each with its own special purpose. Figure 9.18 describes the purpose of some of the most commonly used UML diagrams.

FIGURE 9.18
· · · · · · · · · · · · · · · · ·
UML diagrams

Diagram Type	Description
Class	For each class, shows the name, attributes, and methods of the class, as well as the relationships between the classes in the database.
Use Case	Describes how the system is to behave from the standpoint of the system's users.
State	Shows the possible states of an object. (For example, an order could be in the placed, open, filled, or invoiced states.) Also shows the possible transitions between states (for example, placed→open→filled→invoiced).
Sequence	Shows the sequence of possible interactions between objects over time.
Activity	Shows the sequence of activities that can occur within a use case.
Collaboration	Shows how the elements of the system are to work together.
Component	Complex software systems are usually subdivided into smaller components. This type of diagram shows these components and their relationships with each other.

The type of diagram most relevant to database design is the **class diagram**. Figure 9.19 shows a sample class diagram for the Premiere Products database. A rectangle represents a class. The top portion of a rectangle contains the name of the class, the middle portion contains the attributes, and the bottom portion contains the methods. The lines joining the classes represent the relationships and are called **associations** in UML.

FIGURE 9.19
••••••••••••••••••••
Class diagram for
the Premiere
Products database

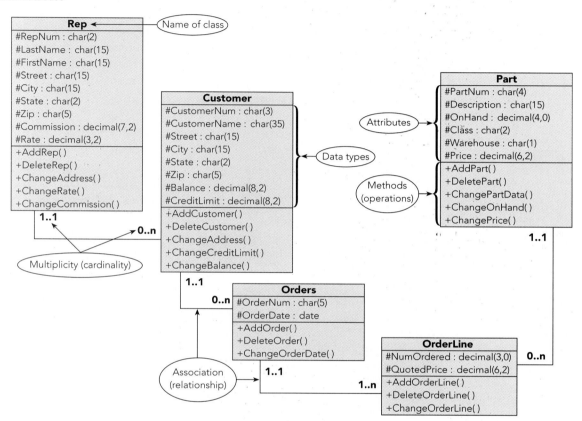

In a class diagram, a visibility symbol precedes each attribute. The **visibility symbol** indicates whether other classes can view or update the value in the attribute. The possible visibility symbols are public visibility (+), protected visibility (#), and private visibility (-). With **public visibility**, any other class can view or update the value. With **protected visibility**, only the class itself or public or protected subclasses of the class can view or update the value. With **private visibility**, only the class itself can view or update the value. The name of the attribute, a colon, and then the data type for the attribute follow the visibility symbol.

At each end of each association is an expression that represents the multiplicity, or cardinality, of the relationship. **Multiplicity** indicates the number of objects that can be related to an individual object at the other end of the relationship. UML provides various alternatives for representing multiplicity. In the alternative shown in Figure 9.19, two periods separate two symbols. The first symbol represents the minimum number of objects and the second symbol represents the maximum number of objects. A second number of *n* indicates that there is no maximum number of objects.

In the association between Customer and Orders, for example, the multiplicity for Customer is *1..1*. This multiplicity indicates that an order must correspond to at least one customer and can correspond to at most one customer. In other words, an order must correspond to *exactly* one customer. The multiplicity for Orders is *0..n*, indicating that a customer can have as few as zero orders (that is, a customer does not have to have any orders currently in the database), and there is no limit on the number of orders a customer can have. In the association between Orders and OrderLine, the multiplicity for OrderLine is *1..n*, rather than *0..n*. This multiplicity indicates that each order must have *at least one* order line, but the number of order lines is unlimited. If, on the other hand, the multiplicity for OrderLine were *1..5*, an order would be required to have anywhere from one to five order lines.

You can also specify constraints, which are restrictions on the data that can be stored in the database. You enter the constraint in the shape shown in Figure 9.20 and then connect the shape to the class to which it applies.

FIGURE 9.20

Class diagram for the Premiere Products database with constraints

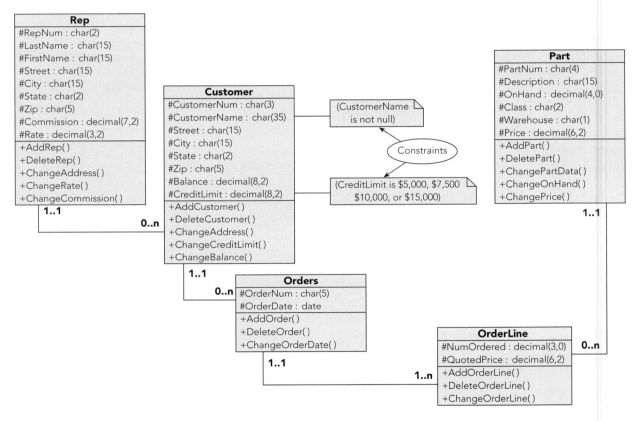

You learned about entity subtypes and how to represent them in entity-relationship (E-R) diagrams. In UML, these entity subtypes are called subclasses. In addition, if one class is a subclass of a second class, you call the second class a **superclass** of the first class. The relationship between a superclass and a subclass is called a **generalization** and is shown in Figure 9.21. This class diagram represents the relationship between the class of students and the subclass of students who live in dorms.

FIGURE 9.21
••••••••••••••••••
Class diagram with
a generalization
and a constraint

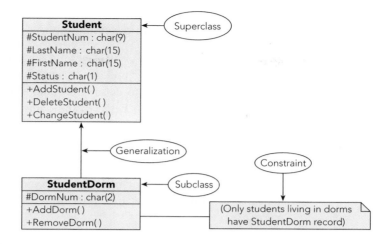

Rules for OODBMSs

Just as there are rules that specify desired characteristics for DDBMSs and OLAP, OODBMSs also have a set of rules. These rules serve as a benchmark against which you can measure object-oriented systems. The rules are as follows:

1. **Complex objects.** An OODBMS must support the creation of complex objects from simple objects such as integers, characters, and so on.

2. **Object identity.** An OODBMS must provide a way to identify objects; that is, the OODBMS must provide a way to distinguish between one object and another.

3. **Encapsulation.** An OODBMS must encapsulate data and associated methods together in the database.

4. **Information hiding.** An OODBMS must hide the details concerning the way data is stored and the actual implementation of the methods from the users of the database.

5. **Types or classes.** You are already familiar with the idea of a class. Types are very similar to classes and correspond to abstract types in programming languages. The differences between the two are subtle and will not be explored here. It is important to know, however, that an OODBMS supports either abstract types or classes (it doesn't matter which).

6. **Inheritance.** An OODBMS must support inheritance.

7. **Late binding.** In this case, **binding** refers to the association of operations to actual program code. With late binding, this association does not happen until runtime, that is, until some user actually invokes the operation. Late binding lets you use the same name for different operations, which is called **polymorphism** in object-oriented systems. For example, an operation to display an object on the screen requires different program code when the object is a picture than when it is text. With late binding, you can use the same name for both operations. At the time a user invokes this "display" operation, the system determines the object being displayed and then binds the operation to the appropriate program code.

8. **Computational completeness.** You can use functions in the language of the OODBMS to perform various computations.

9. **Extensibility.** Any DBMS, object-oriented or not, comes with a set of predefined data types, such as numeric and character. An OODBMS should be **extensible**, meaning that it is possible to define new data types. Furthermore, the OODBMS should make no distinction between the data types provided by the system and these new data types.

10. **Persistence.** In object-oriented programming, **persistence** refers to the ability to have a program *remember* its data from one execution to the next. Although this is unusual in programming languages, it is common in all database systems. After all, one of the fundamental capabilities of any DBMS is its ability to store data for later use.

11. **Performance.** An OODBMS should have sufficient performance capabilities to effectively manage very large databases.

12. **Concurrent update support.** An OODBMS must support concurrent update. (You learned about concurrent update in Chapter 7.)

13. **Recovery support.** An OODBMS must provide recovery services. (You learned about recovery services in Chapter 7.)

14. **Query facility.** An OODBMS must provide query facilities. (You learned about query facilities such as QBE and SQL in Chapters 2 and 3.)

■ Web Access to Databases

Over the past several years, the Internet and the closely related World Wide Web (Web) have shown phenomenal growth, both in numbers of users and the availability of software tools. In particular, you can use a Web browser to access many company databases. Finding information about a company's products and services, paying bills, and ordering merchandise are just a few of the activities you can conduct using the Internet; conducting business activities in this way is called **electronic commerce (e-commerce)**. E-commerce between businesses is called **business to business (B2B)**, and e-commerce between business and consumers is called **business to consumer (B2C)**. In addition, users in many organizations use their company's intranet to perform internal processing.

A three-tier client/server architecture makes Web access to databases possible. If a company uses this architecture, a Web server often replaces the application server to process client requests. A more flexible architecture positions Web servers as an additional tier, as shown in Figure 9.22.

FIGURE 9.22
• • • • • • • • • • • • • • • • • •
Using the Internet
and an intranet to
access a database

As shown in Figure 9.22, external clients use Web browsers to submit requests to the Web server through the Internet, and internal clients use Web browsers to submit requests to the Web server through the company's intranet. For client requests that require access to a database, the Web server sends the request to the application server. The application server performs any necessary business functions and then interacts with the database server, which retrieves the requested data or updates the database. After the database server satisfies the client request, the database server passes its results back to the application server, which assembles a Web page and contacts the Web server. The Web server retrieves the Web page and sends it back to the client.

Many different software languages, products, and standards exist to make e-commerce possible; these languages, products, and standards are constantly evolving with new ones appearing frequently. One language that is becoming prominent for the exchange of data is XML. **XML (Extensible Markup Language)** is a programming language similar in format to HTML, but that is more customizable and is suited to the exchange of data between different programs. Unlike HTML, which uses a fixed set of tags to describe how a Web page should look, you can customize XML code to describe the data it contains and how that data should be structured.

Externally, organizations benefit in two ways from using the Web for database processing. First, they can export and link data from their databases to suppliers, customers, and others outside the company; this provides current information in a timely way to those needing the information. Second, companies can allow customers, for example, to place orders that directly update the organization's database and trigger the processing required to fulfill the orders.

As time passes, organizations will handle more and more database processing through the Internet. Organizations hope that such use will allow them to become more efficient and effective in the way they carry out their business functions.

■ History of Database Management

No study of database management would be complete without an examination of its beginnings. The beginnings of database management coincided with the APOLLO project of the 1960s, which was launched in response to President John F. Kennedy's stated goal of landing a person on the moon by the end of the decade. At that time, no available computer systems could handle the coordination of the vast amounts of data required for the project. North American Rockwell, the prime contractor for the project, asked IBM to develop a system to manage the data. In response, IBM developed the **Generalized Update Access Method (GUAM)**, which went into production in 1964.

IBM realized that this product would be useful in other environments and made the product available to the general public in 1966 under the name **Data Language/I (DL/I)**. DL/I is the data management component of the **Information Management System (IMS)**, which was the dominant DBMS through the 1980s. IBM has enhanced IMS over the years; IMS still is used today by companies in their legacy systems, which are systems developed before PCs became widely used.

Also in the mid-1960s, Charles Bachman led a General Electric team that developed a DBMS named **Integrated Data Store (IDS)**. IDS led to a whole class of DBMSs—the CODASYL systems—that were popular and influential through the late 1980s.

In the late 1960s, the **COnference on DAta SYstems Languages (CODASYL)**, the group responsible for the programming language COBOL, tackled the problem of providing a standard for DBMSs. CODASYL created a task group, the **DataBase Task Group (DBTG)**, to develop specifications for DBMSs. The DBTG did this and in 1971, CODASYL presented these specifications to the **American National Standards Institute (ANSI)** for adoption as a national standard. Although these specifications were not accepted as a standard by ANSI, software vendors developed several systems that followed the CODASYL guidelines. The most widely used CODASYL system was Cullinane's **Integrated Data Management System (IDMS)**.

In 1970, Dr. E.F. Codd presented a paper in which he proposed a new and radically different approach to the management of data: the relational model. Throughout the 1970s, the relational model was the subject of intense research. In addition to purely theoretical research, prototype systems were developed, the most important being a system called **System R**, which was developed by IBM.

It was not until the 1980s that commercial relational DBMSs began to appear. Systems that are at least partly relational now exist in abundance on computers ranging from the smallest personal computer to the largest mainframe. Although System R never became a commercial system, it led to IBM's commercial relational product, **DB2**. Other large relational DBMSs that appeared in the 1980s included **Oracle** and **Sybase**. At the same time, PC-based relational DBMSs, such as **dBASE** and **Paradox**, appeared on the market. Since then, Microsoft **Access** has become the dominant PC-based relational DBMS, and Microsoft **SQL Server** and **MySQL** now compete with Oracle and Sybase.

Beginning in the late 1970s, researchers started working with object-oriented approaches to the development of computer software. In the 1980s, this research led to the development of object-oriented programming languages and OODBMSs. One of the first OODBMSs, **Gemstone** from the then Servio Corporation, appeared in 1987. Dozens of OODBMSs, such as **Objectivity/DB** and **Versant**, have come to the market since then. In addition, some RDBMS vendors have added object-oriented features to their products, creating what are called **object-relational DBMSs (ORDBMSs)**.

■ Hierarchical and Network Databases

The database community categorizes a DBMS by the data model the DBMS follows. There are four **data models**, each of which has two components: structure and operations. The **structure** refers to the way the system structures data or, at least, the way the users of the DBMS feel the data is structured. The **operations** are the facilities given to the users of the DBMS to manipulate data within the database.

The four DBMS data models are the hierarchical model, the network model, the relational model, and the object-oriented model. You learned about the relational and object-oriented models earlier in this text. Because their use is declining rapidly, you'll find the following brief descriptions of the network and hierarchical models to be sufficient.

Network Model

Users perceive a **network model** database as a collection of record types and relationships between these record types. Such a structure is called a network, which is the basis for the model's name. Consider the network database shown in Figure 9.23. The rectangles represent the record types in the database. There is one record type for each of the entities: reps, customers, orders, parts, and order lines. The arrows represent the relationships. In particular, the arrows represent one-to-many relationships. For example, each rep is related to many customers, and each rep is related to exactly one rep.

You manipulate a network model database by essentially following the arrows. This procedure is often referred to as **database navigation**. You can follow arrows in either direction. Suppose you need to print a list of all orders for a given part. You first ask the DBMS to find the part in the Part record. You then repeatedly ask to find the next order line that contains that part until you reach the end of the list for that part. While you're on a given order line, you can ask to find the number of the customer related to the order line. Here you would follow the arrow from the OrderLine record to the Orders record in the reverse direction, and then you would follow the arrow from the Orders record to the Customer record in the reverse direction. If you also wanted to print the customer's name, you could ask the DBMS to find the customer related to the current customer number.

Note that you find the customer to which an order is related by following the arrow, not by looking at a customer number field in an Orders record. There is no need for such a field because the relationships are part of the physical structure of a network model database. In general, you don't need to have fields in network model records to implement the relationships.

Hierarchical Model

Users perceive a **hierarchical model** database as a collection of hierarchies, or trees. Figure 9.24 shows part of a hierarchical database at a typical school, where each department employs many faculty, and each faculty advises many students. At the same time, each student is advised by exactly one faculty, and each faculty is employed in exactly one department.

FIGURE 9.24
....
Hierarchical
structure of part of
a school database

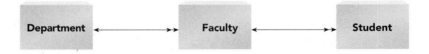

A **hierarchy** is really a network with an added restriction: no rectangle can have more than one "many" arrow entering the rectangle. (It doesn't matter how many "one" arrows leave a rectangle.) A hierarchy is thus a more restrictive structure than a network. Because two "many" arrows—one from the Orders record and a second from the Part record—enter the OrderLine record in the database shown in Figure 9.24, this database is not a hierarchy and cannot be implemented directly in a hierarchical model DBMS.

SUMMARY

- A distributed database is a single logical database that is physically divided among computers at several sites on a network. A user at any site can access data at any other site. A distributed database management system (DDBMS) is a database management system (DBMS) capable of supporting and manipulating distributed databases.

- Computers in a network communicate through messages. Minimizing the number of messages is important for rapid access to distributed databases.

- A homogenous DDBMS is one that has the same local DBMS at each site, whereas a heterogeneous DDBMS is one that does not.

- Location transparency, replication transparency, and fragmentation transparency are important characteristics of DDBMSs.

- DDBMSs permit local control of data, increased database capacity, improved system availability, and added efficiency.

- DDBMSs are more complicated than non-DDBMSs in the areas of the updating of replicated data, the processing of queries, the treatment of concurrent update, the measures for recovery, the management of the data dictionary, and the design of databases.

- The two-phase commit usually uses a coordinator to manage concurrent update.

- C.J. Date presented 12 rules that serve as a benchmark against which you can measure DDBMSs. These rules are local autonomy, no reliance on a central site, continuous operation, location transparency, fragmentation transparency, replication transparency, distributed query processing, distributed transaction management, hardware independence, operating system independence, network independence, and DBMS independence.

- A file server stores the files required by users and sends entire files to the users.

- In a two-tier client/server architecture, a DBMS runs on a file server, and the server sends only the requested data to the clients. The server performs database functions, and the clients perform presentation functions. A fat client can perform the business functions, or the server can perform the business functions in a thin client arrangement.

- In a three-tier client/server architecture, the clients perform the presentation functions, database servers perform the database functions, and application servers perform business functions. A three-tier architecture is more scalable than a two-tier architecture.

- The advantages of client/server systems are lower network traffic; improved processing distribution; thinner clients; greater processing transparency; increased network, hardware, and software transparency; improved security; decreased costs; and increased scalability.

- A trigger is an action that automatically occurs in response to an associated database operation. You can improve the performance of a client/server system by placing triggers in stored procedures.

- Online transaction processing (OLTP) is used with relational database management systems, and online analytical processing (OLAP) is used with data warehouses.

- A data warehouse is a subject-oriented, integrated, time-variant, nonvolatile collection of data in support of management's decision-making process.

- A typical data warehouse data structure is a star schema, consisting of a central fact table surrounded by dimension tables.

- Users perceive the data in a data warehouse as a multidimensional database in the shape of a data cube. OLAP software lets users slice and dice data, drill down data, and roll up data.

- Data mining consists of uncovering new knowledge, patterns, trends, and rules from the data stored in a data warehouse.

- E.F. Codd presented 12 rules that serve as a benchmark against which you can measure OLAP systems. These rules are multidimensional conceptual view; transparency; accessibility; consistent reporting performance; client/server architecture; generic dimensionality; dynamic sparse matrix handling; multiuser support; unrestricted, cross-dimensional operations; intuitive data manipulation; flexible reporting; and unlimited dimensions and aggregation levels.

- Object-oriented DBMSs deal with data as objects. An object is a set of related attributes along with the actions that are associated with the set of attributes. An OODBMS is one in which data and the actions that operate on the data are encapsulated into objects. A domain is the set of values that are permitted for an attribute. The term class refers to the general structure, and the term object refers to a specific occurrence of a class. Methods are the actions defined for a class, and a message is a request to execute a method. A subclass inherits the structure and methods of its superclass.

- The Unified Modeling Language (UML) is an approach to model all the various aspects of software development for object-oriented systems. The class diagram represents the design of an object-oriented database. Relationships are called associations, and visibility symbols indicate whether other classes can view or change the value in an attribute. Multiplicity indicates the number of objects that can be related to an individual object at the other end of the relationship. Generalization is the relationship between a superclass and a subclass.

- Rules that serve as a benchmark against which you can measure object-oriented systems are complex objects, object identity, encapsulation, information hiding, types or classes, inheritance, late binding, computational completeness, extensibility, persistence, performance, concurrent update support, recovery support, and query facility.

- The Internet and Web technology are playing an increasingly important role in an organization's database processing in an electronic commerce (e-commerce) environment—both in business to business (B2B) and business to consumer (B2C) processing. Organizations can place a Web server as an additional tier in a three-tier client/server architecture to handle the client interface.

- Extensible Markup Language (XML) is a customizable programming language suited to the exchange of data between different programs.

- IBM completed its development of Generalized Update Access Method (GUAM), the first DBMS, in 1964 and released it publicly as Data Language/I (DL/I) in 1966. DL/I is the data management component of the Information Management System (IMS). A General Electric team led by Charles Bachman developed the Integrated Data Store (IDS) DBMS, the first of the CODASYL systems, in the mid-1960s. Dr. E.F. Codd originated the relational model for databases, publishing his groundbreaking paper in 1970. IBM's System R was an early prototype DBMS that followed this model. In 1987, Servio Corporation's Gemstone was one of the first object-oriented database management systems (OODBMSs), which are based on object-oriented concepts.

- A data model consists of structure and operations components. DBMSs are characterized by the data model they follow. The four data models are hierarchical, network, relational, and object-oriented.

- A network model database is a collection of record types and relationships between the record types.

- A hierarchical model database is a collection of hierarchies, or trees, where a hierarchy is a network with an additional restriction: each record type can be related on the "many" side in only one one-to-many relationship.

KEY TERMS

Access
access delay
American National Standards Institute (ANSI)
application server
association
back-end machine
back-end processor
binary large objects (BLOBs)
binding
business to business (B2B)
business to consumer (B2C)
class
class diagram
client
client/server
communications network
COnference on DAta SYstems Languages
   (CODASYL)
coordinator
data cube
data fragmentation
Data Language/I (DL/I)
data mining
data model
data warehouse
database navigation
database server
DataBase Task Group (DBTG)
DB2
dBASE
dimension table
distributed database
distributed database management system
   (DDBMS)
domain
drill down
electronic commerce (e-commerce)
encapsulated
extensible
Extensible Markup Language (XML)
fact table
fat client
file server
fragmentation transparency
front-end machine
front-end processor

Gemstone
generalization
Generalized Update Access Method (GUAM)
global deadlock
heterogeneous DDBMS
hierarchical model
hierarchy
homogeneous DDBMS
Information Management System (IMS)
inheritance
Integrated Data Management System (IDMS)
Integrated Data Store (IDS)
local deadlock
local site
location transparency
message
method
multidimensional database
multiplicity
MySQL
network
network model
object
Objectivity/DB
object-oriented database management system
   (OODBMS)
object-relational DBMS (ORDBMS)
online analytical processing (OLAP)
online transaction processing (OLTP)
operations
Oracle
Paradox
persistence
polymorphism
primary copy
private visibility
protected visibility
public visibility
relational DBMS (RDBMS)
remote site
replication transparency
roll up
scalability
server
slice and dice
SQL Server

star schema
stored procedure
structure
subclass
superclass
Sybase
System R
thin client

three-tier architecture
trigger
two-phase commit
two-tier architecture
Unified Modeling Language (UML)
Versant
visibility symbol

■ REVIEW QUESTIONS

1. What is a distributed database? What is a distributed database management system (DDBMS)?

2. What different design decisions do you make to access data rapidly in a centralized database as compared to a distributed database?

3. How does a homogeneous DDBMS differ from a heterogeneous DDBMS? Which is more complex?

4. What is meant by a local site? By a remote site?

5. What is location transparency?

6. What is replication? Why is it used? What benefit is derived from using it? What are the biggest potential problems?

7. What is replication transparency?

8. What is data fragmentation? What purpose does data fragmentation serve?

9. What is fragmentation transparency?

10. Why is local control of data an advantage in a distributed database?

11. Why is the ability to increase system capacity an advantage in a distributed database?

12. Why is system availability an advantage in a distributed database?

13. Why is increased efficiency an advantage in a distributed database?

14. What are two disadvantages of updating replicated data in a distributed database?

15. What causes query processing to be more complex in a distributed database?

16. What is meant by local deadlock? By global deadlock?

17. Describe the two-phase commit process. How does it work? Why is it necessary?

18. Describe three possible approaches to storing data dictionary entries in a distributed system.

19. What additional factors must you consider during the information-level design of a distributed database?

20. What additional factors must you consider during the physical-level design of a distributed database?

21. List and briefly describe the 12 rules against which you can measure DDBMSs.

22. What is the difference between a file server and a client/server system?

23. In a two-tier client/server architecture, what problems occur when you place the business functions on the clients? On the server?

24. What is a fat client? What is a thin client?

25. What is scalability?

26. What is a three-tier architecture?

27. List the advantages of a client/server architecture as compared to a file server.

28. What are triggers? What purpose do they serve?

29. What language do you usually use to create triggers?

30. What are stored procedures? What purpose do they serve?

31. What are the characteristics of online transaction processing (OLTP) systems?

32. What is a data warehouse?

33. What does it mean when a data warehouse is nonvolatile?

34. What is a fact table in a data warehouse?

35. When do you use online analytical processing (OLAP)?

36. What three types of actions do users typically perform when they use OLAP software?

37. What is data mining?

38. What are the 12 rules against which you can measure OLAP systems?

39. What is an object-oriented DBMS?

40. What is a domain?

41. How do classes relate to objects?

42. What is a method? What is a message? How do messages relate to methods?

43. What is inheritance? What are the benefits to inheritance?

44. What is the Unified Modeling Language (UML)?

45. What are relationships called in UML?

46. What is a visibility symbol in UML?

47. What is multiplicity?

48. What is generalization?

49. What are the 14 rules against which you can measure object-oriented systems?

50. What four tiers might you find in a client/server architecture that provides Web access to databases, and what functions does each tier perform?

51. What is XML?

52. Which DBMS is considered to be the first product offered to the public?

53. What is the chief difference between network and hierarchical model databases?

■ Premiere Products Exercises

For the following exercises, you will answer problems and questions from management at Premiere Products. You do not use the Premiere Products database for any exercise.

1. Fragment the Customer table so that customers of rep 20 form a fragment named CustomerRep20, customers of rep 35 form a fragment named CustomerRep35, and customers of rep 65 form a fragment named CustomerRep65. (Include all fields from the Customer table in each fragment.) In addition, you need to fragment the Orders table so that orders are distributed and stored with the customers that placed the orders. For example, fragment OrdersRep20 consists of those orders placed by customers of rep 20. Write the SQL-type statements to create these fragments.

2. Create a class diagram for the Premiere Products database, as shown in Chapter 1, Figure 1.5. If you need to make any assumptions in preparing the diagram, document those assumptions.

3. A user queries the Part table in the Premiere Products database over the company intranet. Assume that the Part table contains 5,000 rows, each row is 1,000 bits long, the access delay is 2.5 seconds, the transmission rate is 50,000 bits per second, and only 20 of the 5,000 rows in the Part table satisfy the query conditions. Calculate the total communication time required for this query based on retrieving all table rows, one row at a time, and then calculate the total communication time required based on retrieving the 20 rows that satisfy the query conditions in a single message.

Henry Books Case

Ray Henry asks you to research improvements he might make to his database processing. To help him, he would like you to complete the following exercises. You do not use the Henry Books database for any exercise.

1. Use computer magazines or the Internet to investigate one of these DBMSs: DB2, SQL Server, MySQL, Oracle, or Sybase. Then prepare a report that explains how that DBMS handles two of the following distributed database functions: deadlock, fragmentation, replication, the data dictionary or log, and distributed queries.

2. Create a class diagram for the Henry Books database, as shown in Chapter 1, Figures 1.17 through 1.20. If you need to make any assumptions in preparing the diagram, document those assumptions.

3. Use computer magazines, books, or the Internet to investigate one of the following Web services: Application Programming Interface (API); Common Gateway Interface (CGI); Simple Object Access Protocol (SOAP); Universal Description, Discovery, and Integration (UDDI); or Web Services Description Language (WSDI). Then prepare a report that defines the Web service and explains its purpose and features.

4. Use computer magazines or the Internet to investigate one of these mobile databases: IBM's DB2 Everyplace, Microsoft SQL Server Windows CE Edition, Oracle Lite, or Sybase's SQL Anywhere. Then prepare a report that describes the features of the mobile database.

APPENDIX A: COMPREHENSIVE DESIGN EXAMPLE: MARVEL COLLEGE

▇ Introduction

Marvel College has decided to computerize its operations. In this appendix, you will design a database that will satisfy a number of user requirements by applying the design techniques that you learned in Chapter 6 to a significant set of requirements.

▇ Marvel College Requirements

Marvel College has provided you with the following requirements that its new system must satisfy. You will use these requirements to design a new database.

General Description

Marvel College is organized by department (math, physics, English, and so on). Most departments offer more than one major; for example, the math department might offer majors in calculus, applied mathematics, and statistics. Each major, however, is offered by only one department. Each faculty member is assigned to a single department. Students can have more than one major, but most students have only one. Each student is assigned a faculty member as an advisor for his or her major; students who have more than one major are assigned a faculty advisor for each major. The faculty member may or may not be assigned to the department offering the major.

A code that has up to three characters (CS for Computer Science, MTH for Mathematics, PHY for Physics, ENG for English, and so on) identifies each department. Each course is identified by the combination of the department code and a three-digit number (CS 162 for Programming I, MTH 201 for Calculus I, ENG 102 for Creative Writing, and so on). The number of credits offered by a particular course does not vary; that is, all students who pass the same course receive the same amount of credit.

A two-character code identifies the semester in which a course is taught (FA for fall, SP for spring, and SU for summer). It is combined with two digits that designate the year (for example, FA07 represents the fall semester of 2007). For a given semester, a department assigns each section of each course a four-digit schedule code (schedule code 1295 for section A of MTH 201, code 1297 for section B of MTH 201, code 1302 for section C of MTH 201, and so on). The schedule codes might vary from semester to semester. The schedule codes are listed in the school's time schedule, and students use them to indicate the sections in which they wish to enroll. (You'll learn more about the enrollment process later in this section.)

After all students have completed the enrollment process for a given semester, each faculty member receives a class list for each section he or she is teaching. In addition to listing the students in each section, the class list provides space to record the grade each student earns in the course. At the end of the semester, the faculty member enters the students' grades in this list and sends a copy of the list to the records office, where the grades are entered into the database. (In the future, the college plans to automate this part of the process.)

After an employee of the records office posts the grades (by entering them into the database), the DBMS generates a report card for each student and then the report cards are mailed to the addresses printed on the report card. The grades earned by a student become part of his or her permanent record and will appear on the student's transcript.

Report Requirements

Employees at Marvel College will require several reports to manage students, classes, schedules, and faculty members; these reports have the following requirements.

Report card: At the end of each semester, the system must produce a report card for each student. A sample report card is shown in Figure A.1.

MARVEL COLLEGE

Department	Course Number	Course Description	Grade	Credits Taken	Credits Earned	Grade Points
Computer Science	CS 162	Programming I	A	4	4	16.0
Mathematics	MTH 201	Calculus I	B+	3	3	9.9

Current Semester Totals

7	7	3.70	25.9
Credits Taken	Credits Earned	GPA	Total Points

Semester: FA07

Cumulative Totals

44	44	3.39	149.2
Credits Taken	Credits Earned	GPA	Total Points

Student Number: 381124188

Student Name & Address	Local Address (IF DIFFERENT)
Brian Connors 686 Franklin Hart, MI 48282	4672 Westchester Trent, MI 48222

Class list: The system must produce a class list for each section of each course; a sample class list is shown in Figure A.2. Note that space is provided for the grades. At the end of the semester, the instructor enters each student's grade and sends a copy of the class list to the records office.

CLASS LIST

Department: CS   Computer Science   Term: FA07
Course: 162   Programming I   (4 CREDITS)
Section: B
Schedule Code: 2366

Time: 1:00 - 1:50 M, T, W, F
PLACE: 118 SCR

Instructor: 462   Diane Johnson

Student Number	Student Name	Class Standing	Grade
381124188	Brian Connors	2	

Grade verification report: After the records office processes the class list, it returns the class list to the instructor with the grades entered into the report. The instructor uses the report to verify that the records office entered the students' grades correctly.

Time schedule: The time schedule shown in Figure A.3 lists all sections of all courses offered during a given semester. Each section has a unique four-digit schedule code. The time schedule lists the schedule code; the department offering the course; the course's number, section letter, and title; the instructor teaching the course; the time at which the course meets; the room in which the course meets; the number of credits generated by the course; and the prerequisites for the course. In addition to the information shown in Figure A.3, the time schedule includes the date the semester begins and ends, the date final exams begin and end, and the last withdrawal date (the last date on which students may withdraw from a course for a refund and without academic penalty).

TIME SCHEDULE   Term: FA07

Course #	Code #	Sect	Time	Room	Faculty
.	.	.	.	.	.
.	.	.	.	.	.
.	.	.	.	.	.

CHEMISTRY (CHM) Office: 341 NSB

111	Chemistry I			4 CREDITS	
	1740	A	10:00-10:50 M, T, W, F	102 WRN	Johnson
	1745	B	12:00-12:50 M, T, W, F	102 WRN	Lawrence

Prerequisite: MTH 110

112	Chemistry II			4 CREDITS	
	1790	A	10:00-11:50 M, W	109 WRN	Adams
	1795	B	12:00-1:50 T, R	102 WRN	Nelson

Prerequisite: CHM 111

114 ....

Registration request form: A sample registration request form is shown in Figure A.4. A student uses this form to request classes for the upcoming semester. Students indicate the sections for which they wish to register by entering each section's schedule code; for each of these sections, they may also enter a code for an alternate section in case the first requested section is full.

REGISTRATION REQUEST FORM

Student Number: 381124188   Term: SP08
Name: Brian Connors
Permanent Address: 686 Franklin   Local Address: 4672 Westchester
City: Hart   City: Trent
State: MI   State: MI
Zip: 48282   Zip: 48222

SCHEDULE CODES

	PRIMARY	ALTERNATE
1.		
2.		
3.		
4.		
5.		
6.		
7.		
8.		
9.		
10.		

Student schedule: After all students have been assigned to sections, the system produces a student schedule form, which is mailed to students so that they know the classes in which they have been enrolled. A sample student schedule form is shown in Figure A.5. This form shows the schedule for an individual student for the indicated semester.

STUDENT SCHEDULE

Student Number: 381124188   Term: SP08
Name: Brian Connors
Permanent Address: 686 Franklin   Local Address: 4672 Westchester
City: Hart   City: Trent
State: MI   State: MI
Zip: 48282   Zip: 48222

Schedule Code	Course Number	Course Description	Section	Credits	Time	Room
2366	CS 253	Programming II	B	4	1:00–1:50 M, T, W, F	118 SCR
.	.	.	.	.		
.	.	.	.	.		
.	.	.	.	.		
		Total Credits:		16		

Full student information report: This report lists complete information about a student, including his or her major(s) and all grades received to date. A sample of a full student information report is shown in Figure A.6.

FULL STUDENT INFORMATION

Student Number: 381124188   Term: FA07
                  Name: Brian Connors
Permanent Address: 686 Franklin   Local Address: 4672 Westchester
                  City: Hart   City: Trent
                  State: MI   State: MI
                  Zip: 48282   Zip: 48222

Major 1: Information Sys.   Department: Computer Science   Advisor: Mark Lawerence
Major 2: Accounting   Department: Business   Advisor: Jill Thomas
Major 3:   Department: Advisor:

Term	Course Number	Course Description	Credits	Grade Earned	Grade Points
SP07	MTH 123	Trigonometry	4	A	16.0
	HST 201	Western Civilization	3	A-	11.1
	ENG 101	American Literature	3	A	12.0
FA07	CS 162	Programming I	4	A	16.0
	MTH 201	Calculus I	4	B+	9.9

Credits Attempted: 44
Credits Earned: 44
Grade Points: 149.2
Grade Point Avg: 3.39
Class Standing: 2

Faculty information report: This report lists all faculty by department and contains each faculty member's ID number, name, address, office location, phone number, current rank (Instructor, Assistant Professor, Associate Professor, or Professor), and starting date of employment. It also lists the number, name, and local and permanent addresses of each of the faculty member's advisees; the code number and description of the major in which the faculty member is advising each advisee; and the code number and description of the department to which this major is assigned. (Remember that this department need not be the one to which the faculty member is assigned.)

Work version of the time schedule: This report is similar to the original time schedule (see Figure A.3), but is designed for the college's internal use. It shows the current enrollments in each section of each course, as well as the maximum enrollment permitted per section. It is more current than the time schedule itself. (When students register for courses, enrollment figures are updated on the work version of the time schedule. When room or faculty assignments are changed, this information also is updated. A new version of this report that reflects the revised figures is printed after being updated.)

Course report: For each course, this report lists the code and name of the department that is offering the course, the course number, the description of the course, and the number of credits awarded. This report also includes the department and course number for each prerequisite course.

332

Update (Transaction) Requirements

In addition to being able to add, change, and delete any of the information in the report requirements, the system must be able to accomplish the following update requirements:

Enrollment: When a student attempts to register for a section of a course, the system must determine whether the student has received credit for all prerequisites to the course. If the student is eligible to enroll in the course and the number of students currently enrolled in the section is less than the maximum enrollment, enroll the student.

Post grades: For each section of each course, the system must post the grades that are indicated on the class list submitted by the instructor and produce a grade verification report. ("Posting the grades" is the formal term for the process of entering them permanently into the students' computerized records.)

Purge: Marvel College retains section information, including grades earned by the students in each section, for two semesters following the end of the semester. Then, the system removes this information from the system. (Grades assigned to students are retained by course, but not by section.)

Marvel College Information-Level Design

You should give some consideration to the overall requirements before you apply the methodology to the individual user requirements. For example, by examining the documents shown in Figures A.1 through A.6, you may have identified the following entities: department, major, faculty member, student, course, and semester.

Note: Your list might include the section and grade entities. On the other hand, you may not have included the semester entity. In the long run, as long as the list is fairly reasonable, it won't make much difference. In fact, you may remember that this step is really not even necessary. The better you do your job now, however, the simpler the process will be later on.

After identifying the entities, you assign a primary key to each one. In general, this step will require some type of consultation with users. You may need to ask users directly for the required information, or you may be able to obtain it from some type of survey form. Let's assume that having had such a consultation, you created a relation for each of these entities and assigned them the following primary keys:

```
Department (DepartmentCode,
Major (MajorNum,
Faculty (FacultyNum,
Student (StudentNum,
Course (DepartmentCode, CourseNum,
Semester (SemesterCode,
```

Note that the primary key for the Course table consists of two attributes, DepartmentCode (such as CS) and CourseNum (such as 153), both of which are required. The database could contain, for example, CS 153 and CS 353. Thus, the department code alone cannot be the primary key. Similarly, the database could contain ART 101 and MUS 101, two courses with the same course number, but with different department codes. Thus, the course number alone cannot be the primary key either.

Now you can begin examining the individual user views as stated in the requirements. You can create relations for these user views, represent any keys, and merge the new user views into the cumulative design. Your first task is to determine the individual user views. In the list of requirements, the term "user view" never appeared. Instead, Marvel College provided a general description of the system, together with a collection of report requirements and another collection of update requirements. How do these requirements relate to user views?

Certainly, you can think of each report requirement and each update requirement as a user view. But what do you do with the general description? Do you think of each paragraph (or perhaps each sentence) in it as representing a user view, or do you use it to furnish additional information about the report and update requirements? Both approaches are acceptable. Because the second approach is often easier, it is the approach you will follow in this text. Think of the report and update requirements as user views, and use the statements in the general description as providing additional information about these user views when needed. You will also consider the general description during the review process to ensure that your final design can satisfy all the functionality it describes.

First, let's consider one of the simpler user views, the course report. (Technically, you can examine user views in any order. Sometimes you take them in the order they are listed. In other cases, you may be able to come up with a better order. Often, examining some of the simpler user views first is a reasonable approach.)

Before you proceed with the design, consider the following methodology. First, with some of the user views, you will attempt to determine the relations involved by carefully determining the entities and relationships between them and using this information when creating the relations. This process means that from the outset, the collection of tables created will be in or close to third normal form. With other user views, you will create a single relation that potentially contains some number of repeating groups. In these cases, as you will see, the normalization process still produces a correct design, but it also involves more work. In practice, the more experience a designer has had, the more likely he or she is to create third normal form relations immediately.

Second, the name of an entity or attribute may vary from one user view to another, and this difference requires resolution. You will attempt to use names that are the same.

User View 1—Course report: For each course, list the code and name of the department that is offering the course, the course number, the course title, and the number of credits awarded. This report also includes the department and course number for each prerequisite course. Forgetting for the moment the requirement to list prerequisite courses, the basic relation necessary to support this report is:

```
Course (DepartmentCode, DepartmentName, CourseNum, CourseTitle,
    NumCredits)
```

The combination of DepartmentCode and CourseNum uniquely determines all the other attributes. In this relation, DepartmentCode determines DepartmentName, and thus the table is not in second normal form (an attribute depends on only a portion of the key). To correct this situation, the table is split into the following two tables:

```
Course (DepartmentCode, CourseNum, CourseTitle, NumCredits)
Department (DepartmentCode, DepartmentName)
```

The DepartmentCode attribute in the first relation is a foreign key identifying the second relation.

To maintain prerequisite information, you need to create the relation Prereq:

```
Prereq (DepartmentCode, CourseNum, DepartmentCode/1,
    CourseNum/1)
```

In this table, the attributes DepartmentCode and CourseNum refer to the course, and the attributes DepartmentCode/1 and CourseNum/1 refer to the prerequisite course. If CS 362 has a prerequisite of MTH 345, for example, there will be a row in the Prereq table in which the DepartmentCode is CS, the CourseNum is 362, the DepartmentCode/1 is MTH, and the CourseNum/1 is 345.

Note: Because there are two attributes named DepartmentCode and two attributes named CourseNum, you must be able to distinguish between them. The software used to produce these diagrams makes the distinction by appending the characters "/1" to one of the names, and that is why these names appear in the Prereq table. In this example, the DepartmentCode/1 and CourseNum/1 attributes represent the department code and course number of the *prerequisite* course. When it is time to implement the design, you will typically assign them names that are more descriptive. For instance, you might name them PrereqDepartmentCode and PrereqCourseNum.

The DBDL version of these tables is shown in Figure A.7.

FIGURE A.7
....
DBDL for
User View 1

```
Department (DepartmentCode, DepartmentName)

Course (DepartmentCode, CourseNum, CourseTitle, NumCredits)
    FK DepartmentCode → Department

Prereq (DepartmentCode, CourseNum, DepartmentCode/1,
    CourseNum/1)
    FK DepartmentCode, CourseNum → Course
    FK DepartmentCode/1, CourseNum/1 → Course
```

The result of merging these relations into the cumulative design appears in the entity-relationship (E-R) diagram shown in Figure A.8. Notice that the Department and Course tables have been merged with the existing Department and Course tables in the cumulative design. In the process, the attribute DepartmentName was added to the Department table, and the attributes CourseTitle and NumCredits were added to the Course table. In addition, the attribute DepartmentCode in the Course table is a foreign key. Because the Prereq table is new, it was added to the cumulative design in its entirety. Notice also that you do not yet have any relationships among the entities Student, Major, Faculty, and Semester.

In Figure A.8, there are two relationships between Course and Prereq. In order to distinguish between them, it is necessary to name the relationships. In the figure, the name for the first relationship is "has prereq" and the name for the second relationship is "is prereq."

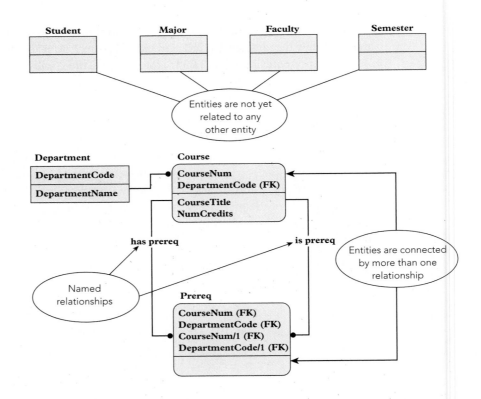

Note: When using a software tool to produce E-R diagrams, the software might reverse the order of the fields that make up the primary key. For example, the E-R diagram in Figure A.8 indicates that the primary key for the Course table is CourseNum and then DepartmentCode, even though you intended it to be DepartmentCode and then CourseNum. This difference is not a problem. Indicating the fields that comprise the primary key is what is significant, not the order in which they appear.

User View 2—Faculty information report: List all faculty by department and each faculty member's ID number, name, address, office location, phone number, current rank (Instructor, Assistant Professor, Associate Professor, or Full Professor), and starting date of employment. In addition, list the number, name, and local and permanent addresses of each of the faculty member's advisees; the code number and description of the major in which the faculty member is advising each advisee; and the code number and description of the department to which this major is assigned. This user view involves three entities (departments, faculty, and advisees), so you can create the following three tables:

```
Department (
Faculty (
Advisee (
```

The next step is to assign a primary key to each table. Before doing so, however, you should briefly examine the tables in the cumulative design and use the same names for any existing tables or attributes. In this case, you would use DepartmentCode as the primary key for the Department table and FacultyNum as the primary key for the Faculty table. There is no Advisee table in the cumulative collection, but there is a Student table. Because advisees and students are the same, rename the Advisee entity to Student and use the StudentNum attribute as the primary key rather than AdvisorNum. Your efforts yield the following tables and primary keys:

```
Department (DepartmentCode,
Faculty (FacultyNum,
Student (StudentNum,
```

Next, add the remaining attributes to the tables:

```
Department (DepartmentCode, DepartmentName)
Faculty (FacultyNum, LastName, FirstName, Street, City, State, Zip,
     OfficeNum, Phone, CurrentRank, StartDate, DepartmentCode)
Student (StudentNum, LastName, FirstName, LocalStreet, LocalCity,
     LocalState, LocalZip, PermStreet, PermCity, PermState,
     PermZip, (MajorNum, Description, DepartmentCode,
     FacultyNum, LastName, FirstName) )
```

The DepartmentCode attribute is included in the Faculty table because there is a one-to-many relationship between departments and faculty. Because a student can have more than one major, the information about majors (number, description, department, and the number and name of the faculty member who advises this student in this major) is a repeating group.

Because the key to the repeating group in the Student table is MajorNum, removing this repeating group yields:

```
Student (StudentNum, LastName, FirstName, LocalStreet, LocalCity,
     LocalState, LocalZip, PermStreet, PermCity, PermState,
     PermZip, MajorNum, Description, DepartmentCode,
     FacultyNum, LastName, FirstName)
```

Converting this relation to second normal form produces the following tables:

```
Student (StudentNum, LastName, FirstName, LocalStreet, LocalCity,
     LocalState, LocalZip, PermStreet, PermCity, PermState,
     PermZip)
Major (MajorNum, Description, DepartmentCode, DepartmentName)
Advises (StudentNum, MajorNum, FacultyNum)
```

In this case, you must remove the following dependencies to create third normal form tables: OfficeNum determines Phone in the Faculty table, and DepartmentCode determines DepartmentName in the Major table. Removing these dependencies produces the following collection of tables:

```
Department (DepartmentCode, DepartmentName)
Faculty (FacultyNum, LastName, FirstName, Street, City, State, Zip,
     OfficeNum, CurrentRank, StartDate, DepartmentCode)
Student (StudentNum, LastName, FirstName, LocalStreet, LocalCity,
     LocalState, LocalZip, PermStreet, PermCity, PermState,
     PermZip)
Advises (StudentNum, MajorNum, FacultyNum)
Office (OfficeNum, Phone)
Major (MajorNum, Description, DepartmentCode)
```

The DBDL representation is shown in Figure A.9.

```
Department (DepartmentCode, DepartmentName)

Student (StudentNum, LastName, FirstName, LocalStreet, LocalCity,
     LocalState, LocalZip, PermStreet, PermCity, PermState, PermZip)

Office (OfficeNum, Phone)

Faculty (FacultyNum, LastName, FirstName, Street, City, State, Zip,
     OfficeNum, CurrentRank, StartDate, DepartmentCode)
     FK OfficeNum → Office
     FK DepartmentCode → Department

Major (MajorNum, Description, DepartmentCode)
     FK DepartmentCode → Department

Advises (StudentNum, MajorNum, FacultyNum)
     FK StudentNum → Student
     FK FacultyNum → Faculty
     FK MajorNum → Major
```

The result of merging these tables into the cumulative design is shown in Figure A.10. The tables Student, Faculty, Major, and Department are merged with the existing tables with the same primary keys and with the same names. Nothing new is added to the Department table in the process, but the other tables receive additional attributes. In addition, the Faculty table also receives two foreign keys, OfficeNum and DepartmentCode. The Major table receives one foreign key, DepartmentCode. The Advises and Office tables are new and thus are added directly to the cumulative design.

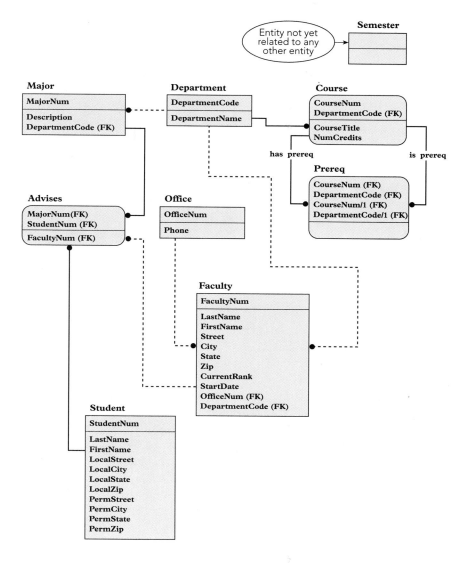

User View 3—Report card: At the end of each semester, the system must produce a report card for each student. Report cards are fairly complicated documents in which the appropriate underlying relations are not immediately apparent. In such a case, it's a good idea to first list all the attributes in the report card and assign them appropriate names, as shown in Figure A.11. After identifying the attributes, you should list the functional dependencies that exist between these attributes. The information necessary to determine functional dependencies must ultimately come from the user, although you can often accurately guess most of them.

FIGURE A.11
· · · · · · · · · · · · · · · · ·
Attributes on a
report card from
Marvel College

```
Department
CourseNum
CourseTitle
Grade
CreditsTaken        (Course)
CreditsEarned       (Course)
GradePoints         (Course)
CreditsTaken        (Semester)
CreditsEarned       (Semester)
GPA                 (Semester)
TotalPoints         (Semester)
CreditsTaken        (Cumulative)
CreditsEarned       (Cumulative)
GPA                 (Cumulative)
TotalPoints         (Cumulative)
SemesterCode
StudentNum
LastName
FirstName
Address
City
State
Zip
LocalAddress
LocalCity
LocalState
LocalZip
```

Note: Notice that there is duplication of names in the list. CreditsEarned, for example, appears three times: once for the course, once for the semester, and once for the cumulative number of credits earned by the student. You could assign these columns different names at this point. The names could be CreditsEarnedCourse, CreditsEarnedSemester, and CreditsEarnedCumulative. Alternatively, you can assign them the same name with an explanation of the purpose of each one in parentheses, as shown in Figure A.11. Of course, after you have determined all the tables and assigned columns to them, you must ensure that the column names within a single table are unique.

Let's assume that the system's users have verified the attributes listed in Figure A.11 and that your work is correct. Figure A.12 shows the functional dependencies among the attributes you identified on the report card. The student number alone determines many of the other attributes.

In addition to the student number, the semester must be identified to determine credits taken and earned, grade point average (GPA), and total points each semester. The combination of a department name (such as Computer Science) and a course number (such as 153) determines a course title and the number of credits.

Finally, the student number, the semester (semester and year), the department, and the course (department and course number) are required to determine an individual grade in a course, the credits earned from the course, and the grade points in a course. (The semester is required because the same course might be offered during more than one semester at Marvel College.)

FIGURE A.12
· · · · · · · · · · · · · · · · · ·
Functional
dependencies
among the
attributes on a
report card

```
StudentNum→
     CreditsTaken          (Cumulative)
     CreditsEarned         (Cumulative)
     GPA                   (Cumulative)
     TotalPoints           (Cumulative)
     LastName
     FirstName
     Address
     City
     State
     Zip
     LocalAddress
     LocalCity
     LocalState
     LocalZip

StudentNum, SemesterCode→
     CreditsTaken          (Semester)
     CreditsEarned         (Semester)
     GPA                   (Semester)
     TotalPoints           (Semester)

DepartmentName, CourseNum→
     CourseTitle
     CreditsTaken          (Course) (Same as NumCredits)

StudentNum, SemesterCode, DepartmentName, CourseNum→
     Grade
     CreditsEarned         (Course)
     GradePoints           (Course)
```

Note: There is a parenthetical comment after CreditsTaken in the section determined by DepartmentName and CourseNum. It indicates that CreditsTaken is the same as NumCredits, which is a column already in the cumulative design. Documenting that the name you have chosen is a synonym for a name already in the cumulative design is a good practice.

The next step is to create a collection of tables that will support this user view. A variety of approaches will work. You could combine all the attributes into a single table, which you then would convert to third normal form. (In such a table, the combination of department, course number, course title, grade, and so on would be a repeating group.) Alternatively, you could use the functional dependencies to determine the following collection of relations:

```
Student (StudentNum, LastName, FirstName, PermStreet, PermCity, Perm
     State, PermZip, LocalStreet, LocalCity, LocalState,
     LocalZip, CreditsTaken, CreditsEarned, GPA, TotalPoints)
StudentSemester (StudentNum, SemesterCode, CreditsTaken,
     CreditsEarned, GPA, TotalPoints)
Course (DepartmentCode, CourseNum, CourseTitle, NumCredits)
StudentGrade (StudentNum, SemesterCode, DepartmentName, CourseNum,
     Grade, CreditsEarned, GradePoints)
```

All these relations are in third normal form. The only change you should make concerns the DepartmentName attribute in the StudentGrade table. In general, if you encounter an attribute for which there exists a determinant that is not in the table, you should add the determinant. In this case, DepartmentCode is a determinant for DepartmentName, but it is not in the table, so you should add DepartmentCode. In the normalization process, DepartmentName will then be removed and placed in another table whose key is DepartmentCode. This other table will be merged with the Department table without the addition of any new attributes. The resulting StudentGrade table is:

```
StudentGrade (StudentNum, SemesterCode, DepartmentCode, CourseNum,
        Grade, CreditsEarned, GradePoints)
```

Before representing this design in DBDL, let's examine the StudentSemester entity. Some of the attributes it contains (CreditsTaken, CreditsEarned, GPA, and TotalPoints) all refer to the current semester and all appear on a report card. Let's assume that after further checking, you find that these attributes are all easily calculated from other fields on the report card. Rather than storing these attributes in the database, you can ensure that the program that produces the report cards performs the necessary calculations. For this reason, you will remove the StudentSemester table from the collection of tables to be documented and merged. (If these attributes are also required by some other user view in which the same computations are not as practical, they might find their way into the database when that user view is analyzed.)

• •

QUESTION

Determine the tables and keys required for User View 3. Merge the result into the cumulative design and draw the E-R diagram for the new cumulative design.

ANSWER

Figure A.13 shows the new cumulative design.

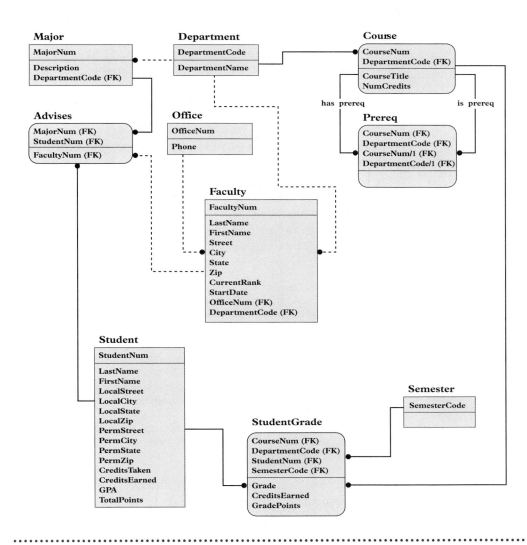

User View 4—Class list: The system must produce a class list for each section of each course. Space is provided for the grades. At the end of the semester, the instructor enters each student's grade and sends a copy of the class list to the records office. Let's assume that after examining the sample class list report (see Figure A.2), you decide to create a single table (actually an unnormalized table) that contains all the attributes on the class list, with the student information (number, name, class standing, and grade) as a repeating group. (Applying the tips for determining the relations to support a given user view would lead more directly to the result, but, for the sake of developing the example, assume you haven't done that yet.) The unnormalized table created by this method would be:

```
ClassList (DepartmentCode, DepartmentName, SemesterCode, CourseNum,
      CourseTitle, NumCredits, SectionLetter, ScheduleCode, Time, Room,
      FacultyNum, FacultyLastName, FacultyFirstName, (StudentNum,
      StudentLastName, StudentFirstName, ClassStanding, Grade) )
```

Note: Because attribute names within a single table must be unique, it is not permissible to assign the attribute name LastName to both the faculty and the student last names. Thus, the attributes that store the last and first names of a faculty member are named FacultyLastName and FacultyFirstName. Similarly, the attributes that store the last and first names of a student are named StudentLastName and StudentFirstName.

Note that you have not yet indicated the primary key. To identify a given class within a particular semester requires either the combination of a department code, course number, and section letter or, more simply, the schedule code. Using the schedule code as the primary key, however, is not quite adequate. Because the information from more than one semester will be on file at the same time and because the same schedule code could be used in two different semesters to represent different courses, the primary key must also contain the semester code. When you remove the repeating group, this primary key expands to contain the key for the repeating group, which, in this case, is the student number. Thus, converting to first normal form yields:

```
ClassList (DepartmentCode, DepartmentName, SemesterCode, CourseNum,
      CourseTitle, NumCredits, SectionLetter, ScheduleCode, Time, Room,
      FacultyNum, FacultyLastName, FacultyFirstName, StudentNum,
      StudentLastName, StudentFirstName, ClassStanding, Grade)
```

Converting to third normal form yields the following collection of tables:

```
Department (DepartmentCode, DepartmentName)
Section (SemesterCode, ScheduleCode, DepartmentCode, CourseNum,
      SectionLetter, Time, Room, FacultyNum)
Faculty (FacultyNum, LastName, FirstName)
StudentClass (SemesterCode, ScheduleCode, StudentNum, Grade)
Student (StudentNum, LastName, FirstName, ClassStanding)
Course (DepartmentCode, CourseNum, CourseTitle, NumCredits)
```

Note: Because the last name of a faculty member is now in a separate table from the last name of a student, it is no longer necessary to have different names. Thus FacultyLastName and StudentLastName have been shortened to LastName. Similarly, FacultyFirstName and StudentFirstName have been shortened to FirstName.

· ·

QUESTION | Why was the grade included?

ANSWER | Although the grade is not actually printed on the class list, it will be entered on the form by the instructor and sent to the records office for posting. The grade verification report differs from the class list only in that the grade is printed. Thus, the grade will ultimately be required and it is appropriate to deal with it here.

QUESTION | Determine the tables and keys required for User View 4. Merge the result into the cumulative design and draw the E-R diagram for the new cumulative design.

ANSWER | Figure A.14 shows the new cumulative design.

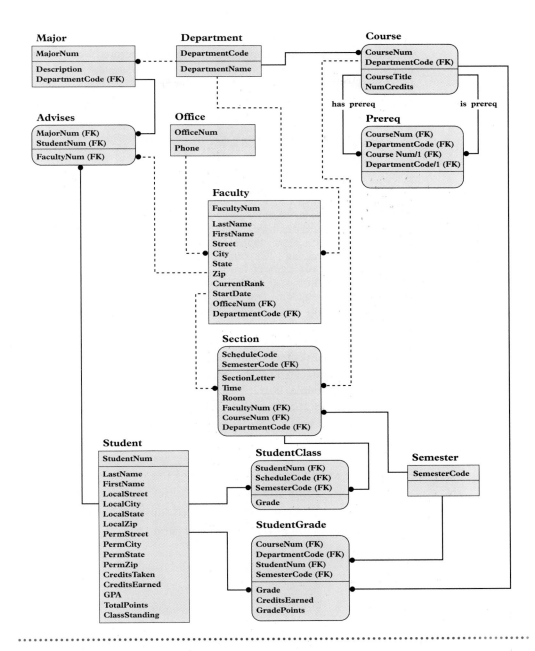

User View 5—Grade verification report: After the records office processes the class list, it returns the class list to the instructor with the grades entered into the report. The instructor uses the report to verify that the records office entered the students' grades correctly. Because the only difference between the class list and the grade verification report is that the grades are printed on the grade verification report, the user views will be quite similar. In fact, because you made a provision for the grade when treating the class list, the views are identical, and no further treatment of this user view is required.

User View 6—Time schedule: List all sections of all courses offered during a given semester. Each section has a unique four-digit schedule code. The time schedule lists the schedule code; the department offering the course; the course's number, section letter, and title; the instructor teaching the course; the time at which the course meets; the room in which the course meets; the number of credits generated by the course; and the prerequisites for the course. In addition to the information shown in the figure, the time schedule includes the date the semester begins and ends, the date final exams begin and end, and the last withdrawal date. The attributes on the time schedule are as follows: term (which is a synonym for semester), department code, department name, location, course number, course title, number of credits, schedule code, section letter, meeting time, meeting place, and instructor name.

You could create a single relation containing all these attributes and then normalize that relation, or you could apply the tips presented in Chapter 8 for determining the collection of relations. In either case, you ultimately create the following collection of relations:

```
Department (DepartmentCode, DepartmentName, Location)
Course (DepartmentCode, CourseNum, CourseTitle, NumCredits)
Section (SemesterCode, ScheduleCode, DepartmentCode, CourseNum,
      SectionLetter, Time, Room, FacultyNum)
Faculty (FacultyNum, LastName, FirstName)
Semester (SemesterCode, StartDate, EndDate, ExamStartDate,
      ExamEndDate, WithdrawalDate)
```

Note: Actually, given the attributes in this user view, the Section relation would contain the instructor's name (LastName and FirstName). There was no mention of instructor number. In general, as you saw earlier, it's a good idea to include determinants for attributes whenever possible. In this example, because FacultyNum determines LastName and FirstName, you add FacultyNum to the Section relation, at which point the Section relation is not in third normal form. Converting to third normal form produces the collection of relations previously shown.

QUESTION Determine the tables and keys required for User View 6. Merge the result into the cumulative design and draw the E-R diagram for the new cumulative design.

ANSWER Figure A.15 shows the new cumulative design.

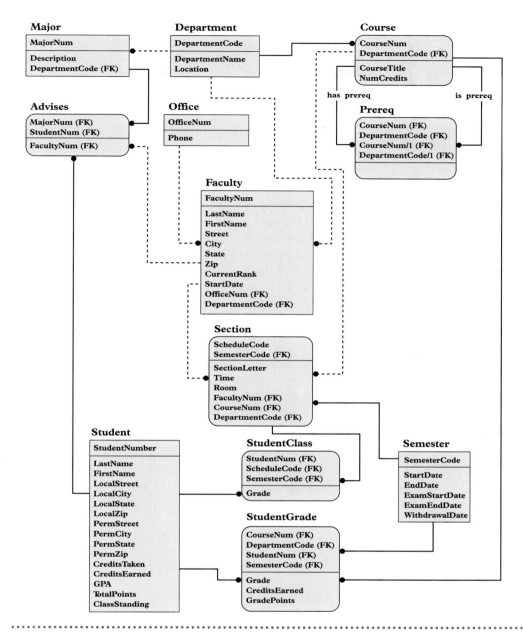

User View 7—Registration request form: A student uses this form to request classes for the upcoming semester. Students indicate the sections for which they want to register by entering the sections' schedule codes; for each section, they may also enter a code for an alternate section in case the requested primary section is full. The collection of tables to support this user view includes a Student table that consists of the primary key, StudentNum, and all the attributes that depend only on StudentNum. These attributes include LastName, FirstName, LocalStreet, and so on. Because all attributes in this table are already in the Student table in the cumulative design, this user view will not add anything new and there is no need for further discussion of it here.

The portion of this user view that is not already present in the cumulative design concerns the primary and alternate schedule codes that students request. A table to support this portion of the user view must contain both a primary and an alternate schedule code. It must also contain the number of the student making the request. Finally, to allow the flexibility of retaining this information for more than a single semester to allow registration for more than a semester at a time, the table must also include the semester in which the request is made. This leads to the following relation:

RegistrationRequest (<u>StudentNum</u>, <u>PrimaryCode</u>, AlternateCode,
    <u>SemesterCode</u>)

For example, if student 381124188 were to request the section with schedule code 2345 and then request the section with schedule code 2396 as an alternate for the FA07 semester, the row (381124188, 2345, 2396, "FA07") would be stored. The student number, the primary schedule code, the alternate schedule code, and the semester code are required to uniquely identify a particular row.

. . . .

QUESTION

Determine the tables and keys required for User View 7. Merge the result into the cumulative design and draw the E-R diagram for the new cumulative design.

ANSWER

Figure A.16 shows the new cumulative design. Notice that there are two relationships joining the Section table to the RegistrationRequest table, so you must name each of them. In this case, you use "primary" and "alternate," indicating that one relationship relates a request to the primary section chosen and that the other relates it to the alternate section, when there is one.

Note: The foreign keys are the combination of PrimaryCode and SemesterCode as well as the combination of AlternateCode and SemesterCode. Because PrimaryCode and AlternateCode are portions of the foreign keys that must match the ScheduleCode in the Section table, they have been renamed ScheduleCode and ScheduleCode/1. Likewise, the second SemesterCode has been renamed SemesterCode/1.

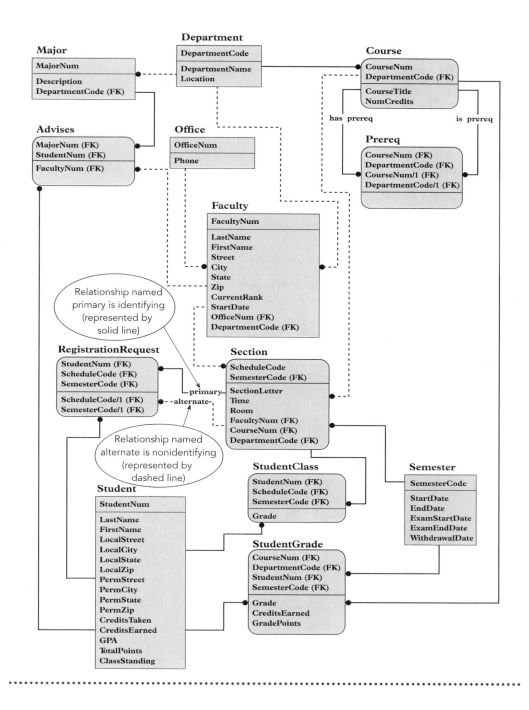

Major

MajorNum
Description
DepartmentCode (FK)

Department

DepartmentCode
DepartmentName
Location

Course

CourseNum
DepartmentCode (FK)
CourseTitle
NumCredits

has prereq   is prereq

Advises

MajorNum (FK)
StudentNum (FK)
FacultyNum (FK)

Office

OfficeNum
Phone

Prereq

CourseNum (FK)
DepartmentCode (FK)
CourseNum/1 (FK)
DepartmentCode/1 (FK)

Faculty

FacultyNum
LastName
FirstName
Street
City
State
Zip
CurrentRank
StartDate
OfficeNum (FK)
DepartmentCode (FK)

Relationship named
primary is identifying
(represented by
solid line)

RegistrationRequest

StudentNum (FK)
ScheduleCode (FK)
SemesterCode (FK)
ScheduleCode/1 (FK)
SemesterCode/1 (FK)

primary
alternate

Section

ScheduleCode
SemesterCode (FK)
SectionLetter
Time
Room
FacultyNum (FK)
CourseNum (FK)
DepartmentCode (FK)

Relationship named
alternate is nonidentifying
(represented by
dashed line)

Student

StudentNum
LastName
FirstName
LocalStreet
LocalCity
LocalState
LocalZip
PermStreet
PermCity
PermState
PermZip
CreditsTaken
CreditsEarned
GPA
TotalPoints
ClassStanding

StudentClass

StudentNum (FK)
ScheduleCode (FK)
SemesterCode (FK)
Grade

Semester

SemesterCode
StartDate
EndDate
ExamStartDate
ExamEndDate
WithdrawalDate

StudentGrade

CourseNum (FK)
DepartmentCode (FK)
StudentNum (FK)
SemesterCode (FK)
Grade
CreditsEarned
GradePoints

User View 8—Student schedule: After all students are assigned to sections, the system produces a student schedule form, which is mailed to students so that they know the classes in which they have been enrolled. Suppose you had created a single unnormalized relation to support the student schedule. This unnormalized relation would contain a repeating group representing the lines in the body of the schedule, as follows:

```
StudentSchedule (StudentNum, SemesterCode, LastName, FirstName, LocalStreet,
    LocalCity, LocalState, LocalZip, PermStreet, PermCity,
    PermState, PermZip, (ScheduleCode, DepartmentName,
    CourseNum, CourseTitle, SectionLetter, NumCredits,
    Time, Room) )
```

At this point, you remove the repeating group to convert to first normal form, yielding the following:

```
StudentSchedule (StudentNum, SemesterCode, LastName, FirstName, LocalStreet,
    LocalCity, LocalState, LocalZip, PermStreet, PermCity,
    PermState, PermZip, ScheduleCode, DepartmentCode,
    CourseNum, CourseTitle, SectionLetter, NumCredits,
    Time, Room)
```

Note that the primary key expands to include ScheduleCode, which is the key to the repeating group. Converting this table to second normal form produces:

```
Student (StudentNum, LastName, FirstName, LocalStreet, LocalCity,
    LocalState, LocalZip, PermStreet, PermCity, PermState,
    PermZip)
StudentSchedule (StudentNum, SemesterCode, ScheduleCode)
Section (SemesterCode, ScheduleCode, DepartmentCode, CourseNum,
    CourseTitle, SectionLetter, NumCredits, Time, Room)
Course (DepartmentCode, CourseNum, CourseTitle, NumCredits)
```

Removing the attributes that depend on the determinant of DepartmentCode and CourseNum from the Section table and converting this collection of tables to third normal form produces:

```
Student (StudentNum, LastName, FirstName, LocalStreet, LocalCity,
    LocalState, LocalZip, PermStreet, PermCity, PermState,
    PermZip)
StudentSchedule (StudentNum, SemesterCode, ScheduleCode)
Section (SemesterCode, ScheduleCode, DepartmentCode, CourseNum,
    SectionLetter, Time, Room)
Course (DepartmentCode, CourseNum, CourseTitle, NumCredits)
```

Merging this collection into the cumulative design does not add anything new. In the process, you can merge the StudentSchedule table with the StudentClass table.

User View 9—Full student information report: List complete information about a student, including his or her majors and all grades received to date. Suppose you attempted to place all the attributes on the full student information report into a single unnormalized relation. The table has two separate repeating groups: one for the different majors a student might have, and the other for all the courses the student has taken.

Note: There are also several attributes, such as name, address, and so on, that would not be in the repeating groups. All these attributes are already in the cumulative design, however, and are not addressed here.

The table with repeating groups is:

```
Student (StudentNum, (MajorNum, DepartmentCode, LastName, FirstName),
     (SemesterCode, DepartmentCode, CourseNum, CourseTitle,
     NumCredits, Grade, GradePoints) )
```

Recall from Chapter 5 that you should separate repeating groups when a relation has more than one. If you don't do so, you will typically have problems with fourth normal form. Separating the repeating groups in this example produces:

```
StudentMajor (StudentNum, (MajorNum, DepartmentCode, LastName,
     FirstName) )
StudentCourse (StudentNum, (SemesterCode, DepartmentCode,
     CourseNum, CourseTitle, NumCredits, Grade, GradePoints) )
```

Converting these tables to first normal form and including FacultyNum, which is a determinant for LastName and FirstName, produces:

```
StudentMajor (StudentNum, MajorNum, DepartmentCode, FacultyNum,
     LastName, FirstName)
StudentCourse (StudentNum, SemesterCode, DepartmentCode, CourseNum,
     CourseTitle, NumCredits, Grade, Grade Points)
```

The StudentCourse table is not in second normal form because CourseTitle and NumCredits depend only on the DepartmentCode, CourseNum combination. The StudentMajor table is not in second normal form either because DepartmentCode depends on MajorNum. Removing these dependencies produces the following tables:

```
StudentMajor (StudentNum, MajorNum, FacultyNum, LastName,
     FirstName)
Major (MajorNum, DepartmentCode)
StudentCourse (StudentNum, SemesterCode, DepartmentCode, CourseNum,
     Grade, GradePoints)
Course (DepartmentCode, CourseNum, CourseTitle, NumCredits)
```

Other than the StudentMajor table, all these relations are in third normal form. Converting the StudentMajor table to third normal form produces the following tables:

```
StudentMajor (StudentNum, MajorNum, FacultyNum)
Faculty (FacultyNum, LastName, FirstName)
```

Merging this collection into the cumulative design does not add anything new. (You can merge the StudentMajor table with the Advises table without adding any new attributes.)

User View 10—Work version of the time schedule: This report is similar to the original time schedule (see Figure A.3), but it is designed for the college's internal use. It shows the current enrollments in each section of each course, as well as each section's maximum enrollment. The only difference between the work version of the time schedule and the time schedule itself (see User View 6) is the addition of two attributes for each section: current enrollment and maximum enrollment. Because these two attributes depend only on the combination of the semester code and the schedule code, you would place them in the Section table of User View 6 and, after the merge, they would be in the Section table in the cumulative design. The cumulative design thus far is shown in Figure A.17.

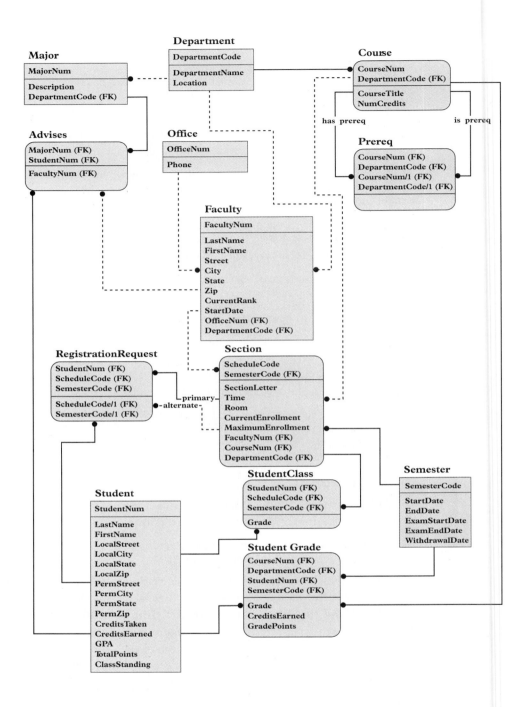

Because the process of determining whether a student has had the prerequisites for a given course involves examining the grades (if any) received in these prior courses, it makes sense to analyze the user view that involves grades before treating the user view that involves enrollment.

User View 11—Post grades: For each section of each course, the system must post the grades that are indicated on the class list submitted by the instructor and produce a grade verification report. There is a slight problem with posting grades—grades must somehow be posted by section to produce the grade report (in other words, you must record the fact that student 381124188 received an A in the section of CS 162 whose schedule code was 2366 during the fall 2007 semester). On the other hand, for the full student information report, there is no need to have any of the grades related to an actual section of a course. Further, because section information, including these grades, is kept for only two semesters, grades would be lost after two semesters if they were kept only by section because section information would be purged at that time.

A viable alternative is to post two copies of the grade: one copy will be associated with the student, the term, and the section; and the other copy will be associated with only the student and the term. The first copy would be used for the grade verification report and the second for the full student information report. Report cards would probably utilize the second copy, although not necessarily.

Thus, you would have the following two grade tables:

```
GradeSection (StudentNum, DepartmentCode, CourseNum, ScheduleCode,
     SemesterCode, Grade)
GradeStudent (StudentNum, DepartmentCode, CourseNum, SemesterCode,
     Grade)
```

Because the DepartmentCode and CourseNum in the GradeSection table depend only on the concatenation of ScheduleCode and SemesterCode, they will be removed from the GradeSection table during the normalization process and placed in a table whose primary key is the concatenation of ScheduleCode and SemesterCode. This table will be combined with the Section table in the cumulative design without adding new fields. The GradeSection table that remains will be merged with the StudentClass table without adding new fields. Finally, the GradeStudent table will be combined with the StudentGrade table in the cumulative design without adding any new fields. Thus, treatment of this user view does not change the cumulative design.

User View 12—Enrollment: When a student attempts to register for a section of a course, you must determine whether the student has received credit for all prerequisites to the course. If the student is eligible to enroll in the course and the number of students currently enrolled in the section is less than the maximum enrollment, enroll the student. With the data already in place in the cumulative design, you can determine what courses a student has taken. You can also determine the prerequisites for a given course. The only remaining issue is the ability to enroll a student in a course. Because the system must retain information for more than one semester, you must include the semester code in the table. (You must have the information that student 381124188 enrolled in section 2345 in SP08 rather than in FA07, for example.) The additional table is as follows:

```
Enroll (StudentNum, SemesterCode, ScheduleCode)
```

The primary key of this table matches the primary key of the StudentClass table in the cumulative design. The fields occur in a different order here, but that makes no difference. Thus, this table will be merged with the StudentClass table. There are no new fields to be added, so the cumulative design remains unchanged.

User View 13—Purge: Marvel College retains section information, including grades earned by the students in each section, for two semesters following the end of the semester, at which time the system removes this information from the system. Periodically, certain information that is more than two terms old is removed from the database. This includes all information concerning sections of courses, such as the time, room, and instructor, as well as information about the students in the sections and their grades. The grade each student received will remain in the database by course, but not by section. For example, you will always retain the fact that student 381124188 received an A in CS 162 during the fall semester of 2007, but once the data for that term is purged, you will no longer know the precise section of CS 162 that awarded this grade.

If you examine the current collection of tables, you see that all the data to be purged is already included in the cumulative design and that you don't need to add anything new at this point.

■ Final Information-Level Design

Now that you are finished examining the user views, Marvel College can review the cumulative design to ensure that all user views have been met. You should conduct this review on your own to make certain that you understand how the requirements of each user can be satisfied. You will assume that this review has taken place and that no changes have been made. Therefore Figure A.17 represents the final information-level design.

At this point, Marvel College is ready to move on to the physical-level design process. In this process, the appropriate team members will use the information-level design you produced to create the design for the specific DBMS that Marvel College selects. After they have done so, they will be able to create the database, load the data, and create the forms, reports, queries, and programs necessary to satisfy their requirements.

EXERCISES

1. Discuss the effect of the following changes on the design for the Marvel College requirements:

    a. More than one instructor might teach a given section of a course, and each instructor must be listed on the time schedule.

    b. Each department offers only a single major.

    c. Each department offers only a single major, and each faculty member can advise students only in the major that is offered by the department to which the faculty member is assigned.

    d. Each department offers only a single major, and each faculty member can advise students only in the major that is offered by the department to which the faculty member is assigned. In addition, a student can have only one major.

    e. There is an additional transaction requirement: Given a student's name, find the student's number.

    f. More than one faculty member can be assigned to one office.

    g. The number of credits earned in a particular course cannot vary from student to student or from semester to semester.

    h. Instead of a course number, course codes are used to uniquely identify courses (in other words, department numbers are no longer required for this purpose). However, it is still important to know which courses are offered by which departments.

    i. On the registration request form, a student may designate a number of alternates along with his or her primary choice. These alternates are listed in a priority order, with the first one being the most desired and the last one being the least desired.

2. Complete an information-level design for Holt Distributors.

    General description. Holt Distributors is a distributor that buys products from its vendors and sells these products to its customers. The Holt Distributors operation is divided into territories. Each customer is represented by a single sales rep, who must be assigned to the territory in which the customer is located. Although each sales rep is assigned to a single territory, more than one sales rep can be assigned to the same territory.

    When a customer places an order, the computer assigns the order the next available order number. The data entry clerk enters the customer number, the customer purchase order (PO) number, and the date. (Customers can place orders by submitting a purchase order, in which case, a PO number is recorded.) For each part that is ordered, the clerk enters the part number, quantity, and quoted price. (When it is time for the clerk to enter the quoted price, the computer displays the price from the master price list on the screen. If the quoted price is the same as the actual price, the clerk takes no special action. If not, the clerk enters the quoted price.)

    When the clerk completes the order, the system prints the order acknowledgment/picking list form shown in Figure A.18 and sends it to the customer for confirmation and payment. When Holt Distributors is ready to ship the customer's order, this same form is used to "pick" the merchandise in the warehouse and to prepare it for delivery.

FIGURE A.18
....
Order acknow-
ledgment/picking
list for Holt
Distributions

ORDER ACKNOWLEDGMENT / PICKING LIST

10/12/2007   Order 12424

HOLT DISTRIBUTORS
146 NELSON PLACE
BRONSTON, MI 49802

SOLD   SHIP
  TO: Smith Rentals   TO: A & B Supplies
        153 Main St.   2180 Halton Pl.
        Suite 102   Arendville, MI 49232
        Grandville, MI 49494

Customer	P.O. No.	Order Date	Sales Rep
1354	PO335	10/02/2007	10-Brown, Sam

Quantity						
Order	**Ship**	**B/O**	**Item Number**	**Description**	**Price**	**Amount**
6			AT414	Lounge Chair	$42.00	$252.00
4			BT222	Arm Chair	$51.00	$204.00

An order that hasn't been shipped (filled) is called an open order; an order that has been shipped is called a released order. When orders are released, the system prints an invoice, sends it to the customer, and then increases the customer's balance by the invoice amount. Some orders are completely filled; others are only partially filled, meaning that only part of the customer's order was shipped. In either case, when an entire order or a partial order has been shipped, the order is considered to have been filled and is no longer considered an open order. (Another possibility is to allow back orders when the order cannot be completely filled. In this case, the order would remain open but only for the back-ordered portion.) When the system generates an invoice, it removes the order from the open orders file. The system stores summary information about the invoice (number, date, customer, invoice total, and freight) until the end of the month. A sample invoice is shown in Figure A.19.

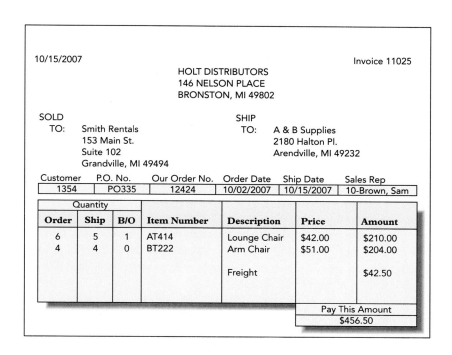

Most companies use one of two methods for accepting payments from customers: open items and balance forward. In the open-item method, customers make payments on specific invoices. An invoice remains on file until the customer pays it in full. In the balance-forward method, customers have balances. When the system generates an invoice, the customer's balance is increased by the amount of the invoice. When a customer makes a payment, the system decreases the customer's balance by the payment amount. Holt Distributors uses the balance-forward method.

At the end of each month, the system updates and ages customers' accounts. (You'll learn about month-end processing requirements and the update and aging processes in the following sections.) The system prints customer statements, an aged trial balance (described in the report requirements section), a monthly cash receipts journal, a monthly invoice register, and a sales rep commission report. The system then removes cash receipts and invoice summary records from the database and sets month-to-date (MTD) fields to zero. When the system processes the monthly data for December, it also sets the year-to-date (YTD) fields to zero.

Transaction requirements. The following transaction requirements are required by Holt Distributors:

a. Enter and edit territories (territory number and name).

b. Enter and edit sales reps (sales rep number, name, address, city, state, zip, MTD sales, YTD sales, MTD commission, YTD commission, and commission rate). Each sales rep represents a single territory.

c. Enter and edit customers (customer number, name, first line of address, second line of address, city, state, zip, MTD sales, YTD sales, current balance, and credit limit). A customer can have a different name and address to which goods are shipped (called the "ship-to" address). Each customer has a single sales rep and is located in a single territory. The sales rep must represent the territory in which the customer is located.

d. Enter and edit parts (part number, description, price, MTD and YTD sales, units on hand, units allocated, and reorder point). Units allocated is the number of units that are currently present on some open orders. The reorder point is the lowest value acceptable for units on hand without reordering the product. On the stock status report, which will be described later, an asterisk indicates any part for which the number of units on hand is less than the reorder point.

e. Enter and edit vendors (vendor number, name, address, city, state, and zip). In addition, for each part supplied by the vendor, enter and edit the part number, the price the vendor charges for the part, the minimum order quantity that the vendor will accept for this part, and the expected lead time for delivery of this part from this vendor.

f. Order entry (order number, date, customer, customer PO number, and the order detail lines). An order detail line consists of a part number, a description, the number ordered, and the quoted price. Each order detail line includes a sequence number that is entered by the user. Detail lines on an order must print in the order of this sequence number. The system should calculate and display the order total. After all orders for the day have been entered, the system prints order acknowledgment/picking list reports (see Figure A.18). In addition, for each part ordered, the system must increase the units allocated for the part by the number of units that the customer ordered.

g. The invoicing system has the following requirements:

1. Enter the numbers of the orders to be released. For each order, enter the ship date for invoicing and the freight amount. Indicate whether the order is to be shipped in full or in part. If an order is to be partially shipped, enter the number shipped for each order detail line. The system will generate a unique invoice number for this invoice.

2. Print invoices for each of the released orders. (A sample invoice is shown in Figure A.19.)

3. Update files with information from the printed invoices. For each invoice, the system adds the invoice total to the current invoice total. It also adds the current balance and the MTD and YTD sales for the customer that placed the order. The system also adds the total to the MTD and YTD sales for the sales rep who represents the customer. In addition, the system multiplies the total by the sales rep's commission rate and adds this amount to the MTD commission earned and the YTD commission earned. For each part shipped, the system decreases units on hand and units allocated by the number of units of the part(s) that were shipped. The system also increases the MTD and YTD sales of the part by the amount of the number of units shipped multiplied by the quoted price.

4. Create an invoice summary record for each invoice printed. These records contain the invoice number, date, customer, sales rep, invoice total, and freight.

5. Delete the released orders.

h. Receive payments on account (customer number, date, and amount). The system assigns each payment a number, adds the payment amount to the total of current payments for the customer, and subtracts the payment amount from the current balance of the customer.

Report requirements. The following is a list of the reports required by Holt Distributors:

a. **Territory List:** For each territory, list the number and name of the territory; the number, name, and address of each sales rep in the territory; and the number, name, and address of each customer represented by these sales reps.

b. **Customer Master List:** For each customer, list the customer number, the bill-to address, and the ship-to address. Also list the number, name, address, city, state, and zip of the sales rep who represents the customer, and the number and name of the territory in which the customer is located.

c. **Customer Open Order Report:** This report lists open orders organized by customer. It is shown in Figure A.20.

	HOLT DISTRIBUTORS	
10/16/2007	CUSTOMER OPEN ORDER REPORT	PAGE 1

Order Number	Item Number	Item Description	Order Date	Order Qty	Quoted Price
Customer 1354 - Smith Rentals					
12424	AT414	Lounge Chair	10/02/2007	1	$42.00
Customer 1358 - • • • • • • • •					

d. **Item Open Order Report:** This report lists open orders organized by item and is shown in Figure A.21.

	HOLT DISTRIBUTORS	
10/16/2007	ITEM OPEN ORDER REPORT	PAGE 1

Item Number	Item Description	Customer Number	Customer Name	Order Number	Order Date	Order Qty	Quoted Price
AT414	Lounge Chair	1354	Smith Rentals	12424	10/02/2007	1	$42.00
		1358	Kayland Enterprises	12489	10/03/2007	8	$42.00
			Total on order -			9	
BT222	Arm Chair	1358	Kayland Enterprises	12424	10/03/2007	3	$51.00

e. **Daily Invoice Register:** For each invoice produced on a given day, list the invoice number, invoice date, customer number, customer name, sales amount, freight, and invoice total. A sample of this report is shown in Figure A.22.

	HOLT DISTRIBUTORS	
10/16/2007	DAILY INVOICE REGISTER FOR 10/15/2007	PAGE 1

Invoice Number	Invoice Date	Customer Number	Customer Name	Sales Amount	Freight	Invoice Amount
11025	10/15/2007	1354	Smith Rentals	$414.00	$42.50	$456.50
				$2,840.50	$238.20	$3,078.70

f. **Monthly Invoice Register:** The monthly invoice register has the same format as the daily invoice register, but it includes data for all invoices that occurred during the selected month.

g. **Stock Status Report:** For each part, list the part number, description, price, MTD and YTD sales, units on hand, units allocated, and reorder point. For each part for which the number of units on hand is less than the reorder point, an asterisk should appear at the far right of the report.

h. **Reorder Point List:** This report has the same format as the stock status report. Other than the title, the only difference is that parts for which the number of units on hand is greater than or equal to the reorder point will not appear on this report.

i. **Vendor Report:** For each vendor, list the vendor number, name, address, city, state, and zip. In addition, for each part supplied by the vendor, list the part number, the description, the price the vendor charges for the part, the minimum order quantity that the vendor will accept for this part, and the expected lead time for delivery of this part from this vendor.

j. **Daily Cash Receipts Journal:** For each payment received on a given day, list the number and name of the customer that made the payment and the payment amount. A sample report is shown in Figure A.23.

FIGURE A.23
· · · · · · · · · · · · · · · · ·
Daily cash receipts
journal

10/05/2007	**HOLT DISTRIBUTORS**		PAGE 1
	DAILY CASH RECEIPTS JOURNAL		

Payment Number	Customer Number	Customer Name	Payment Amount
:	:	:	:
:	:	:	:
5807	1354	Smith Rentals	$1,000.00
:	:	:	:
			$12,235.50

k. **Monthly Cash Receipts Journal:** The monthly cash receipts journal has the same format as the daily cash receipts journal, but it includes all cash receipts for the month.

l. **Customer Mailing Labels:** A sample of the three-across mailing labels printed by the system is shown in Figure A.24.

FIGURE A.24
· · · · · · · · · · · · · · · · ·
Customer
mailing labels

Smith Rentals
153 Main St.
Suite 102
Grandville, MI 49494

Kayland Enterprises
267 29th St
Wyoming, MI 48222

John & Sons, Inc.
5563 Crestview
Ada, MI 49292

m. **Statements:** The system must produce a monthly statement for each active customer. A sample statement is shown in Figure A.25.

11/01/2007	**HOLT DISTRIBUTORS**
	146 NELSON PLACE
	BRONSTON, MI 49802

Smith Rentals   Customer Number: 1354
153 Main St.   Sales Rep: 10 - Brown, Sam
Suite 102
Grandville, MI 49494   Limit: $5,000.00

Invoice Number	Date	Description	Total Amount
10945	10/02/2007	Invoice	$1,230.00
	10/05/2007	Payment	$1,000.00CR
11025	10/15/2007	Invoice	$456.50
	10/22/2007	Payment	$500.00CR

Over 90	Over 60		
$0.00	$198.50		
Over 30	Current	Total Due >>>>>>	$2,325.20
$490.20	$1,686.50		
Previous Balance	Current Invoices	Current Payments	
$2,138.70	$1,686.50	$1,500.00	

n. **Monthly Sales Rep Commission Report:** For each sales rep, list his or her number, name, address, MTD sales, YTD sales, MTD commission earned, YTD commission earned, and commission rate.

o. **Aged Trial Balance:** The aged trial balance report contains the same information that is printed on each customer's statement.

Month-end processing. Month-end processing consists of the following actions that occur at the end of each month:

a. Update customer account information. In addition to the customer's actual balance, the system must maintain the following records: current debt, debt incurred within the last 30 days, debt that is more than 30 days past due but less than 60 days past due, debt that is 60 or more days past due but less than 90 days past due, and debt that is 90 or more days past due. The system updates the actual balance, the current invoice total, and the current payment total when it produces a new invoice or receives a payment; however, the system updates these aging figures only at the end of the month. The actual update process is as follows:

1. The system processes payments received within the last month and credits these payments to the past due amount for 90 or more days. The system then credits any additional payment to the 60 or more days past due amount, then to the more than 30 days past due amount, and then to the current debt amount (less than 30 days).

2. The system "rolls" the amounts by adding the 60 or more days past due amount to the 90 or more days past due amount and by adding the more than 30 days past due amount to the 60 or more days past due amount. The current amount becomes the new more than 30 days past due amount. Finally, the current month's invoice total becomes the new current amount.

3. The system prints the statements and the aged trial balances.
4. The system sets the current invoice total to zero, sets the current payment total to zero, and sets the previous balance to the current balance in preparation for the next month.

To illustrate, let's assume that before the update begins, the amounts for customer 1354 are as follows:

```
Current Balance:   $2,375.20   Previous Balance: $2,138.70
Current Invoices: $1,686.50              Current:   $490.20
Current Payments: $1,500.00              Over 30:   $298.50
                                         Over 60:   $710.00
                                         Over 90:   $690.00
```

The system subtracts the current payments ($1,500.00) from the over 90 amount ($690.00), reduces the over 90 amount to zero, and calculates an excess payment of $810.00. The system subtracts this excess payment from the over 60 amount ($710.00), reduces the over 60 amount to zero, and calculates an excess payment of $100.00. The system then subtracts the excess payment from the over 30 amount ($298.50) and reduces this amount to $198.50. At this point, the system rolls the amounts and sets the current amount to the current invoice total, producing the following:

```
Current Balance:   $2,375.20   Previous Balance:  $2,138.70
Current Invoices: $1,686.50              Current: $1,686.50
Current Payments: $1,500.00              Over 30:   $490.20
                                         Over 60:   $198.50
                                         Over 90:     $0.00
```

The system then produces statements and the aged trial balance and updates the Previous Balance, Current Invoices, and Current Payments amounts, yielding:

```
Current Balance: $2,375.20   Previous Balance:  $2,375.20
Current Invoices:    $0.00              Current: $1,686.50
Current Payments:    $0.00              Over 30:   $490.20
                                        Over 60:   $198.50
                                        Over 90:     $0.00
```

b. Print the monthly invoice register and the monthly cash receipts journal.
c. Print a monthly sales rep commission report.
d. Set all MTD fields to zero. If necessary, set all YTD fields to zero.
e. Remove all cash receipts and invoice summary records. (In practice, such records would be moved to a historical type of database for future reference. For the purposes of this assignment, you will omit this step.)

APPENDIX B: ADDITIONAL CASE PROBLEM: ALEXAMARA MARINA GROUP

Alexamara Marina Group offers in-water storage to boat owners by providing boat slips that boat owners can rent on an annual basis. Alexamara owns two marinas: Alexamara East and Alexamara Central. In addition to boat slips, Alexamara also provides a variety of boat repair and maintenance services to the boat owners who rent the slips. Alexamara stores the data it needs to manage its operations in a relational database containing the tables described in the following sections.

Alexamara stores information about its two marinas in the Marina table shown in Figure B.1. A marina number uniquely identifies each marina. The table also includes the marina name, address, city, state, and zip code.

FIGURE B.1
....
Marina table

Marina

MarinaNum	Name	Address	City	State	Zip
1	Alexamara East	108 2nd Ave.	Brinman	FL	32273
2	Alexamara Central	283 Branston	W. Brinman	FL	32274

Alexamara stores information about the boat owners to whom it rents slips in the Owner table shown in Figure B.2. An owner number that consists of two uppercase letters followed by a two-digit number uniquely identifies each owner. For each owner, the table also includes the last name, first name, address, city, state, and zip code.

FIGURE B.2
....
Owner table

Owner

OwnerNum	LastName	FirstName	Address	City	State	Zip
AD57	Adney	Bruce and Jean	208 Citrus	Bowton	FL	31313
AN75	Anderson	Bill	18 Wilcox	Glander Bay	FL	31044
BL72	Blake	Mary	2672 Commodore	Bowton	FL	31313
EL25	Elend	Sandy and Bill	462 Riverside	Rivard	FL	31062
FE82	Feenstra	Daniel	7822 Coventry	Kaleva	FL	32521
JU92	Juarez	Maria	8922 Oak	Rivard	FL	31062
KE22	Kelly	Alyssa	5271 Waters	Bowton	FL	31313
NO27	Norton	Peter	2811 Lakewood	Lewiston	FL	32765
SM72	Smeltz	Becky and Dave	922 Garland	Glander Bay	FL	31044
TR72	Trent	Ashton	922 Crest	Bay Shores	FL	30992

Each of the marinas contains slips that are identified by slip numbers. Marina 1 (Alexamara East) has two sections named A and B. Slips are numbered within each section. Thus, slip numbers at marina 1 consist of the letter A or B followed by a number (for example, A3 or B2). At marina 2 (Alexamara Central), a number (1, 2, 3) identifies each slip.

Information about the slips in the marinas is contained in the MarinaSlip table shown in Figure B.3. The table contains the marina number and slip number, the length of the slip (in feet), the annual rental fee, the name of the boat currently occupying the slip, the type of boat, and the boat owner's number.

MarinaSlip

SlipID	MarinaNum	SlipNum	Length	RentalFee	BoatName	BoatType	OwnerNum
1	1	A1	40	$3,800.00	Anderson II	Sprite 4000	AN75
2	1	A2	40	$3,800.00	Our Toy	Ray 4025	EL25
3	1	A3	40	$3,600.00	Escape	Sprite 4000	KE22
4	1	B1	30	$2,400.00	Gypsy	Dolphin 28	JU92
5	1	B2	30	$2,600.00	Anderson III	Sprite 3000	AN75
6	2	1	25	$1,800.00	Bravo	Dolphin 25	AD57
7	2	2	25	$1,800.00	Chinook	Dolphin 22	FE82
8	2	3	25	$2,000.00	Listy	Dolphin 25	SM72
9	2	4	30	$2,500.00	Mermaid	Dolphin 28	BL72
10	2	5	40	$4,200.00	Axxon II	Dolphin 40	NO27
11	2	6	40	$4,200.00	Karvel	Ray 4025	TR72

Alexamara provides boat maintenance service at its two marinas. The types of service provided are stored in the ServiceCategory table shown in Figure B.4. A category number uniquely identifies each service that Alexamara performs. The table also contains a description of the category.

ServiceCategory

CategoryNum	CategoryDescription
1	Routine engine maintenance
2	Engine repair
3	Air conditioning
4	Electrical systems
5	Fiberglass repair
6	Canvas installation
7	Canvas repair
8	Electronic systems (radar, GPS, autopilots, etc.)

Information about the services requested by owners is stored in the ServiceRequest table shown in Figure B.5. A ServiceID number uniquely identifies each row in the table. The combination of the marina number and slip number identifies the location of each boat to be serviced. For example, the SlipID on the second row is 5. As indicated in the MarinaSlip table, the SlipID 5 identifies the boat in marina 1 and slip number B2. The ServiceRequest table also contains the category number of the service to be performed, a description of the specific service to be performed, and a description of the current status of the service. It also contains the estimated number of hours required to complete the service. For completed jobs, the table contains the actual number of hours it took to complete the service. If another appointment is required to complete additional service, the appointment date appears in the NextServiceDate column.

ServiceRequest

Service ID	Slip ID	Category Num	Description	Status	Est Hours	Spent Hours	Next ServiceDate
1	1	3	Air conditioner periodically stops with code indicating low coolant level. Diagnose and repair.	Technician has verified the problem. Air conditioning specialist has been called.	4	2	7/12/2007
2	5	4	Fuse on port motor blown on two occasions. Diagnose and repair.	Open	2	0	7/12/2007
3	4	1	Oil change and general routine maintenance (check fluid levels, clean sea strainers, etc.).	Service call has been scheduled.	1	0	7/16/2007
4	1	2	Engine oil level has been dropping drastically. Diagnose and repair.	Open	2	0	7/13/2007
5	3	5	Open pockets at base of two stantions.	Technician has completed the initial filling of the open pockets. Will complete the job after the initial fill has had sufficient time to dry.	4	2	7/13/2007
6	11	4	Electric-flush system periodically stops functioning. Diagnose and repair.	Open	3	0	
7	6	2	Engine overheating. Loss of coolant. Diagnose and repair.	Open	2	0	7/13/2007

(continued)

ServiceRequest *(continued)*

Service ID	Slip ID	Category Num	Description	Status	Est Hours	Spent Hours	Next ServiceDate
8	6	2	Heat exchanger not operating correctly.	Technician has determined that the exchanger is faulty. New exchanger has been ordered.	4	1	7/17/2007
9	7	6	Canvas severely damaged in windstorm. Order and install new canvas.	Open	8	0	7/16/2007
10	2	8	Install new GPS and chart plotter.	Scheduled	7	0	7/17/2007
11	2	3	Air conditioning unit shuts down with "HHH" showing on the control panel.	Technician not able to replicate the problem. Air conditioning unit ran fine through multiple tests. Owner to notify technician if the problem recurs.	1	1	
12	4	8	Both speed and depth readings on data unit are significantly less than the owner thinks they should be.	Technician has scheduled appointment with owner to attempt to verify the problem.	2	0	7/16/2007
13	8	2	Customer describes engine as making a "clattering" sound.	Technician suspects problem with either propeller or shaft and has scheduled the boat to be pulled from the water for further investigation.	5	2	7/12/2007

(continued)

ServiceRequest (continued)

Service ID	Slip ID	Category Num	Description	Status	Est Hours	Spent Hours	Next ServiceDate
14	7	5	Owner accident caused damage to forward portion of port side.	Technician has scheduled repair.	6	0	7/13/2007
15	11	7	Canvas leaks around zippers in heavy rain. Install overlap around zippers to prevent leaks.	Overlap has been created. Installation has been scheduled.	8	3	7/17/2007

An entity-relationship (E-R) diagram of the Alexamara database is shown in Figure B.6.

FIGURE B.6

E-R diagram for the Alexamara database

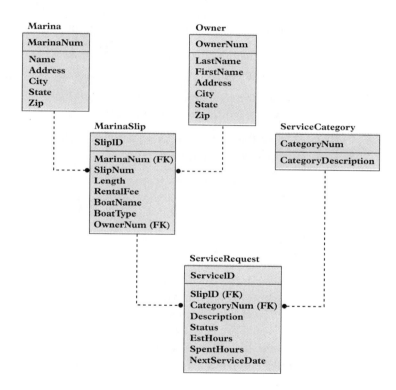

■ Alexamara Marina Group Exercises

Chapter 1

Answer each of the following questions using the Alexamara Marina Group data shown in Figures B.1 through B.5. No computer work is required.

1. List the names of all boats that are stored in a slip with a length of less than 30 feet.

2. List the last name, first name, and street address of every owner located in Glander Bay.

3. List the last name, first name, and city of every owner who has more than one boat stored at the marina.

4. List the last name, first name, and city of every owner with a boat located in a slip whose rental fee is greater than $2,500.00.

5. List the boat name and boat type of all boats that are stored at the Alexamara East marina.

6. List the last name and first name of the owner and the boat name of all boats that have a completed or open service request.

7. List the boat name and boat type of all boats that have two or more completed or open service requests.

8. List the boat name and boat type of all boats that have a completed or open service request for a fiberglass or canvas-related service.

Chapter 2

In the following exercises, you will use the data in the Alexamara Marina Group database shown in Figures B.1 through B.5. (If you use a computer to complete these exercises, use a copy of the Alexamara Marina Group database so you will still have the original data when you complete Chapter 3.) In each step, use Query-By-Example (QBE) to obtain the desired results. You can use the query feature in a database management system (DBMS) to complete the exercises using a computer or you can simply write a description of how you would complete the task. Check with your instructor if you are uncertain of which approach to take.

1. List the owner number, last name, and first name of every boat owner.

2. List the complete Marina table (all rows and all columns).

3. List the last name and first name of every owner located in Bowton.

4. List the last name and first name of every owner not located in Bowton.

5. List the marina number and slip number for every slip whose length is equal to or less than 30 feet.

6. List the marina number and slip number for every boat with the type Dolphin 28.

7. List the slip number for every boat with the type Dolphin 28 that is located in marina 1.

8. List the boat name for each boat located in a slip whose length is between 25 and 30 feet.

9. List the slip number for every slip in marina 1 whose annual rental fee is less than $3,000.00.

10. Labor is billed at the rate of $60.00 per hour. List the slip ID, category number, estimated hours, and estimated labor cost for every service request. To obtain the estimated labor cost, multiply the estimated hours by 60. Use the column name "EstimatedCost" for the estimated labor cost.

11. List the marina number and slip number for all slips containing a boat with the type Sprite 4000, Sprite 3000, or Ray 4025.

12. List the marina number, slip number, and boat name for all boats. Sort the results by boat name within the marina number.

13. How many Dolphin 25 boats are stored at both marinas?

14. Calculate the total rental fees Alexamara receives each year based on the length of the slip.

15. For every boat, list the marina number, the slip number, the boat name, the owner number, the owner's first name, and the owner's last name.

16. For every completed or open service request for routine engine maintenance, list the slip ID, the description, and the status.

17. For every service request for routine engine maintenance, list the slip ID, marina number, slip number, estimated hours, spent hours, owner number, and the owner's last name.

18. Create a new table named LargeSlip using the data in the MarinaNum, SlipNum, RentalFee, BoatName, and OwnerNum columns in the MarinaSlip table for slips with lengths of 40 feet.

19. Use an update query to change the rental fee of any slip in the LargeSlip table whose fee is currently $3,800.00 to $3,900.00.

20. Use a delete query to delete slips in the LargeSlip table whose rental fee is $3,600.00.

Chapter 3

In the following exercises, you will use the data in the Alexamara Marina Group database shown in Figures B.1 through B.5. (If you use a computer to complete these exercises, use a copy of the original Alexamara Marina Group database so you will still have the original data when you complete Chapter 4.) In each step, use structured query language (SQL) to obtain the desired results. You can use a DBMS to complete the exercises using a computer, or you can simply write the SQL command to complete each step. Check with your instructor if you are uncertain of which approach to take.

1. List the owner number, last name, and first name of every boat owner.

2. List the complete Marina table (all rows and all columns).

3. List the last name and first name of every owner located in Bowton.

4. List the last name and first name of every owner not located in Bowton.

5. List the marina number and slip number for every slip whose length is equal to or less than 30 feet.

6. List the marina number and slip number for every boat with the type Dolphin 28.

7. List the slip number for every boat with the type Dolphin 28 that is located in marina 1.

8. List the boat name for each boat located in a slip whose length is between 25 and 30 feet.

9. List the slip number for every slip in marina 1 whose annual rental fee is less than $3,000.00.

10. Labor is billed at the rate of $60.00 per hour. List the slip ID, category number, estimated hours, and estimated labor cost for every service request. To obtain the estimated labor cost, multiply the estimated hours by 60. Use the column name "EstimatedCost" for the estimated labor cost.

11. List the marina number and slip number for all slips containing a boat with the type Sprite 4000, Sprite 3000, or Ray 4025.

12. List the marina number, slip number, and boat name for all boats. Sort the results by boat name within the marina number.

13. How many Dolphin 25 boats are stored at both marinas?

14. Calculate the total rental fees Alexamara receives each year based on the length of the slip.

15. For every boat, list the marina number, the slip number, the boat name, the owner number, the owner's first name, and the owner's last name.

16. For every completed or open service request for routine engine maintenance, list the slip ID, the description, and the status.

17. For every service request for routine engine maintenance, list the slip ID, marina number, slip number, estimated hours, spent hours, owner number, and the owner's last name.

18. Create a new table named LargeSlip using the data in the MarinaNum, SlipNum, RentalFee, BoatName, and OwnerNum columns in the MarinaSlip table for slips with lengths of 40 feet.

19. Use an update query to change the rental fee of any slip in the LargeSlip table whose fee is currently $3,800.00 to $3,900.00.

20. Use a delete query to delete slips in the LargeSlip table whose rental fee is $3,600.00.

Chapter 4

In the following exercises, you will use the data in the Alexamara Marina Group database shown in Figures B.1 through B.5. (If you use a computer to complete these exercises, use a copy of the original Alexamara Marina Group database so your data will not reflect the changes you made in Chapter 3.) If you have access to a DBMS, use the DBMS to perform the tasks and explain the steps you used in the process. If not, explain how you would use SQL to obtain the desired results. Check with your instructor if you are not certain about which approach to take.

1. Create a view named LargeSlip using the data in the MarinaNum, SlipNum, RentalFee, BoatName, and OwnerNum columns in the MarinaSlip table for those slips with lengths of 40 feet. Display the data in the view.

2. Create a view named InitialService using the slip ID, category number, category description, and estimated hours for every service request for which the spent hours is zero. Display the data in the view.

3. Create a view named TypesOfBoats using the boat type and a count of all boats of each type. Display the data in the view.

4. Create the following indexes. If it is necessary to name the index in your DBMS, use the indicated name.

   a. Create an index named BoatIndex on the BoatName field in the MarinaSlip table.
   b. Create an index named BoatIndex2 on the OwnerNum field in the MarinaSlip table.
   c. Create an index named BoatIndex3 on the Length and BoatName fields in the MarinaSlip table and list the lengths in descending order.

5. Drop the BoatIndex3 index.

6. Assume that the MarinaSlip table has been created, but that there are no integrity constraints. Create the necessary integrity constraints so that the rental fee must be less than $5,000.00 and the slip length must be 25, 30, or 40.

7. Add the following foreign keys (that is, specify referential integrity) in the Alexamara database:

   a. MarinaNum is a foreign key in the MarinaSlip table.
   b. OwnerNum is a foreign key in the MarinaSlip table.
   c. CategoryNum is a foreign key in the ServiceRequest table.
   d. SlipID is a foreign key in the SerivceRequest table.

8. Add to the MarinaSlip table a new character field named FeePaid that is one character in length. Change the value for the FeePaid field on all records to Y.

9. Change the FeePaid field in the MarinaSlip table to N for the slip whose slip ID is 4.

10. Change the length of the BoatName field in the MarinaSlip table to 60.

Chapter 5

The following exercises are based on the Alexamara Marina Group database shown in Figures B.1 through B.5.

1. Using the types of entities found in the Alexamara Marina Group database (marinas, owners, boat slips, categories, and service requests), create an example of a table that is in first normal form but not in second normal form and an example of a table that is in second normal form but not in third normal form. In each case, justify your answer and show how to convert to the higher forms.

2. Determine the functional dependencies that exist in the following table, and then convert this table to an equivalent collection of tables that are in third normal form:

   ```
   Marina (MarinaNum, Name, (SlipNum, Length, RentalFee, BoatName) )
   ```

3. Determine the functional dependencies that exist in the following table, and then convert this table to an equivalent collection of tables that are in third normal form:

   ```
   MarinaSlip (SlipID, MarinaNum, SlipNum, Length, RentalFee, BoatName,
        BoatType, OwnerNum, LastName, FirstName)
   ```

Chapter 6

Complete the following tasks:

1. Design a database to produce the following reports. Do not use any surrogate keys in your design.

   a. For each marina, list the marina number, name, address, city, state, and zip code.

   b. For each boat owner, list the owner number, last name, first name, address, city, state, and zip code.

   c. For each marina, list all the slips in the marina. For each slip, list the length of the slip, the annual rental fee, the name and type of the boat occupying the slip, and the boat owner's number, last name, and first name.

   d. For each service category, list the category number and description. In addition, for each service request in each category, list the marina number, the slip number of the boat receiving the service, the estimated hours for the service, the hours already spent on the service, and the next scheduled service date.

   e. For each service request, list the marina number, slip number, category description, description of the particular service, and a description of the current status of the service.

2. Expand the database design you created in Exercise 1 so that it will also support the following situation. A specific technician handles each service request. Along with all the details concerning a service request listed in Exercise 1, list the number, last name, and first name of the technician assigned to handle the request.

APPENDIX C: SQL REFERENCE

You can use this appendix to obtain details concerning important components and syntax of the structured query language (SQL) language. Items are arranged alphabetically. Each item contains a description and, where appropriate, both an example and a description of the query results. Some SQL commands also include a description of the clauses associated with them. For each clause, there is a brief description and an indication of whether the clause is required or optional.

ALTER TABLE

Use the ALTER TABLE command to change a table's structure. As shown in Figure C.1, you type the ALTER TABLE command, followed by the table name, and then the alteration to perform. (**Note:** In Access you would usually make these changes to a table in Design view rather than using ALTER TABLE.)

Clause	Description	Required?
ALTER TABLE *table name*	Indicates the name of table to be altered.	Yes
alteration	Indicates the type of alteration to be performed.	Yes

The following command alters the Customer table by adding a new column named CustType:

```
ALTER TABLE Customer
ADD CustType CHAR(1)
;
```

The following command alters the Customer table by changing the length of the CustomerName column:

```
ALTER TABLE Customer
CHANGE COLUMN CustomerName TO CHAR(50)
;
```

The following command alters the Part table by deleting the Warehouse column:

```
ALTER TABLE Part
DELETE Warehouse
;
```

■ Column or Expression List (SELECT Clause)

To select columns, use a SELECT clause with the list of columns, separated by commas.

The following SELECT clause selects the CustomerNum, CustomerName, and Balance columns:

```
SELECT CustomerNum, CustomerName, Balance
```

Use an asterisk in a SELECT clause to select all columns in the table. The following SELECT command selects all columns in the Part table:

```
SELECT *
FROM Part
;
```

Computed Fields

You can use a computation in place of a field by typing the computation. For readability, you can type the computation in parentheses, although it is not necessary to do so.

The following SELECT clause selects the CustomerNum and CustomerName columns as well as the results of subtracting the Balance column from the CreditLimit column:

```
SELECT CustomerNum, CustomerName, CreditLimit-Balance
```

Functions

You can use aggregate functions in a SELECT clause. The most commonly used functions are AVG (to calculate an average), COUNT (to count the number of rows), MAX (to determine the maximum value), MIN (to determine the minimum value), and SUM (to calculate a total).

The following SELECT clause calculates the average balance:

```
SELECT AVG(Balance)
```

■ Conditions

A condition is an expression that can be evaluated as either true or false. When you use a condition in a WHERE clause, the results of the query contain those rows for which the condition is true. You can create simple conditions and compound conditions using the BETWEEN, LIKE, and IN operators, as described in the following sections.

Simple Conditions

A simple condition has the form: column name, comparison operator, and then either another column name or a value. The available comparison operators are = (equal to), < (less than), > (greater than), <= (less than or equal to), >= (greater than or equal to), and < > (not equal to).

The following WHERE clause uses a condition to select rows in which the balance is greater than the credit limit:

```
WHERE Balance>CreditLimit
```

Compound Conditions

Compound conditions are formed by connecting two or more simple conditions using the AND, OR, and NOT operators. When simple conditions are connected by the AND operator, all of the simple conditions must be true in order for the compound condition to be true. When simple conditions are connected by the OR operator, the compound condition will be true whenever any one of the simple conditions is true. Preceding a condition by the NOT operator reverses the truth of the original condition.

The following WHERE clause is true if the warehouse number is equal to 3 *or* the units on hand is greater than 20, *or* both:

```
WHERE Warehouse='3'
OR OnHand>20
```

The following WHERE clause is true if the warehouse number is equal to 3 and the units on hand is greater than 20:

```
WHERE Warehouse='3'
AND OnHand>20
```

The following WHERE clause is true if the warehouse number is not equal to 3:

```
WHERE NOT (Warehouse='3')
```

BETWEEN Conditions

You can use the BETWEEN operator to determine if a value is within a range of values.

The following WHERE clause is true if the balance is between 1,000 and 5,000:

```
WHERE Balance BETWEEN 1000 AND 5000
```

LIKE Conditions

LIKE conditions use wildcards to select rows. Use the percent sign (%) to represent any collection of characters. The condition LIKE '%Oxford%' will be true for data consisting of any character or characters, followed by the letters "Oxford," followed by any other character or characters. Another wildcard is the underscore character (_), which represents any individual character. For example, "T_m" represents the letter "T," followed by any single character, followed by the letter "m," and would be true for a collection of characters such as Tim, Tom, or T3m.

Note: Access uses different wildcard symbols. The symbol for any collection of characters is the asterisk (*), and the symbol for an individual character is the question mark (?).

The following WHERE clause is true if the value in the Street column is Ox, ford, Oxford, or any other value that contains "Oxford:"

```
WHERE Street LIKE '%Oxford%'
```

Access version:

```
WHERE Street LIKE '*Oxford*'
```

IN Conditions

You can use the IN operator to determine whether a value is in some specific collection of values.

The following WHERE clause is true if the credit limit is 7,500, 10,000, or 15,000:

```
WHERE CreditLimit IN (7500, 10000, 15000)
```

The following WHERE clause is true if the part number is in the collection of part numbers located in warehouse 3:

```
WHERE PartNum IN
(SELECT PartNum
FROM Part
WHERE Warehouse='3')
```

■ CREATE INDEX

Use the CREATE INDEX command to create an index for a table. Figure C.2 describes the CREATE INDEX command.

The following CREATE INDEX command creates an index named RepBal for the

FIGURE C.2
. . . .
CREATE INDEX
command

Clause	Description	Required?
CREATE INDEX *index name*	Indicates the name of the index.	Yes
ON *table name*	Indicates the table for which the index is to be created.	Yes
column list	Indicates the column or columns on which the index is to be based.	Yes

Customer table on the combination of the RepNum and Balance columns:

```
CREATE INDEX RepBal
ON Customer (RepNum, Balance)
;
```

■ CREATE TABLE

Use the CREATE TABLE command to define the structure of a new table. Figure C.3 describes the CREATE TABLE command.

FIGURE C.3
•••••••••••••••••••
CREATE TABLE
command

Clause	Description	Required?
CREATE TABLE *table name*	Indicates the name of the table to be created.	Yes
(column and data type list)	Indicates the columns that comprise the table along with their corresponding data types (see the "Data Types" section).	Yes

The following CREATE TABLE command creates the Rep table and its associated columns and data types.

```
CREATE TABLE Rep
(RepNum  CHAR(2),
LastName  CHAR(15),
FirstName  CHAR(15),
Street  CHAR(15),
City  CHAR(15),
State  CHAR(2),
Zip  CHAR(5),
Commission  DECIMAL(7,2),
Rate  DECIMAL(3,2) )
;
```

Access version:

```
CREATE TABLE Rep
(RepNum  CHAR(2),
LastName  CHAR(15),
FirstName  CHAR(15),
Street  CHAR(15),
City  CHAR(15),
State  CHAR(2),
Zip  CHAR(5),
Commission  CURRENCY,
Rate  NUMBER )
;
```

Note: Access doesn't have a DECIMAL data type. To create numbers with decimals, you must use either the CURRENCY or NUMBER data type. Use the CURRENCY data type for fields that will contain currency values; use the NUMBER data type for all other numeric fields.

CREATE VIEW

Use the CREATE VIEW command to create a view. Figure C.4 describes the CREATE VIEW command.

FIGURE C.4
....
CREATE VIEW
command

Clause	Description	Required?
CREATE VIEW *view name* AS	Indicates the name of the view to be created.	Yes
query	Indicates the defining query for the view.	Yes

The following CREATE VIEW command creates a view named Housewares, which consists of the part number, description, on hand, and price for all rows in the Part table on which the Class is HW:

```
CREATE VIEW Housewares AS
SELECT PartNum, Description, OnHand, Price
FROM Part
WHERE Class = 'HW'
;
```

Data Types

Figure C.5 describes the data types that you can use in a CREATE TABLE command.

FIGURE C.5
....
Data types

Data Type	Description
CHAR (*n*)	Stores a character string *n* characters long. You use the CHAR type for fields that contain letters and other special characters, and for fields that contain numbers that will not be used for arithmetic. Because neither sales rep numbers nor customer numbers will be used in any arithmetic operations, for example, they both are assigned CHAR as the data type. [Some database management systems (DBMSs) use TEXT rather than CHAR, but they mean the same thing.]
DATE	Stores dates in the form DD-MON-YYYY or MM/DD/YYYY. For example, May 12, 2007, could be stored as 12-MAY-2007 or 5/12/2007. (**Note:** The specific format in which the dates are stored varies from one SQL implementation to another.)
DECIMAL *(p,q)*	Stores a decimal number *p* digits long with *q* of these digits being decimal places. For example, DECIMAL(5,2) represents a number with three places to the left and two places to the right of the decimal. You can use the contents of DECIMAL fields for arithmetic. (**Note:** The specific meaning of DECIMAL varies from one SQL implementation to another. In some implementations, the decimal point counts as one of the places, and in other implementations it does not. Likewise, in some implementations a minus sign counts as one of the places, but in others it does not.)

Data Type	Description
INTEGER	Stores integers, which are numbers without a decimal part. The valid data range is −2147483648 to 2147483647. You can use the contents of INTEGER fields for arithmetic.
SMALLINT	Stores integers, but uses less space than the INTEGER data type. The valid data range is −32768 to 32767. SMALLINT is a better choice than INTEGER when you are certain that the field will store numbers within the indicated range. You can use the contents of SMALLINT fields for arithmetic.

DELETE Rows

Use the DELETE command to delete one or more rows from a table. Figure C.6 describes the DELETE command.

FIGURE C.6
··················
DELETE command

Clause	Description	Required?
DELETE FROM *table name*	Indicates the table from which the row or rows are to be deleted.	Yes
WHERE *condition*	Indicates a condition. Those rows for which the condition is true will be retrieved and deleted.	No (If you omit the WHERE clause, all rows will be deleted.)

The following DELETE command deletes any row from the OrderLine table on which the part number is BV06.

```
DELETE
FROM OrderLine
WHERE PartNum='BV06'
;
```

DROP INDEX

Use the DROP INDEX command to delete an index, as shown in Figure C.7.

FIGURE C.7
··················
DROP INDEX
command

Clause	Description	Required?
DROP INDEX *index name*	Indicates the name of the index to be dropped.	Yes

The following DROP INDEX command deletes the index named RepBal:

```
DROP INDEX RepBal
;
```

■ DROP TABLE

Use the DROP TABLE command to delete a table, as shown in Figure C.8.

FIGURE C.8
• • • • • • • • • • • • • • • • •
DROP TABLE
command

Clause	Description	Required?
DROP TABLE *table name*	Indicates the name of the table to be dropped.	Yes

The following DROP TABLE command deletes the table named SmallCust:

```
DROP TABLE SmallCust
;
```

■ GRANT

Use the GRANT command to grant privileges to a user. Figure C.9 describes the GRANT command.

FIGURE C.9
• • • • • • • • • • • • • • • • •
GRANT command

Clause	Description	Required?
GRANT *privilege*	Indicates the type of privilege(s) to be granted.	Yes
ON *database object*	Indicates the database object(s) to which the privilege(s) pertain.	Yes
TO *user name*	Indicates the user(s) to whom the privilege(s) are to be granted.	Yes

The following GRANT command grants the user named Johnson the privilege of selecting rows from the Rep table:

```
GRANT SELECT ON Rep TO Johnson
;
```

INSERT

Use the INSERT command and the VALUES clause to insert a row into a table by specifying the values for each of the columns. As shown in Figure C.10, you must indicate the table into which to insert the values, and then list the values to insert in parentheses.

FIGURE C.10
••••••••••••••••••
INSERT command

Clause	Description	Required?
INSERT INTO *table name*	Indicates the name of the table into which the row will be inserted.	Yes
VALUES *(values list)*	Indicates the values for each of the columns on the new row.	Yes

The following INSERT command inserts the values shown in parentheses as a new row in the Rep table:

```
INSERT INTO Rep
VALUES
('16','Rands','Sharon','826 Raymond','Altonville','FL','32543',0.00,0.05)
;
```

Integrity

You can use the ALTER TABLE command with an appropriate CHECK, PRIMARY KEY, or FOREIGN KEY clause to specify integrity. Figure C.11 describes the ALTER TABLE command for specifying integrity.

FIGURE C.11
••••••••••••••••••
Integrity options

Clause	Description	Required?
ALTER TABLE *table name*	Indicates the table for which integrity is being specified.	Yes
integrity clause	CHECK, PRIMARY KEY, or FOREIGN KEY	Yes

The following ALTER TABLE command changes the Part table so that the only legal values for the Class column are AP, HW, and SG:

```
ALTER TABLE Part
CHECK (Class IN ('AP','HW','SG') )
;
```

The following ALTER TABLE command changes the Rep table so that the RepNum column is the table's primary key:

```
ALTER TABLE Rep
ADD PRIMARY KEY(RepNum)
;
```

The following ALTER TABLE command changes the Customer table so that the RepNum column in the Customer table is a foreign key referencing the primary key of the Rep table:

```
ALTER TABLE Customer
ADD FOREIGN KEY (RepNum) REFERENCES Rep
;
```

■ REVOKE

Use the REVOKE statement to revoke privileges from a user. Figure C.12 describes the REVOKE statement.

FIGURE C.12
••••••••••••••••••••
REVOKE statement

Clause	Description	Required?
REVOKE *privilege*	Indicates the type of privilege(s) to be revoked.	Yes
ON *database object*	Indicates the database object(s) to which the privilege pertains.	Yes
FROM *user name*	Indicates the user name(s) from whom the privilege(s) are to be revoked.	Yes

The following REVOKE statement revokes the SELECT privilege for the Rep table from the user named Johnson:

```
REVOKE SELECT ON Rep FROM Johnson
;
```

■ SELECT

Use the SELECT command to retrieve data from a table or from multiple tables. Figure C.13 describes the SELECT command.

FIGURE C.13
· · · · · · · · · · · · · · · ·
SELECT command

Clause	Description	Required?
SELECT *column or expression list*	Indicates the column(s) and/or expression(s) to be retrieved.	Yes
FROM *table list*	Indicates the table(s) required for the query.	Yes
WHERE *condition*	Indicates one or more conditions. Only the rows for which the condition(s) are true will be retrieved.	No (If you omit the WHERE clause, all rows will be retrieved.)
GROUP BY *column list*	Indicates the column(s) on which rows are to be grouped.	No (If you omit the GROUP BY clause, no grouping will occur.)
HAVING *condition involving groups*	Indicates a condition for groups. Only groups for which the condition is true will be included in query results. Use the HAVING clause only if the query output is grouped.	No (If you omit the HAVING clause, all groups will be included.)
ORDER BY *column or expression list*	Indicates the column(s) on which the query output is to be sorted.	No (If you omit the ORDER BY clause, no sorting will occur.)

The following SELECT command groups and orders rows by rep number. It displays the rep number, the count of the number of customers having this rep, and the average balance of these customers. It renames the count as NumCustomers and the average balance as AverageBalance. The HAVING clause restricts the reps to be displayed to only those having fewer than four customers.

```
SELECT RepNum, COUNT(*) AS NumCustomers, AVG(Balance)
     AS AverageBalance
FROM Customer
GROUP BY RepNum
HAVING COUNT(*)<4
ORDER BY RepNum
;
```

▇ SELECT INTO

Use the SELECT command with an INTO clause to insert the rows retrieved by a query into a table. As shown in Figure C.14, you must indicate the name of the table into which the row(s) will be inserted and the query whose results will be inserted into the named table.

FIGURE C.14
· · · · · · · · · · · · · · · ·
SELECT command
with INTO clause

Clause	Description	Required?
SELECT *field list*	Indicates the list of fields to be selected.	Yes
INTO *table name*	Indicates the name of the table into which the row(s) will be inserted.	Yes
remainder of query	Indicates the remainder of the query (for example, FROM clause and WHERE clause) whose results will be inserted into the table.	Yes

The following SELECT command with an INTO clause inserts rows selected by a query into the SmallCust table:

```
SELECT *
INTO SmallCust
FROM Customer
WHERE CreditLimit<=7500
;
```

Subqueries

You can use nested queries in SQL. The inner query is called a subquery and it is evaluated first. The outer query is evaluated next.

The following command contains a subquery that produces a list of part numbers located in Warehouse 3. The outer query then produces those order numbers in the OrderLine table that are on any rows containing a part number in the list:

```
SELECT OrderNum
FROM OrderLine
WHERE PartNum IN
(SELECT PartNum
FROM Part
WHERE Warehouse='3')
;
```

UNION

Connecting two SELECT commands with the UNION operator produces all the rows that would be in the results of the first command, the second command, or both.

The following query displays the customer number, last name, and first name of all customers that are represented by sales rep 65, *or* that have orders, *or* both:

```
SELECT CustomerNum, CustomerName
FROM Customer
WHERE RepNum='35'
UNION
SELECT Customer.CustomerNum, CustomerName
FROM Customer, Orders
WHERE Customer.CustomerNum=Orders.CustomerNum
;
```

■ UPDATE

Use the UPDATE command to change the contents of one or more rows in a table. Figure C.15 describes the UPDATE command.

FIGURE C.15
················
UPDATE command

Clause	Description	Required?
UPDATE *table name*	Indicates the table whose contents will be changed.	Yes
SET *column = expression*	Indicates the column to be changed, along with an expression that provides the new value.	Yes
WHERE *condition*	Indicates a condition. The change will occur only on those rows for which the condition is true.	No (If you omit the WHERE clause, all rows will be updated.)

The following UPDATE command changes the street address on the row in the Customer table on which the customer number is 524 to 1445 Rivard:

```
UPDATE Customer
SET Street='1445 Rivard'
WHERE CustomerNum='524'
;
```

APPENDIX D: "HOW DO I" REFERENCE

This appendix answers frequently asked questions about how to accomplish a variety of tasks using structured query langauge (SQL). Use the second column to locate the correct section in Appendix C that answers your question.

How do I?	Review the Named Section(s) in Appendix C
Add columns to an existing table?	ALTER TABLE
Add rows?	INSERT
Calculate a statistic (sum, average, maximum, minimum, or count)?	1. SELECT 2. Column or Expression List (SELECT Clause) (Use the appropriate function in the query.)
Change rows?	UPDATE
Create a data type for a column?	1. Data Types 2. CREATE TABLE
Create a table?	CREATE TABLE
Create a view?	CREATE VIEW
Create an index?	CREATE INDEX
Delete a table?	DROP TABLE
Delete an index?	DROP INDEX
Delete rows?	DELETE Rows
Drop a table?	DROP TABLE
Drop an index?	DROP INDEX
Grant a privilege?	GRANT
Group data in a query?	SELECT (Use a GROUP BY clause.)
Insert rows using a query?	SELECT INTO
Insert rows?	INSERT
Join tables?	Conditions (Include a WHERE clause to relate the tables.)
Order query results?	SELECT (Use an ORDER BY clause.)
Remove a privilege?	REVOKE
Remove rows?	DELETE Rows
Retrieve all columns?	1. SELECT 2. Column or Expression List (SELECT Clause) (Type * in the SELECT clause.)
Retrieve all rows?	SELECT (Omit the WHERE clause.)
Retrieve only certain columns?	1. SELECT 2. Column or Expression List (SELECT Clause) (Type the list of columns in the SELECT clause.)
Revoke a privilege?	REVOKE

(continued)

How do I?	Review the Named Section(s) in Appendix C
Select all columns?	1. SELECT 2. Column or Expression List (SELECT Clause) (Type * in the SELECT clause.)
Select all rows?	SELECT (Omit the WHERE clause.)
Select only certain columns?	1. SELECT 2. Column or Expression List (SELECT Clause) (Type the list of columns in the SELECT clause.)
Select only certain rows?	1. SELECT 2. Conditions (Use a WHERE clause.)
Sort query results?	SELECT (Use an ORDER BY clause.)
Specify a foreign key?	Integrity (Use a FOREIGN KEY clause in an ALTER TABLE command.)
Specify a primary key?	Integrity (Use a PRIMARY KEY clause in an ALTER TABLE command.)
Specify a privilege?	GRANT
Specify integrity?	Integrity (Use a CHECK clause in an ALTER TABLE command.)
Specify legal values?	Integrity (Use a CHECK clause in an ALTER TABLE command.)
Update rows?	UPDATE
Use a calculated field?	1. SELECT 2. Column or Expression List (SELECT Clause) (Enter a calculation in the query.)
Use a compound condition in a query?	Conditions
Use a compound condition?	1. SELECT 2. Conditions (Use simple conditions connected by AND, OR, or NOT in a WHERE clause.)
Use a condition in a query?	1. SELECT 2. Conditions (Use a WHERE clause.)
Use a subquery?	Subqueries
Use a wildcard?	1. SELECT 2. Conditions (Use LIKE and a wildcard in a WHERE clause.)
Use UNION operation?	UNION (Connect two SELECT commands with UNION.)

APPENDIX E: ANSWERS TO ODD-NUMBERED REVIEW QUESTIONS

■ Chapter 1—Introduction to Database Management

1. Redundancy is the duplication of data or the storing of the same data in more than one place. Redundancy wastes space, makes the updating of data more cumbersome and time-consuming, and can lead to inconsistencies.

3. An entity is a person, place, object, event, or idea for which you want to store and process data. An attribute, which is also called a field or column in many database systems, is a characteristic or property of an entity.

5. A database is a structure that can store information about multiple types of entities, the attributes of those entities, and the relationships among the entities.

7. An E-R diagram represents a database in a visual way by using a rectangle for each entity, using a line to connect two entities that have a relationship, and placing a dot at the end of a line to indicate the "many" part of a one-to-many relationship.

9. Database design is the process of determining the table structure of the desired database.

11. A switchboard system is a set of special forms used to provide controlled access to the data, forms, reports, and other objects in a database.

13. Sharing of data means that several users can have access to the same piece of data and use it in a variety of ways.

15. Multiple copies of the same data in an organization leads to inconsistency because each piece of data can have different values. Controlling redundancy is the result of eliminating, or at least reducing, the multiple copies. Improved consistency is the result of this controlled redundancy.

17. Security is the prevention of access to the database by unauthorized users. A DBMS provides security features such as passwords. As additional security, the DBA can assign users to a group and restrict each group to certain data and to certain types of access.

19. In a database environment, file size is a disadvantage because the DBMS is a large program that occupies a great amount of disk space and internal memory. Also, because all the data that the database manages for you is stored in one file, the database file itself requires a large amount of disk space and internal memory.

21. In a nondatabase environment, each user has a completely separate system; the failure of any single user's system does not necessarily affect any other user. On the other hand, if several users are sharing the same database, a failure on the part of any one user that damages the database in some way might affect all the other users.

Chapter 2—The Relational Model 1: Introduction, QBE, and Relational Algebra

1. A relation is a two-dimensional table in which (1) the entries in the table are single-valued; (2) each column has a distinct name; (3) all of the values in a column are values of the same attribute; (4) the order of the columns is immaterial; (5) each row is distinct; and (6) the order of the rows is immaterial.

3. An unnormalized relation is a structure that satisfies all the properties of a relation except the restriction that entries must be single-valued. It is not a relation.

5. In the shorthand representation, each table is listed, and after each table, all the columns of the table are listed in parentheses. Primary keys are underlined.

   ```
   Branch (BranchNum, BranchName, BranchLocation, NumEmployees)
   Publisher (PublisherCode, PublisherName, City)
   Author (AuthorNum, AuthorLast, AuthorFirst)
   Book (BookCode, Title, PublisherCode, Type, Price, Paperback)
   Wrote (BookCode, AuthorNum, SequenceNum)
   Inventory (BookCode, BranchNum, OnHand)
   ```

7. The primary key is the column or collection of columns that uniquely identifies a given row. The primary key of the Branch table is BranchNum. The primary key of the Publisher table is PublisherCode. The primary key of the Author table is AuthorNum. The primary key of the Book table is BookCode. The primary key of the Wrote table is the concatenation (combination) of BookCode and AuthorNum. The primary key of the Inventory table is the concatenation of BookCode and BranchNum.

9. Enter the criteria in the Criteria row for the appropriate column name.

11. Type the computation instead of a field name in the design grid. Alternatively, you can type the computation in the Zoom dialog box.

13. Indicate the appropriate sort order (Ascending or Descending) in the Sort row of the design grid.

15. Include field lists from both tables in the query. Provided they have matching fields, a join line will connect the tables. Include the desired fields from either table in the design grid.

17. Use a delete query when you want to delete all rows satisfying some criteria.

19. Relational algebra is a theoretical way of manipulating a relational database. Relational algebra includes operations that act on existing tables to produce new tables, similar to the way the operations of addition and subtraction act on numbers to produce new numbers in the mathematical algebra with which you are familiar.

21. The PROJECT command selects only the specified columns.

23. The UNION command selects all rows that are in the first table, in the second table, or both.

25. The INTERSECT command selects all rows that are in both tables.

27. The PRODUCT command (mathematically called the Cartesian product) is the table obtained by concatenating every row in the first table with every row in the second table.

■ Chapter 3—The Relational Model 2: SQL

1. To create a table in SQL, use a CREATE TABLE command that includes the name of the table followed by the names and data types of the columns in the table. INTEGER allows the field to contain integers. SMALLINT allows the field to contain "small" integers (integers less than 32,767). DECIMAL allows the field to contain numbers that have a decimal part. CHAR allows the field to contain character strings. Date allows the field to contain dates.

3. Compound conditions are formed by connecting simple conditions using the operators AND, OR, or NOT. A compound condition using AND is true only if all the simple conditions it contains are true. A compound condition using OR is true if one (or more) of the simple conditions it contains are true. Preceding a simple or compound condition with NOT reverses the truth or falsity of the original condition.

5. To use the LIKE or IN operators in an SQL query, include them in the WHERE clause. A character string containing one or more wildcards is included after the word LIKE. The word IN is followed by a list of values.

7. Use a SQL built-in function (COUNT, SUM, AVG, MAX, and MIN) by including it in the SELECT clause followed by the name of the field to which it applies.

9. To group data in SQL, include the words GROUP BY followed by the field or fields on which the data is to be grouped in the query results. If you group data, you can only include the fields on which you are grouping or statistics in the SELECT clause.

11. To qualify a field in an SQL query, precede the field with the name of the table to which it belongs, followed by a period. It is necessary to qualify a field if the field name occurs in more than one of the tables listed in the FROM clause.

13. The update commands in SQL are INSERT, which inserts new rows in a table; UPDATE, which changes all the rows in a table that satisfy some condition; and DELETE, which deletes all the rows in a table that satisfy some condition.

■ Chapter 4—The Relational Model 3: Advanced Topics

1. A view is an individual user's picture of the database. It is defined using a defining query. The data in the view never actually exists in the form described in the view. Rather, when a user accesses the view, his or her query is merged with the defining query of the view to form a query that pertains to the whole database.

3. a.

```
CREATE VIEW PartOrder AS
SELECT Part.PartNum, Description, Price, OrderNum, OrderDate,
      NumOrdered, QuotedPrice
FROM Part, OrderLine
WHERE Part.PartNum = OrderLine.PartNum;
```

b.

```
SELECT PartNum, Description, OrderNum, QuotedPrice
FROM PartOrder
WHERE QuotedPrice > 100;
```

c.

```
SELECT Part.PartNum, Description, OrderNum, QuotedPrice
FROM Part, OrderLine
WHERE Part.PartNum = OrderLine.PartNum
AND QuotedPrice > 100;
```

5. The GRANT statement is used to assign privileges to users of a database. It relates to security because a user who does not have the privilege of accessing a certain portion of a database cannot access that portion of the database. The privileges that can be assigned include the privilege of selecting rows from a table, inserting new rows, and updating existing rows. The REVOKE command is used to revoke privileges.

7. `REVOKE SELECT ON Part FROM Stillwell;`

9. a.

```
SELECT Name
FROM Systables
WHERE Creator = 'your name';
```

b.

```
SELECT Colname, Coltype
FROM Syscolumns
WHERE Tbname = 'Customer';
```

c.

```
SELECT Tbname
FROM Syscolumns
WHERE Colname = 'PartNum';
```

11. Null is a special value that represents missing information. Nulls are used when a value is either unknown or inapplicable. The primary key cannot accept null values. With a null value in the primary key, the primary key could not fulfill its main purpose of being the unique identifier for records in a table.

13. Adding an order to the Orders table on which the customer number does not match a customer number in the Customer table would violate referential integrity. In addition, changing the customer number on a record in the Orders table to a number that does not match a customer number in the Customer table would also violate referential integrity. If deletes do not cascade, deleting a customer that has orders would violate referential integrity. If deletes cascade, such a customer can be deleted, in which case all orders for that customer will automatically be deleted.

■ Chapter 5—Database Design 1: Normalization

1. A column (attribute) B is functionally dependent on another column A (or possibly a collection of columns) if each value for A in the database is associated with exactly one value of B.

3. Column A (or a collection of columns) is the primary key for a relation (table) R, if: Property 1—all columns in R are functionally dependent on A; and Property 2—no subcollection of the columns in A (assuming A is a collection of columns and not just a single column) also has Property 1.

5. A table is in first normal form if it does not contain a repeating group.

7. A table is in third normal form if it is in second normal form and if the only determinants it contains are candidate keys. If a table is not in third normal form, redundant data will cause wasted space and update problems. Inconsistent data might also be a problem.

9. An interrelation constraint is a condition that involves two or more relations. Requiring the value of a RepNum on a row in the Customer relation to match a value of RepNum in the Rep relation is an example of an interrelation constraint.

11.

```
Patient (PatientNum, HouseholdNum, PatientName)
Household (HouseholdNum, HouseholdName, Street, City, State,
     Zip, Balance)
Service (ServiceCode, Description, Fee)
Appointment (PatientNum, ServiceCode, Date)
```

13.

```
StudentNum → StudentName
ActivityNum → ActivityName
CourseNum → Description
StudentNum →→ ActivityNum
StudentNum →→ CourseNum

Student (StudentNum, StudentName)
Activity (ActivityNum, ActivityName)
Course (CourseNum, Description)
StudentActivity (StudentNum, ActivityNum)
StudentCourse (StudentNum, CourseNum)
```

■ Chapter 6—Database Design 2: Design Methodology

1. A user view is the view of data that is necessary to support the operations of a particular user. By considering individual user views rather than the complete design problem, the database design process is greatly simplified.

3. If the design problem were extremely simple, the overall design might not have to be broken down into a consideration of individual user views.

5. The primary key is the column or columns that uniquely identify a given row and that furnish the main mechanism for directly accessing a row in the table. An alternate key is a column or combination of columns that could have functioned as the primary key but was not chosen to do so. A secondary key is a column or combination of columns that is not any other type of key but is of interest for purposes of retrieval. A foreign key is a column or combination of columns in one table whose values are required to match the primary key in another table. Foreign keys furnish the mechanism through which relationships are made explicit.

7.

```
Department (DepartmentNum, DepartmentName)
Advisor (AdvisorNum, LastName, FirstName, DepartmentNum)
FK    DepartmentNum → Department

Course (CourseCode, Description)
Student (StudentNum, LastName, FirstName, AdvisorNum)
FK    AdvisorNum → Advisor

StudentCourse (StudentNum, CourseCode, Grade)
FK    StudentNum → Student
FK    CourseCode → Course
```

9. a. No change is necessary.
   b. Store both the AdvisorNum and DepartmentNum columns in the Student table.

11. The methodology presented in this text is bottom-up; that is, an approach in which specific user requirements are synthesized into a design. By initially reviewing the requirements and determining a possible list of entities prior to following the steps in this methodology, you can gain the advantages to both top-down and bottom-up approaches.

13. Many answers are possible. Be sure the functional dependencies you represent are based on reasonable assumptions and that the tables you create are in third normal form based on these dependencies.

15. a. There is a many-to-many-to-many relationship between students, courses, and faculty members.
    b. There is a many-to-many relationship between students and courses. A faculty number functionally determines a course number; that is, each faculty member is associated with exactly one course.
    c. There are separate many-to-many relationships between students and courses; courses and faculty members; and students and faculty members.
    d. There is a many-to-many relationship between students and courses. A given student number-course number combination uniquely determines a faculty number. That is, when a student takes a course, a single faculty member teaches the course.
    e. There is a many-to-many relationship between students and courses. There is also a many-to-many relationship between courses and faculty members. A student number functionally determines a faculty number; that is, each student is associated with exactly one faculty member.

17. If one 3NF relation contains a column that is a determinant for a column in another 3NF relation with the same primary key, merging the relations will produce a relation that is not in third normal form. The following is an example of two such relations:

```
Student1 (StudentNum, LastName, FirstName, AdvisorNum)
Student2 (StudentNum, LastName, FirstName, AdvisorLast,
         AdvisorFirst)
```

The following is the result of merging them:

```
Student (StudentNum, LastName, FirstName, AdvisorNum,
        AdvisorLast, AdvisorFirst)
```

This table is not in third normal form because

```
AdvisorNum → AdvisorLast, AdvisorFirst
```

Chapter 7—DBMS Functions

1. In updating and retrieving data, users do not need to know how data is physically structured on disk nor what processes the DBMS uses to manipulate the data. These structures and manipulations are solely the responsibility of the DBMS.

3. Most PC-based DBMSs contain a catalog that maintains metadata about tables, fields, table relationships, views, indexes, users, privileges, and replicated data. Large, expensive DBMSs usually contain a data dictionary that serves as a super catalog containing the same metadata as a catalog and additional metadata such as that needed to track fragmented data.

5. A lost update could occur in a concurrent update situation when two users attempt to update the same data at the same time, and the DBMS does not does not have concurrency update features such as two-phase locking or timestamping.

7. A transaction is a set of steps completed by a DBMS that accomplishes a single user task; the DBMS must successfully complete all transaction steps or none at all for the database to remain in a correct state.

9. Deadlock occurs in a concurrent update situation when the first user is waiting for data that has been locked by a second user, and the second user is waiting for data that has been locked by the first user. Unless outside action occurs, each user could wait for the other's data forever. Deadlock occurs when each of the two users is attempting to access data that is locked by the other user.

11. Recovery is the process of returning the database to a state that is known to be correct from a state known to be incorrect.

13. A DBA uses forward recovery when a catastrophe destroys a database. Forward recovery consists of two steps. First, the DBA copies the most recent database backup over the live database. Second, the DBMS forward recovery feature uses the log to apply after images for committed transactions.

15. Security is the prevention of unauthorized access, either intentional or accidental, to a database.

17. Authentication refers to techniques for identifying the person who is attempting to access the DBMS. Three types of authentication are passwords, biometrics, and database passwords. A password is a string of characters assigned by the DBA to a user that the user must enter to access to the database. Biometrics identify users by physical characteristics such as fingerprints, voiceprints, handwritten signatures, and facial characteristics. A database password is a string of characters assigned by the DBA to a database that users must enter before they can access the database.

19. Permissions specify what kind of access a user has to objects in a database. A workgroup is a group of users, and a DBA usually assigns appropriate permissions to workgroups.

21. Privacy refers to the right of individuals to have certain information about them kept confidential. Privacy and security are related because it is only through appropriate security measures that privacy can be ensured.

23. Data independence is a property that lets you change a database structure without requiring you to change the programs that access the database.

25. Some utility services that a DBMS should provide include: services that let you change the database structure; services that let you add new indexes and delete indexes that are no longer needed; facilities that let you use the services available from your operating system; services that let you export data to and import data from other software products; services that provide support for easy-to-use edit and query capabilities, screen generators, report generators, and so on; support for both procedural and nonprocedural languages; and support for an easy-to-use menu-driven or switchboard-driven interface that allows users to tap into the power of the DBMS without having to resort to a complicated set of commands.

■ Chapter 8—Database Administration

1. DBA is database administrator, the group or person responsible for the database. DBA is necessary because its responsibilities are critical to success in a database environment, especially if the database is shared among many users.

3. After DBA determines the access privileges for each user, DBA creates security policies and procedures, obtains management approval of the policies and procedures, and then distributes them to authorized users. DBA uses the DBMS's security features, such as encryption, authentication, authorizations, and views, and uses special security programs, if necessary. Finally, DBA monitors database usage to detect potential security violations and takes corrective action if a violation occurs.

5. Certain data, although no longer needed in the production database, must be kept for future reference. A data archive is a place for storing this type of data. The use of data archives lets an organization keep data indefinitely, without causing the database to become unnecessarily large. Data can be removed from the database and placed in the data archive, instead of just being deleted.

7. A shared lock permits other users to read the data. An exclusive lock prevents other users from accessing the data in any way.

9. Context-sensitive help means that if a user is having trouble and asks for help, the DBMS will provide assistance for the particular feature being used at the time the user asks for it.

11. DBA installs the DBMS, makes any changes to its configuration when they are required, determines whether it is appropriate to install a new version of the DBMS when it becomes available, applies any vendor fixes to problems, coordinates problem resolution, and coordinates the users so that their needs are satisfied without unduly affecting other users.

13. DBA does some of the training of computer users. DBA coordinates other training, such as that which is provided by software vendors.

15. The production system is the hardware, software, and database for the users; the test system is a separate system that programmers use to develop new programs and modify existing programs.

17. If users access only certain fields in a table, splitting the table results in smaller tables than the original; the lower amount of data moves faster between disk and memory and across a network.

■ Chapter 9—Database Management Approaches

1. A distributed database is a single logical database that is physically divided among computers at several sites on a network. A distributed database management system (DDBMS) is a DBMS capable of supporting and manipulating distributed databases.

3. A homogeneous DDDBS is one that has the same local DBMS at each site. A heterogeneous DDBMS is one that does not; there are at least two sites at which the local DBMSs are different. Heterogeneous systems are more complex.

5. Location transparency is the characteristic of a DDBMS that states that users do not need to be aware of the location of data in a distributed database. Data should be accessible at a remote site just as easily as it is at a local site; the only difference should be the response time.

7. Replication transparency is the characteristic that users do not need to be aware of any replication that has taken place in a distributed database. The DDBMS should handle updates to all copies of the data without users being aware of the steps taken by the DDBMS.

9. Fragmentation transparency is the characteristic that users do not need to be aware of any data fragmentation (splitting of data) in a distributed database. Users should feel as if they are using a single central database, even if data is stored at different sites.

11. In a well-designed distributed database, you can often increase its capacity by increasing the capacity at only one site. Also, you can increase capacity through the addition of new sites to the network.

13. Increased efficiency is an advantage in a distributed database because data available locally can be retrieved much more rapidly than data stored on a remote, centralized system.

15. Query processing is more complex in a distributed environment because of the difference between the time it takes to send messages between sites and the time it takes to access a disk.

17. With a two-phase commit, one site (often the site initiating the update) acts as coordinator. In the first phase, the coordinator sends messages to all other sites requesting they prepare to update the database (or acquire all necessary locks). Once the coordinator receives positive replies from all sites, the coordinator sends a message to each site to commit the update. At this point, each site must proceed with the commit process, and all sites must abort if any reply is negative. Two-phase commit guarantees consistency of the database.

19. The information-level design process is unaffected by the fact that the database is distributed.

21. The 12 rules against which you can measure DDBMSs are:

- Local autonomy: No site should depend on another site to perform its functions.

- No reliance on a central site: A DDBMS should not need to rely on one site more than any other site.

- Continuous operation: Performing any function should not shut down the entire distributed database.

- Location transparency: Users should feel as if the entire database is stored at their location.

- Fragmentation transparency: Users should feel as if they are using a single central database.

- Replication transparency: Users should not be aware of any data replication.

- Distributed query processing: A DDBMS must process queries as rapidly as possible even though the data is distributed.

- Distributed transaction management: A DDBMS must effectively manage transaction updates at multiple sites.

- Hardware independence: A DDBMS must be able to run on different types of hardware.

- Operating system independence: A DDBMS must be able to run on different operating systems.

- Network independence: A DDBMS must be able to run on different types of networks.

- DBMS independence: A DDBMS must be heterogeneous.

23. Placing the business functions on the client causes a client maintenance problem. Whenever programmers change the business functions, they must place the updated business functions on every client. Placing the business functions on the server causes a scalability problem and might cause the server to become a bottleneck and degrade the system's responsiveness to clients.

25. Scalability is the ability of a computer system to continue to function well even as utilization of the system increases.

27. Compared to a file server, the advantages of a client/server architecture are: lower network traffic; improved processing distribution; thinner clients; greater processing transparency; increased network, hardware, and software transparency; improved security; decreased costs; and increased scalability.

29. Triggers are usually created with SQL.

31. Online transaction processing (OLTP) systems typically deal with a small number of rows from the tables in a database in a highly structured, repetitive, and predetermined way.

33. Nonvolatile means that the data in a data warehouse is read-only; data is loaded periodically into a data warehouse, but users cannot update the data directly.

35. You use OLAP to work efficiently with data warehouses.

37. Data mining consists of uncovering new knowledge, patterns, trends, and rules from the data stored in a data warehouse.

39. An object-oriented DBMS is one in which data and the actions that operate on the data are encapsulated into objects.

41. The term class refers to the general structure of an entity. The term object refers to a specific occurrence of a class.

43. When one class is a subclass of another, the subclass inherits the structure of the class as well as the methods that apply to the class. The benefit to inheritance is that you need to define only the additional attributes and methods that were not inherited from the class.

45. Relationships are called associations in UML.

47. Multiplicity indicates the number of objects that can be related to an individual object at the other end of the relationship.

49. You can measure object-oriented systems against the following 14 rules: an OODBMS must support the creation of complex objects; an OODBMS must provide a way to identify objects; an OODBMS must encapsulate data and associated methods together in the database; an OODBMS must hide the details concerning the way data is stored and the actual implementation of the methods from the users of the database; an OODBMS must support either types or classes; an OODBMS must support inheritance; an OODBMS must support late binding; you can use functions in the language of the OODBMS to perform various computations; an OODBMS should let you define new data types; an OODBMS should be able to remember its data from one execution to the next; an OODBMS should have sufficient performance capabilities to effectively manage very large databases; an OODBMS must support concurrent update; an OODBMS must provide recovery services; and an OODBMS must provide query facilities.

51. XML (Extensible Markup Language) is a programming language similar in format to HTML that is more customizable and is suited to the exchange of data between different programs.

53. The chief difference between network and hierarchical model databases is that in the hierarchical model no rectangle can have more than one "many" arrow entering the rectangle. A hierarchy is thus a more restrictive structure than a network.

GLOSSARY

Access   A PC-based relational DBMS first marketed by Microsoft in the 1990s.

Access delay   A fixed amount of time required for every message sent over a network.

After image   A record that the DBMS places in the journal or log that shows what the data in a row looked liked in the database after a transaction update.

Aggregate function   A function to calculate the number of entries, the sum or average of all the entries in a given column, or the largest or smallest of the entries in a given column; also called *built-in function*.

Alternate key   A candidate key that was not chosen to be the primary key.

American National Standards Institute (ANSI)   An organization that in 1971 considered, but did not adopt, the CODASYL DBMS specifications as a national standard.

AND criterion   Combination of criteria in which both criteria must be true.

ANSI   See *American National Standards Institute*.

Application server   In a three-tier client/server architecture, a computer that performs the business functions and serves as an interface between clients and the database server.

Archive   See *data archive*.

Artificial key   A column created for an entity to serve solely as the primary key and that is visible to users.

Association   A relationship in UML.

Attribute   A characteristic or property of an entity; also called a *field* or *column*.

Authentication   A technique for identifying the person who is attempting to access a DBMS.

Authorization rule   A rule that specifies which user has what type of access to which data in a database.

B2B   See *business to business*.

B2C   See *business to consumer*.

Back-end machine   See *server*.

Back-end processor   See *server*.

Backup   A copy of a database made periodically; the backup is used to recover the database when it has been damaged or destroyed; also called a *save*.

Backward recovery   A process used to recover a database to a valid state by reading the log for problem transactions and applying the before images to undo their updates; also called *rollback*.

Batch processing   The processing of a transaction file that contains a group, or "batch," of records to update a database or another file.

Before image   A record that the DBMS places in the journal or log that shows what the data in a row looked liked in the database before a transaction update.

Binary large object (BLOB)   A generic term for a special data type used by relational DBMSs to store complex objects.

Binding   The association of operations to actual program code.

Biometrics   A technique to identify users of a database or other resource by physical characteristics such as fingerprints, voiceprints, handwritten signatures, and facial characteristics.

BLOB   See *binary large object*.

Bottom-up design methodology   A design methodology in which specific user requirements are synthesized into a design.

Boyce-Codd normal form (BCNF)   A relation is in Boyce-Codd normal form if it is in second normal form and the only determinants it contains are candidate keys; also called *third normal form* in this text.

Built-in function   A function to calculate the number of entries in a given column, the sum or average of all the entries in a given column, or the largest or smallest of the entries in a given column; also called *aggregate function*.

Business to business (B2B)   E-commerce between businesses.

Business to consumer (B2C)   E-commerce between business and consumers.

Calculated field   A field whose value may be computed from other fields in the database; also called *computed field*.

Candidate key   A minimal collection of columns (attributes) in a table on which all columns are functionally dependent but that has not necessarily been chosen as the primary key.

Cardinality   The number of items that must be included in a relationship; also called *multiplicity*.

Cartesian product   The table obtained by concatenating every row in the first table with every row in the second table.

Cascade delete   A delete option in which related records are automatically deleted.

Cascade update   An update option in which related records are automatically updated.

Catalog   A source of data, usually stored in hidden database tables, about the types of entities, attributes, and relationships in a database.

Category   The IDEF1X name for an entity subtype.

CHAR(*n*)   The SQL data type for character data.

Class   The general structure and actions of an object in an object-oriented system.

Class diagram   A UML diagram that for each class, shows the name, attributes, and methods of the class, as well as the relationships between the classes in a database.

Client   A computer that is connected to a network and that people use to access data stored on a server in a client/server system; also called a *front-end machine* or a *front-end processor*.

Client/server   A networked system in which a special site on the network, called the *server*, provides services to the other sites, called the *clients*. Clients send requests for the specific services they need. Software, often including a DBMS, running on the server then processes the requests and sends only the appropriate data and other results back to the clients.

CODASYL   See *COnference on DAta SYstems Languages*.

Column   A characteristic or property of an entity; also called an *attribute* or a *field*.

Command   An instruction by a user that directs a database to perform a certain function.

Commit   A special record in a database journal or log that indicates the successful completion of a transaction.

Communications network   Several computers configured in such a way that data can be sent from any one computer on the network to any other.

Comparison operator   An operator used to compare values. The valid operators are =, <, >, <=, <>, and !=. Also called a *relational operator*.

Complete category   In IDEF1X, a collection of subtypes with the property that every element of the supertype is an element of at least one subtype.

Composite entity   An entity in the entity-relationship model used to implement a many-to-many relationship.

Compound condition   See *compound criteria*.

Compound criteria   Two simple criteria (conditions) in a query that are combined with the AND or OR operators.

Computed field   A field whose value can be computed from other fields in the database; also called a *calculated field*.

Concatenation   The combination of columns. To say a primary key is a concatenation of two columns means that a combination of values of both columns is required to uniquely identify a given row.

Concurrent update   A situation when multiple users make updates to the same database at the same time.

Condition   See *criterion*.

COnference on DAta SYstems Languages (CODASYL)   The group responsible for the programming language COBOL and that proposed the CODASYL model for database management. The CODASYL model led to the development of several DBMSs that were popular and influential through the late 1980s.

Context-sensitive help   The assistance a DBMS provides for the particular feature being used at the time a user asks for help.

Coordinator   In a distributed network, the site that directs the update to the database for a transaction. Often, it is the site that initiates the transaction.

CREATE TABLE command   The SQL command used to describe the layout of a table. The word TABLE is followed by the name of the table to be created and then by the names and data types of the columns (fields) that comprise the table.

Criteria   The plural version of the word "criterion".

Criterion   A statement that can be either true or false. In queries, only records for which the statement is true will be included; also called a *condition*.

Cumulative design   A design that supports all the user views encountered thus far in a design process.

Data archive   A place where a record of certain corporate data is kept; also called an *archive*. Data that is no longer needed in a corporate database but must be retained for future reference is removed from the database and placed in the archive.

Data cube   The perceived shape by a user of a multidimensional database in a data warehouse.

Data dictionary   A catalog, usually found in large, expensive DBMSs, that stores data about the entities, attributes, relationships, programs, and other objects in a database.

Data file   A file used to store data about a single entity. It's the computer counterpart to an ordinary paper file you might keep in a file cabinet, an accounting ledger, and so on. Such a file can be thought of as a table.

Data fragmentation   The process of dividing a logical object, such as the records in a table, among the various locations in a distributed database.

Data independence   The property that lets you change the structure of a database without requiring you to change the programs that access the database; examples of these programs are the forms you use to interact with the database and the reports that provide information from the database.

Data Language/I (DL/I)   An early DBMS product developed by IBM and first offered publicly in 1966.

Data mining   The uncovering of new knowledge, patterns, trends, and rules from the data stored in a data warehouse.

Data model   One of the four categories of DBMS (hierarchical, network, relational, object-oriented), each characterized by the way the users of the DBMS feel the data is structured. A data model has two components: structure and operations.

Data warehouse   A subject-oriented, integrated, time-variant, nonvolatile collection of data in support of management's decision-making process.

Database   A structure that can store information about multiple types of entities, the attributes of these entities, and the relationships among the entities.

Database administration (DBA)   The individual or group that is responsible for a database.

Database administrator   The individual who is responsible for a database, or the head of database administration.

Database design   The process of determining the content and structure of data in a database in order to support some activity on behalf of a user or group of users.

Database Design Language (DBDL)   A relational-like language that is used to represent the result of the database design process.

Database management system (DBMS)   A program, or a collection of programs, through which users interact with a database. DBMSs let you create forms, reports, and switchboards quickly and easily, as well as obtain answers to questions about the data stored in a database.

Database navigation   The procedure used by programs and users to manipulate a network-structured database.

Database password   A string of characters assigned by DBA to a database that users must enter before they can access a database.

Database server   In a three-tier client/server architecture and in other architectures, a computer that performs the database functions such as storing and retrieving data in a database.

DataBase Task Group (DBTG)   The group that was created by CODASYL to develop DBMS specifications, which CODASYL presented to ANSI in 1971.

DATE   The SQL data type for date data.

DB2   IBM's commercial relational DBMS.

DBA   See *database administration*. (Sometimes the acronym stands for *database administrator*.)

dBASE   A PC-based relational DBMS first marketed in the 1980s.

DBDL   See *Database Design Language*.

DBMS   See *database management system*.

DBTG   See *DataBase Task Group*.

DDBMS   See *distributed database management system*.

Deadlock   A situation in which two or more database users are each waiting to use resources that are held by the other(s); also called *deadly embrace*.

Deadly embrace   See *deadlock*.

DECIMAL(*p,q*)   The SQL data type for decimal data.

Decryption   A process that reverses the encryption of a database.

Defining query   The query that is used to define the structure of a view.

DELETE   The SQL command used to delete a table. The word DELETE is followed by a FROM clause identifying the table. Use a WHERE clause to specify a condition. Any records satisfying the condition will be deleted.

Delete query   A query that deletes all records that satisfy some criterion.

Denormalizing   The conversion of a table that is in third normal form to a table that is no longer in third normal form. Denormalizing introduces anomaly problems but can decrease the number of disk accesses required by certain types of transactions, thus improving performance.

Dependency diagram   A diagram that indicates the dependencies among the columns in a table.

Dependent entity   An entity that requires a relationship to another entity for identification.

Design grid   The portion of the Query Design screen in Microsoft Access where you enter fields, criteria, sort orders, and so on.

Determinant   A column in a table that determines at least one other column.

Difference   When comparing tables, the set of all rows that are in the first table but that are not in the second table.

Dimension table   A table in a data warehouse that contains a single-part primary key, serving as an index into the central fact table, and other fields associated with the primary key value.

Disaster recovery plan   A plan that specifies the ongoing and emergency actions and procedures required to ensure data availability, even if a disaster occurs.

Distributed database   A single logical database that is physically divided among computers at several sites on a computer network.

Distributed database management system (DDBMS)   A DBMS capable of supporting and manipulating distributed databases.

Division   The relational algebra command that combines tables and searches for rows in the first table that match all rows in the second table.

DL/I   See *Data Language/I.*

Documenter   An Access tool that prints documentation about the objects in a database.

Domain   The set of values that are permitted for an attribute.

Drill down   The process of viewing and analyzing lower levels of aggregation, or a more detailed view of the data.

. . . .

E-commerce   See *electronic commerce.*

Electronic commerce (e-commerce)   Business conducted on the Internet and Web.

Encapsulated   In an object-oriented system, defining an object to contain both data and its associated actions.

Encryption   A security measure that converts the data in a database to a format that's indecipherable to normal programs. The DBMS decrypts, or decodes, the data to its original form for any legitimate user who accesses the database.

Entity   A person, place, object, event, or idea for which you want to store and process data.

Entity integrity   The rule that no column (attribute) that is part of the primary key may accept null values.

Entity subtype   Entity A is a subtype of entity B if every occurrence of entity A is also an occurrence of entity B.

Entity-relationship (E-R) diagram   A graphic model for database design in which entities are represented as rectangles and relationships are represented as either arrows or diamonds connected to the entities they relate.

Entity-relationship (E-R) model   An approach to representing data in a database that uses E-R diagrams exclusively as the tool for representing entities, attributes, and relationships.

Exclusive lock   A lock that prevents other users from accessing the locked data in any way.

Existence dependency   A relationship in which the existence of one entity depends on the existence of another related entity,

Extensible Markup Language   See *XML.*

Extensible   The capability of defining new data types in an OODBMS.

. . . .

Fact table   The central table in a data warehouse that consists of rows that contain consolidated and summarized data.

Fat client   In a two-tier client/server architecture, a client that performs presentation functions and business functions.

Field   A characteristic or property of an entity; also called an *attribute* or a *column.*

File server   A networked system in which a special site on the network stores files for users at other sites. When a user needs a file, the file server sends the entire file to the user.

First normal form (1NF)   A table is in first normal form if it does not contain repeating groups. (This is part of the definition of a table.)

Foreign key   An column (attribute) or collection of columns in a table whose value is required either to match the value of a primary key in a table or to be null.

Form   A screen object you use to maintain, view, and print data from a database.

Forward recovery   A process used to recover a database by reading the log and applying the after images of committed transactions to bring the database up to date.

Fourth normal form (4NF)   A table is in fourth normal form if it is in third normal form and there are no multi-valued dependencies.

Fragmentation transparency   The characteristic that users do not need to be aware of any data fragmentation (splitting of data) that has taken place in a distributed database.

FROM clause   The part of an SQL SELECT command that indicates the tables in the query.

Front-end machine   See *client.*

Front-end processor   See *client.*

Functional dependence   See *functionally dependent.*

Functionally dependent   Column B is functionally dependent on column A (or on a collection of columns) if a value for A determines a single value for B at any one time.

Functionally determine   Column A functionally determines column B if B is functionally dependent on A.

∙∙∙

Gemstone   One of the first OODBMSs from Servio Corporation that appeared on the market in 1987.

Generalization   In UML, the relationship between a superclass and a subclass.

Generalized Update Access Method (GUAM)   An early DBMS that was produced for use in the moon launch program and that was capable of handling vast amounts of data.

Global deadlock   In a distributed database, deadlock that cannot be detected solely at any individual site.

GROUP BY clause   The part of an SQL SELECT command that indicates grouping.

Grouping   The process of creating collections of records that share some common characteristic.

Growing phase   A phase during a database update in which the DBMS locks all the data needed for a transaction and releases none of the locks.

GUAM   See *Generalized Update Access Method*.

∙∙∙

HAVING clause   The part of an SQL SELECT command that restricts the groups to be displayed.

Heterogeneous DDBMS   A distributed DBMS in which at least two of the local DBMSs are different from each other.

Hierarchical model   A database model that is perceived by users as a collection of hierarchies, or trees.

Hierarchy   A network with an added restriction that each record type cannot be related on the "many" side in more than one one-to-many relationship.

Homogeneous DDBMS   A distributed DBMS in which all the local DBMSs are the same.

∙∙∙

IDEF1X   A type of E-R diagram; or, technically, a language in the IDEF (Integrated Definition) family of languages that is used for data modeling.

Identifying relationship   A relationship that is necessary for identification of an entity.

IDMS   See *Integrated Data Management System*.

IDS   See *Integrated Data Store*.

IMS   See *Information Management System*.

Incomplete category   In IDEF1X, a collection of subtypes with the property that there are elements of the supertype that are not elements of any subtype.

Independent entity   An entity that does not require a relationship to another entity for identification.

Index   A file that relates key values to records that contain those key values.

Index key   The field or fields on which an index is built.

Information Management System (IMS)   An early, dominant DBMS developed by IBM and still in use in some legacy systems.

Information-level design   The step during database design in which the goal is to create a clean, DBMS-independent design that will support all user requirements.

Inheritance   The property that a subclass inherits the structure of the class as well as its methods.

INSERT   The SQL command to add new data to a table. After the words INSERT INTO, you list the name of the table, followed by the word VALUES. Then you list the values for each of the columns in parentheses.

INTEGER   The SQL data type for integer data.

Integrated Data Management System (IDMS)   The most widely used CODASYL DBMS that was created by Cullinane.

Integrated Data Store (IDS)   An early DBMS developed by a team at General Electric.

Integrity   A database has integrity if the data in it satisfies all established integrity constraints.

Integrity constraint   A rule that must be followed by data in a database.

Integrity rules   See *entity integrity, legal-values integrity,* and *referential integrity*.

Intelligent key   A primary key that consists of a column or collection of columns that is an inherent characteristic of the entity.

Interrelation constraint   A constraint that involves more than one relation.

INTERSECT   The relational algebra command for performing the intersection of two tables.

Intersection   When comparing tables, an intersection is a new table containing all rows that are in both original tables.

INTO clause   The SQL clause that inserts values into a table. An INTO clause consists of the word INTO followed by the name of the table to be created.

Intranet   An internal company network that uses software tools typically used on the Internet and the World Wide Web.

∙∙∙

Join   In relational algebra, the operation in which two tables are connected on the basis of common data.

Join column   The column on which two tables are joined. Also see *join*.

Join line   In an Access query, the line drawn between tables to indicate how they are related.

Journal   A file that contains a record of all the updates made to a database; also called a *log*. The DBMS uses the journal to recover a database that has been damaged or destroyed.

Journaling   Maintaining a journal or log of all updates to a database.

Key   The field on which data will be sorted; also called a *sort key*.

LAN   See *local area network*.

Legal-values integrity   The property that no record can exist in the database with a value in a field other than a legal value.

Live system   The hardware, software, and database for the users; also called a *production system*.

Local area network (LAN)   A configuration of several computers connected together that allows users to share a variety of hardware and software resources.

Local deadlock   In a distributed database, deadlock that occurs at a single site.

Local site   From a user's perspective, the site in a distributed system at which the user is working.

Location transparency   The property that users do not need to be aware of the location of data in a distributed database.

Locking   A DBMS's denial of access by other users to data while the DBMS processes one user's updates to the database.

Log   A file that contains a record of all the updates made to a database; also called a *journal*. The DBMS uses the log to recover a database that has been damaged or destroyed.

Logical key   A primary key that consists of a column or collection of columns that is an inherent characteristic of the entity.

Major sort key   When sorting on two fields, the more important field; also called a *primary sort key*.

Make-table query   An Access query that creates a table containing the results of a query.

Mandatory role   The role in a relationship played by an entity with a minimum cardinality of 1 (that is, there must be at least one occurrence of the entity).

Many-to-many relationship   A relationship between two entities in which each occurrence of each entity can be related to many occurrences of the other entity.

Many-to-many-to-many relationship   A relationship between three entities in which each occurrence of each entity can be related to many occurrences of each of the other entities.

Message   A request to execute a method. Also, data, requests, or responses sent from one computer to another computer on a network.

Metadata   Data about the data in a database.

Method   An action defined for a object class.

Minor sort key   When sorting on two fields, the less important field; also called a *secondary sort key*.

Multidependent   In a table with columns A, B, and C, B is multidependent on A if each value for A is associated with a specific collection of values for B and, further, this collection is independent of any values for C.

Multidetermines   In a table with columns A, B, and C, A multidetermines B if each value for A is associated with a specific collection of values for B and, further, this collection is independent of any values for C.

Multidimensional database   The perceived structure by users of the data in a data warehouse.

Multiple-column index   See *multiple-field index*.

Multiple-field index   An index built on more than one field (column).

Multiplicity   In UML, the number of objects that can be related to an individual object on the other side of a relationship; also called *cardinality*.

Multivalued dependence   In a table with columns A, B, and C, there is a multivalued dependence of column B on column A (also read as "B is multidependent on A" or "A multidetermines B"), if each value for A is associated with a specific collection of values for B and, furthermore, this collection is independent of any values for C.

MySQL   A large relational DBMS first marketed in the 1990s.

Natural join   The most common form of a join.

Natural key   A primary key that consists of a column or collection of columns that is an inherent characteristic of the entity.

Network   The logical structure of records and relationships for a network model database. Also see *communications network*.

Network model   A database model in which users perceive a database as a collection of record types and relationships between these record types.

Nonidentifying relationship   A relationship that is not necessary for identification.

Nonkey attribute An attribute (column) that is not part of the primary key.

Nonkey column An attribute (column) that is not part of the primary key.

Nonprocedural language A language in which a user describes the task that is to be accomplished by the computer rather than the steps that are required to accomplish it.

Normal form See *first normal form*, *second normal form*, *third normal form*, and *fourth normal form*.

Normalization process The process of removing repeating groups to produce a first normal form table. Sometimes refers to the process of creating a third normal form table.

Null A data value meaning "unknown" or "not applicable."

Object A unit of data (set of related attributes) along with the actions that are associated with that data.

Objectivity/DB An OODBMS.

Object-oriented database management system (OODBMS) A DBMS in which data and the methods that operate on that data are encapsulated into objects.

Object-relational DBMS (ORDBMS) A relational DBMS that has added object-oriented features.

OLAP See *online analytical processing*.

OLTP See *online transaction processing*.

One-to-many relationship A relationship between two entities in which each occurrence of the first entity is related to many occurrences of the second entity and each occurrence of the second entity is related to only one occurrence of the first entity.

One-to-one relationship A relationship between two entities in which each occurrence of the first entity is related to one occurrence of the second entity and each occurrence of the second entity is related to one occurrence of the first entity.

Online analytical processing (OLAP) Software that is optimized to work efficiently with multidimensional databases in a data warehouse environment.

Online transaction processing (OLTP) A system that processes a transaction by dealing with a small number of rows in a relational database in a highly structured, repetitive, and predetermined way.

OODBMS See *object-oriented database management system*.

Operations One of the two components of a data model; the facilities given to the users of a DBMS to manipulate data within the database.

Optional role The role in a relationship played by an entity with a minimum cardinality of 0 (that is, there need not be any occurrences of the entity).

OR criterion A combination of criteria in which at least one of the criteria must be true.

Oracle A large relational DBMS first marketed in the 1980s.

ORDBMS See *object-relational DBMS*.

ORDER BY clause The part of an SQL SELECT command that indicates a sort order.

Outer join The form of a join in which all records appear, even if they don't match.

Paradox A PC-based relational DBMS first marketed in the 1980s.

Partial dependency A dependency of a column on only a portion of the primary key.

Password A string of characters assigned by DBA to a user that the user must enter to access a database.

Permission The specification of the kind of access a user has to the objects in a database.

Persistence The ability to have a program "remember" its data from one execution to the next.

Physical-level design The step during database design in which a design for a given DBMS is produced from the final information-level design.

Polymorphism The use of the same name for different operations in an object-oriented system.

Primary copy In a distributed database with replicated data, the copy of the database that must be updated in order for the update to be deemed complete.

Primary key A minimal collection of columns (attributes) in a table on which all columns are functionally dependent and that is chosen as the main direct-access vehicle to individual rows. Also see *candidate key*.

Primary sort key When sorting on two fields, the more important field; also called a *major sort key*.

Privacy The right of individuals to have certain information about them kept confidential.

Private visibility In UML, an indication that only the class itself can view or update the attribute value.

Procedural language A language in which a user specifies the steps that are required for accomplishing a task instead of merely describing the task itself.

Product The table obtained by concatenating every row in the first table with every row in the second table.

Production system The hardware, software, and database for the users; also called *live system*.

PROJECT The relational algebra command to select columns from a table.

Protected visibility In UML, an indication that only the class itself or public or protected subclasses of the class can view or update the attribute value.

Public visibility In UML, an indication that any class can view or update the attribute value.

QBE See *Query-By-Example.*

Qualify To indicate the table (relation) of which a given column (attribute) is a part by preceding the column name with the table name. For example, *Customer.Address* indicates the column named Address within the table named Customer.

Query A question, the answer to which is found in the database; also used to refer to a command in a nonprocedural language such as SQL that is used to obtain the answer to such a question.

Query-By-Example (QBE) A data manipulation language for relational databases in which users indicate the action to be taken by completing on-screen forms.

RDBMS See *relational DBMS.*

Record A collection of related fields; can be thought of as a row in a table.

Recovery The process of returning a database to a state that is known to be correct from a state known to be incorrect.

Redundancy Duplication of data, or the storing of the same data in more than one place.

Referential integrity The rule that if a table A contains a foreign key that matches the primary key of table B, then the value of this foreign key must either match the value of the primary key for some row in table B or be null.

Relation A two-dimensional table-style collection of data in which all entries are single-valued, each column has a distinct name, all the values in a column are values of the attribute that is identified by the column name, the order of columns is immaterial, each row is distinct, and the order of rows is immaterial.

Relational algebra A relational data manipulation language in which new tables are created from existing tables through the use of a set of operations.

Relational database A collection of relations (tables).

Relational DBMS (RDBMS) A DBMS that supports and manipulates a relational database.

Relational operator An operator used to compare values. Valid operators are =, <, >, <=, >=, <>, and !=. Also called a *comparison operator.*

Relationship An association between entities.

Remote site From a user's perspective, any site other than the one at which the user is working.

Repeating group Several entries at a single location in a table.

Replica A copy of the data in a database that a user can access at a remote site.

Replicate A duplicate of the data in a database that a user can access at a remote site.

Replication transparency The property that users do not need to be aware of any replication that has taken place in a distributed database.

Reserved word A word that is part of the SQL language.

Roll up View and analyze higher levels of aggregation.

Rollback A process to recover a database to a valid state by reading the log for problem transactions and applying the before images to undo their updates; also called *backward recovery.*

Row-and-column subset view A view that consists of a subset of the rows and columns in a table.

Save A copy of a database made periodically; the backup is used to recover the database when it has been damaged or destroyed; also called a *backup.*

Scalability The ability of a computer system to continue to function well as utilization of the system increases.

Second normal form (2NF) A relation is in second normal form if it is in first normal form and no nonkey attribute is dependent on only a portion of the primary key.

Secondary key A column (attribute) or collection of columns that is of interest for retrieval purposes (and that is not already designated as some other type of key).

Secondary sort key When sorting on two fields, the less important field; also called *minor sort key.*

Security The prevention of access to the database by unauthorized users.

SELECT The relational algebra command to select rows from a table. Also, the retrieval command in SQL.

SELECT clause The part of an SQL SELECT command that indicates the columns to be included in the query results.

Server A computer that provides services to the clients in a client/server system; also called a *back-end processor* or a *back-end machine.*

Shared lock A lock that lets other users read the locked data.

Shrinking phase A phase during a database update in which the DBMS releases all the locks previously acquired for a transaction and acquires no new locks.

Simple condition A condition that involves only a single field and a single value.

Single-column index See *single-field index*.

Single-field index An index built on a single field (column).

Slice and dice In a data warehouse, selecting portions of the available data, or reducing the data cube.

SMALLINT The SQL data type for integer data for small integers.

Sort The process of arranging rows in a table or results of a query in a particular order.

Sort key The field on which data are sorted; also called a *key*.

SQL See *Structured Query Language*.

SQL Server A large relational DBMS first marketed by Microsoft in the 1990s.

Star schema A multidimensional database whose conceptual shape resembles a star.

Statistic A mathematical function supported by Access; statistics include Count, Sum, Avg (average), Max (largest value), Min (smallest value), StDev (standard deviation), Var (variance), First, and Last; also called an *aggregate function*.

Stored procedure A file containing a collection of compiled and optimized SQL statements that are available for future use.

Structure One of the two components of a data model: the manner in which the system structures data or, at least, the manner in which the users perceive that the data is structured.

Structured Query Language (SQL) A very popular relational data definition and manipulation language that is used in many relational DBMSs.

Subclass A class that inherits the structure and methods of another class and for which you can define additional attributes and methods.

Subquery In SQL, a query that appears within another query.

SUBTRACT The relational algebra command for performing the difference of two tables.

Superclass In UML, a class that has subclasses.

Surrogate key A system-generated primary key that is usually hidden from users.

Switchboard A set of special forms used to provide controlled access to the data, forms, reports, and other objects in a database.

Sybase A large relational DBMS first marketed in the 1980s.

Synchronization The periodic exchange by a DBMS of all updated data between two databases in a replica set.

Synthetic key A system-generated primary key that is usually hidden from users.

Syscolumns The portion of the system catalog that contains column information.

Sysindexes The portion of the system catalog that contains index information.

Systables The portion of the system catalog that contains table information.

System catalog A structure that contains information about the objects (tables, columns, indexes, views, and so on) in a database.

System R A theoretical research, prototype relational DBMS developed by IBM in the 1970s that led to its commercial relational product, DB2.

Sysviews The portion of the system catalog that contains view information.

Table See *relation*.

Test system The hardware, software, and database that programmers use to develop new programs and modify existing programs.

Thin client In a client/server architecture, a client that performs only presentation functions.

Third normal form (3NF) A table is in third normal form if it is in second normal form and if the only determinants it contains are candidate keys.

Three-tier architecture A client/server architecture in which the clients perform the presentation functions, a database server performs the database functions, and the application servers perform the business functions and serve as an interface between clients and the database server.

Timestamp The unique time when the DBMS starts a transaction update to a database.

Timestamping The process of using timestamps to avoid the need to lock rows in a database and to eliminate the processing time needed to apply and release locks and to detect and resolve deadlocks.

Top-down design methodology A design methodology that begins with a general database design that models the overall enterprise and then repeatedly refines the model to achieve a design that supports all necessary applications.

Transaction A set of steps completed by a DBMS to accomplish a single user task.

Trigger An action that automatically occurs in response to an associated database operation.

Tuning The process of changing the database design to improve performance.

Tuple The formal name for a row in a table.

Two-phase commit An approach to the commit process in distributed systems in which there are two phases. In the first phase, each site is instructed to prepare to commit and must indicate whether the commit will be possible. After each site has responded, the second phase begins. If every site has replied in the affirmative, all sites must commit. If any site has replied in the negative, all sites must abort the transaction.

Two-phase locking An approach to locking that is used to manage concurrent update in which there are two phases: a growing phase, in which the DBMS locks more rows and releases none of the locks, and a shrinking phase, in which the DBMS releases all the locks and acquires no new locks.

Two-tier architecture A client/server architecture in which the clients perform the presentation functions, and a database server performs the database functions. In a fat client configuration, the clients perform the business functions, whereas in a thin client configuration, the database server performs the business functions.

UML See *Unified Modeling Language.*

Unified Modeling Language (UML) An approach to model all the various aspects of software development for object-oriented systems.

Union A combination of two tables consisting of all records that are in either table.

Union compatible Two tables are union compatible if they have the same number of fields and if their corresponding fields have identical data types.

Unnormalized relation A structure that satisfies the properties required to be a relation (table) with one exception: repeating groups are allowed; that is, the entries in the table do not have to be single-valued.

UPDATE The SQL command to make changes to existing table data. After the word UPDATE, you indicate the table to be updated. After the word SET, you indicate the field to be changed, followed by an equals sign and the new value. Finally, you can include a condition in the WHERE clause, in which case, only the records that satisfy the condition will be changed.

Update anomaly An update problem that can occur in a database as a result of a faulty design.

Update query In Access, a query that updates the contents of a table.

User view The view of data that is necessary to support the operations of a particular user.

User-level security A feature that DBA can use in Access to implement authorization rules granting users specific permissions to the objects in a database.

Utility services DBMS-supplied services that assist in the general maintenance of a database.

Validation rule In Access, a rule that data entered in a field must satisfy.

Validation text In Access, a message that is displayed when a validation rule is violated.

Versant An OODBMS.

Victim In a deadlock situation, the deadlocked user's transaction that the DBMS chooses to abort to break the deadlock.

View An application program's or an individual user's picture of a database.

Visibility symbol In UML, a symbol preceding an attribute in a class diagram to indicate whether other classes can view or change the value in the attribute. The possible visibility symbols are public visibility (+), protected visibility (#), and private visibility (-). With public visibility, any other class can view or change the value. With protected visibility, only the class itself or public or protected subclasses of the class can view or change the value. With private visibility, only the class itself can view or change the value.

Weak entity An entity that depends on another entity for its own existence.

WHERE clause The part of an SQL SELECT command that indicates a condition the rows to be displayed must satisfy.

Workgroup In Access, a group of users who are assigned the same permissions to various objects in a database.

Workgroup information file In Access, a file that contains group, user, and group membership information.

XML (Extensible Markup Language) A programming language similar in format to HTML that is more customizable and is suited to the exchange of data between different programs. You can customize XML code to describe the data it contains and how that data should be structured.

INDEX

C

C++ computer language, DBMS support for, 256

calculated fields. *See* computed fields

calculating

    averages, 41–42

    statistics in databases, 40

candidate keys, 146

cardinality in E-R diagram, 219

Cartesian product, 59

cascade delete, update option and referential integrity, 125

catalog, system, 131, 135

catalogs and hidden tables, 228

categories and entity subtypes, 211–213

centralized vs. distributed data processing, 286

changing

    field characteristics, 129

    relationships, 253–254

chapter review questions

    answers to, 389–400

    database administration, 282

    database concepts, 27

    database design methodology, 221–223

    database management approaches, 324

    DBMS features, 258–259

    normalization, 169

    relational databases, advanced topics, 136

    relational databases and QBE, 62–63

    SQL (structured query language), 105

chapter summaries

    database administration, 281

    database concepts, 26

    database design methodology, 220

    database management approaches, 321

    DBMS functions, 257

    normalization, 168

    relational databases, advanced topics, 135

    relational databases and QBE, 61

    SQL (structured query language), 104

CHAR(n) data type, 69

CHECK clause

    described, 135

    and legal-values integrity, 127

Chen, Peter, 215

classes

    inheritance, 312

    in OODBMSs, 310–312, 315

clients, fat and thin, 298

client/server systems

    advantages of, 299–300

    and distributed DBMSs, 296–300

COBOL computer language, DBMS support for, 256

CODASYL (COnference on DAta SYstems Languages), 318

Codd, Dr. E.F., 122, 307, 322

Codd, S.B., 307

column or expression list (SELECT clause), 374–376

columns

    in databases described, 6

    and functional dependence, 142

    functional dependence of, 168

    join, 54

    nonkey, 149

    in relational databases, 31, 32

    system catalog information about, 131–132

commands. *See also specific commands*

    'How Do I' reference (table), 387–388

    SQL, 68, 97–98, 373–385

committing transactions, 241

communications networks described, 286–287

comparison operators, 36, 73

complete categories, 213

composite entities, 217

compound criteria, conditions, 37

computed fields

    described, using, 39, 79

    query that uses (fig.), 40

    SQL Reference, 374

computers

    in client/server architecture, 297–298

    locking on PC-based database management systems, 238

    messages, 287

concatenation described, 54, 143

concurrent updates

    avoiding lost updates, 230–234

    and database complexity, 240

    described, 229

    in distributed databases, 293

    guidelines for programs performing, 239

    in OODBMSs, 316

    problem, and two-phase locking, 234–237

conditions

    compound, 75–78, 375

    simple, 73, 374–376

    SQL Reference, 374–376

COnference on DAta SYstems Languages (CODASYL), 318

consistency, controlling with database processing, 17

constraint enforcement, 251

context-sensitive help, 274

S

Salley, C.T., 307
saves and data recovery, 240
saving queries, 113
scalability
    of client/server systems, 300
    of computer systems, 298
second normal form
    described, 168
    in normalization process, 148–151
secondary sort key, 43, 177
security
    in client/server systems, 300
    database, generally, 121–122
    DBA responsibilities, 265–267
    DBMS evaluation, 273
    DBMS features, 257
    encryption. *See* encryption
    improvements with database processing, 17
    and privacy, 250
    splitting table for, 279
    user-level, 247
SELECT clauses
    in defining query, 112
    in simple retrievals, 70, 72
    SQL Reference, 374–376
    in table joins, 91–94
SELECT command, 61
    with INTO clause, 383–384
    in relational algebra, 53
    SQL Reference, 382–383
semicolons (;) in SQL commands, 70
SEQUEL. *See also* SQL
    hardware product, 68
servers
    application, described, 298
    database, described, 298
    file, and networks, 296
sharing data, database processing advantage,
    16
shrinking phase, locking, 237
simple condition, 73
simple queries, 34–35
simple retrievals, 70–75
single-column index, 120
single-field index, 120
sites, remote and local in networks, 288
SMALLINT data type, 69
sort keys, 43
sorting
    alphabetically, 44
    database records, 43

    on multiple keys, 44–46
    query results, 82–84, 104
spaces in table and column names, 68–69
splitting tables, 207–214, 278–279
spreadsheets
    redundancy and data access in, 4
    sample orders (fig.), 3
SQL Server, 13, 319
SQL (structured query language)
    command summary, reference, 97–98, 373–385
    data updating methods, 95–96
    frequently asked questions (table), 378–388
    nested queries in, 384
    programming language described, 68
    reserved words, 70
    simple retrievals, 70–75
    versions, differences between, 71
square brackets ([ ]) in relational database fields,
    40
star schema, data warehouse structure and access,
    302–303
statistics, calculating with queries, 40–41
stored procedures and triggers in client/server systems,
    300
structured query language. *See* SQL (structured query
    language)
structures, data model, 319–320
subclasses in OODBMSs, 312
subject-oriented data, 301
subqueries
    described, 87, 104
    SQL Reference, 384
SUBSTRACT command, 57–58, 61
SUM function
    described, 84
    SQL Reference, 374
    vs. COUNT, 85
sums. *See* totals
superclasses, 314
surrogate keys, 178
survey forms, gathering database design information
    from, 220
switchboard systems, 14–15
Sybase, 319
synchronization data, 255
synthetic keys. *See* surrogate keys
Syscolumns, 131, 133
Sysindexes, Systables, 131
system catalog, 131, 135
System R, 319
Sysviews, 131

WHERE clauses
    conditions in, 374
    in defining query, 112
    and HAVING clause, 90
    in join operation, 53
    simple condition, 73
    in simple retrievals, 70–73
    in table joins, 90–94
    vs. HAVING clause, 89

wildcard symbols
    Access, 375
    with LIKE operator, 80
workgroups
    assigning user to (fig.), 249
    described, restricting database access, 247

X

XML (Extensible Markup Language), 318, 322